ASHEVILLE-BUNCOMBE TECHNICAL INSTITUTE NORTH CAROLINA
                                        ST   ON
                                        DEP   GES
                                                 LIBRARIES

Asheville-Buncombe Technical Institute
LIBRARY
340 Victoria Road
Asheville, North Carolina 28801

DISCARDED

JUN 2 5 2025

Fundamental Electronics

Fundamental Electronics

James F. Turner
Instructor, West Valley College, Campbell, California
and
Staff Engineer, IBM Corporation

CHARLES E. MERRILL PUBLISHING COMPANY
A Bell & Howell Company
Columbus, Ohio

Published by
Charles E. Merrill Publishing Co.
A Bell & Howell Company
Columbus, Ohio 43216

Copyright © 1972 by Bell & Howell Company. All rights reserved. No part of this book may be reproduced in any form, electronic or mechanical, including photocopy, recording, or any information storage and retrieval system, without permission in writing from the publisher.

International Standard Book Number: 0-675-09175-6
Library of Congress Catalog Card Number: 72-169985

1 2 3 4 5 6 7 8 – 77 76 75 74 73 72
Printed in the United States of America

**To Pigeon,
my wife and partner**

Merrill's International Series in Electrical and Electronics Technology

SAMUEL L. OPPENHEIMER, *Consulting Editor*
BEATY   *Basic ET Laboratory — An Audio-Tutorial Program*
BOYLESTAD   *Introductory Circuit Analysis, Second Edition*
FABER   *Introduction to Electronic Amplifiers*
HARTKEMEIER *Fortran Programming of Electronic Computers*
HERRICK   *Introduction to Electronic Communications*
HERRICK   *Mathematics for Electronics*
LENERT   *Semiconductor Physics, Devices, and Circuits*
OPPENHEIMER *Semiconductor Logic and Switching Circuits*
PRISE   *Electronic Circuit Packaging*
SCHWARTZ   *Survey of Electronics*
TOCCI   *Fundamentals of Electronic Devices*
TOCCI   *Fundamentals of Pulse and Digital Circuits*
TURNER   *Digital Computer Analysis*
TURNER   *Fundamental Electronics*

Preface

Fundamental Electronics has been specifically designed to prepare students thoroughly for continuing courses in electronics. The text assumes a high school education, and is intended for a first two-semester course in electronics.

The first two chapters cover basic physics and magnetism. Then, Ohm's law and the various combinations of resistance are studied. AC, dc, and pulsating dc are introduced prior to capacitance, inductance, and impedance. Frequency effects and resonant circuits complete the study of the basic circuits. Next, mathematical approaches to circuit analysis include loop equations, Thevenin's and Norton's theorems, and complex numbers.

The text progresses into vacuum tubes and vacuum tube circuits. Semiconductor physics, diodes, transistors, and semiconductor circuits are covered in the next chapter, followed by special semiconductor devices, field effect transistors, and integrated circuits. The final chapter discusses power supplies and regulator circuits.

Appendixes are organized to provide useful information pertaining to electronic terms, symbols, mathematic relationships and tables, and conversion factors. Each chapter is provided with an ample number of questions and problems. The odd numbered questions are answered in the Appendix as an aid to better understanding of the material. An index has over 1,000 entries for ease of reference.

Summarizing, the text is concise and complete, offering the additional coverage of semiconductor devices and integrated circuits not often found in a basic text of this type.

March, 1972 *James F. Turner*

Contents

1 Physics — 1
    1.1 Basic Atoms 1
    1.2 Interaction of Atoms 4
    1.3 Excitation of Atoms 5
    1.4 Electrical Charges 6
    1.5 Potential Difference 8
    1.6 Resistivity 9
    1.7 Resistance 10
    1.8 Conductance 11
    Questions 12

2 Magnetism — 14
    2.1 Magnets 14
    2.2 Magnetic Theory 17
    2.3 Magnetic Flux 18
    2.4 Magnetic Induction 20
    2.5 Permeability 21
    2.6 Magnetic Materials and Types of Magnets 22
    2.7 Electric Field 23

2.8 Field Intensity 26
2.9 Reluctance 29
2.10 B and H Curve 30
2.11 Hysteresis 31
2.12 Field Force on a Conductor 33
2.13 Induced Current 34
2.14 Induced Voltage 36
Questions 39

3 Resistance 41

3.1 Definitions 41
3.2 Basic Ohm's Law 42
3.3 Finding the Current 44
3.4 Finding the Resistance 45
3.5 Finding the Voltage 45
3.6 Finding the Power 46
3.7 Examples of Problems 47
3.8 Ohm's Law with Two Power Sources 50
Questions 55

4 Series Circuits 56

4.1 Kirchhoff's Law 56
4.2 DC Current 57
4.3 Voltage Drops 59
4.4 Power Dissipation 63
4.5 Determining an Unknown Quantity 64
Questions 67

5 Parallel Circuits 69

5.1 Kirchhoff's Law 69
5.2 DC Currents 70
5.3 Circuit Resistance 72
5.4 DC Voltage 79
5.5 Power Dissipation 82
5.6 Determining an Unknown Quantity 84
Questions 87

6 Series/Parallel Circuits 89

6.1 Kirchhoff's Law 89
6.2 Series/Parallel 90

Contents   xi

    6.3 Parallel/Series 98
    6.4 Combination Circuits 101
    Questions 105

7 Alternating Voltage, Alternating Current, and Pulsating DC   107

    7.1 Alternating Voltage 107
    7.2 Angular Rotation 109
    7.3 AC Current 111
    7.4 Amplitude Measurements 113
    7.5 Phase Angle 116
    7.6 Frequency 117
    7.7 Pulsating DC 121
    Questions 125

8 Capacitance   127

    8.1 Charging a Capacitor 127
    8.2 Measurement of Capacitance 131
    8.3 DC and AC Effects 134
    8.4 Capacitors in Series 138
    8.5 Capacitors in Parallel 142
    8.6 Capacitors in Series/Parallel 145
    8.7 Capacitive Reactance 146
    8.8 Phase Angle of Voltage and Current 152
    Questions 154

9 Inductance   157

    9.1 Induction 157
    9.2 Inductance 159
    9.3 Transformer Action 163
    9.4 Transformer Cores 171
    9.5 Series and Parallel Inductance 173
    9.6 Inductive Reactance 178
    9.7 Series/Parallel Inductive Reactance 180
    9.8 Phase Angle of Voltage and Current 183
    Questions 185

10 Impedance   189

    10.1 Series Resistance and Capacitive Reactance 189
    10.2 Parallel Resistance and Capacitive Reactance 194

10.3 *RC* Time Constant   197
10.4 Series Resistance and Inductive Reactance   200
10.5 Parallel Resistance and Inductive Reactance   205
10.6 *LR* Time Constant   208
10.7 Series Inductive and Capacitive Reactance   211
10.8 Parallel Inductive and Capacitive Reactance   213
10.9 Series Resistance and Reactance   215
10.10 Parallel Resistance and Reactance   217
10.11 Series/Parallel Resistance and Reactance   220
Questions   222

11 Frequency Considerations   225

11.1 Series Resonance   225
11.2 Parallel Resonance   230
11.3 *Q* Effects   233
11.4 Bandwidth   236
11.5 Filters   238
11.6 Transient Circuits   245
11.7 Non-Linear Circuits   249
Questions 253

12 Circuit Analysis   257

12.1 Loop Equations   257
12.2 Thévenin's Theorem   262
12.3 Norton's Theorem   266
12.4 Superposition Theorem   269
12.5 *j* Factors   272
12.6 Complex Numbers in Polar Form   278
12.7 Applications of Complex Numbers   280
Questions   286

13 Vacuum Tubes   290

13.1 Diodes   291
13.2 Triodes   298
13.3 Characteristic Curves   304
13.4 Tetrodes and Pentodes   310
13.5 Specifications   316
13.6 Special Amplifier Configurations   318
13.7 Special Purpose Tubes   321
Questions   326

Contents   xiii

14 Semiconductors   328

14.1 Semiconductor Theory   329
14.2 P-N Junction   333
14.3 Diodes   335
14.4 PNP/NPN Junction   337
14.5 Bias Effects   341
14.6 Basic Transistor Configurations   345
14.7 Characteristic Curves   347
14.8 Specifications   350
14.9 Basic Transistor Circuits   352
Questions   358

15 Special Devices and Integrated Circuits   361

15.1 Zener Diodes   361
15.2 Photo Devices   367
15.3 Field Effect Transistors   372
15.4 Integrated Circuits   379
Questions   386

16 Power Supplies   388

16.1 Vacuum Tube Power Supplies   388
16.2 Power Supply Filters   392
16.3 Solid State Power Supplies   394
16.4 Voltage Regulation   396
16.5 Current Regulation   401
Questions   403

Appendixes

A Electronic Units and Symbols   405
B Mathematic Functions and Tables   410
C Conversion Tables and Graphs   415
D References   420
E Answers to the Odd Numbered Questions   421

Index   431

1 Physics

A basic study of physics is necessary in order to understand the fundamentals of electronics. This chapter is devoted to specific areas associated with electricity and electronics, from the atomic structure of matter to the flow of electricity.

1.1 Basic Atoms

The atom is considered the basis of matter. Various types and combinations of atoms in motion determine the physical nature of matter and its characteristics. A useful concept for studying the motion or planetary action of atoms is the **Bohr** model. The basic model is illustrated in Figure 1-1. The center of the atom is called the **nucleus**, which contains **protons** and **neutrons**. The model shows one proton and one neutron. Protons have an electrically positive charge and neturons are electrically neutral (no charge). A neutron has a slightly greater mass than a proton. An electron (which has a negative charge) is shown orbiting the nucleus.

Elementary physics explains the idea of charges. Like charges (both positive or both negative) repel, and unlike charges (one positive and one negative) attract. In the model, Figure 1-1, the nucleus contains a positive proton that weighs about 1840 times as much as the electron. The negatively charged electron is held in orbit by its electrostatic attraction to the

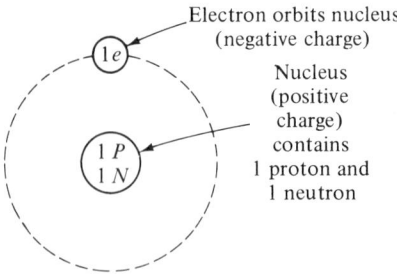

FIGURE 1-1 *Basic Atom—Bohr Model*

large nucleus. The theory is likened to the earth's rotation around the sun or the moon's rotation around the earth.

Electrons are the smallest units of electrical charge that have a negative polarity. Protons are the basic particles that have positive polarity. Both have equal charges of opposite polarity, even though the protons are heavier. The nucleus also has a neutron (in the model). A neutron does not have a charge and weighs approximately the same as a proton. The number of neutrons in a nucleus does not always match the number of protons.

In Figure 1-2(a), a hydrogen atom is illustrated. The electron orbits in a circular path called a **shell** or **band**, labeled K. Since it has one proton and one electron, its atomic number is 1. A carbon atom is shown in drawing (b). Its atomic number is 6 because there are six protons in the nucleus and six electrons in orbit. Notice that there are two orbital paths labeled K and L. K has two electrons, and L has four electrons. The K band can hold a maximum of 2 electrons, and the L band can hold a maximum of 8. The L band in the carbon atom is only half filled. Since the K band is closest, the nucleus exerts more force on it than on the outer band. The inner band is more stable, whereas the outer band can be more easily upset by external forces; that is, it can lose or gain electrons. If the outer band had its maximum allotment of eight electrons, it would be considered fairly stable because it cannot gain electrons, and losing electrons is unlikely since the atom has become stabilized.

Figure 1-2(c) illustrates a silicon atom with 14 protons and 14 electrons and an atomic number of 14. There are three orbital paths labeled K, L, and M. The first two are filled with two and eight respectively. The maximum number of electrons an outside band may hold is always eight. Band M is therefore half full. Most atoms follow the rule of $2n^2$, where n is the band 1, 2, etc., to determine the number of electrons in a given band. The rule does not apply to the outermost band which is always eight or less. Also, several atoms take exception and have fewer electrons than the rule would indicate.

[1.1] Physics 3

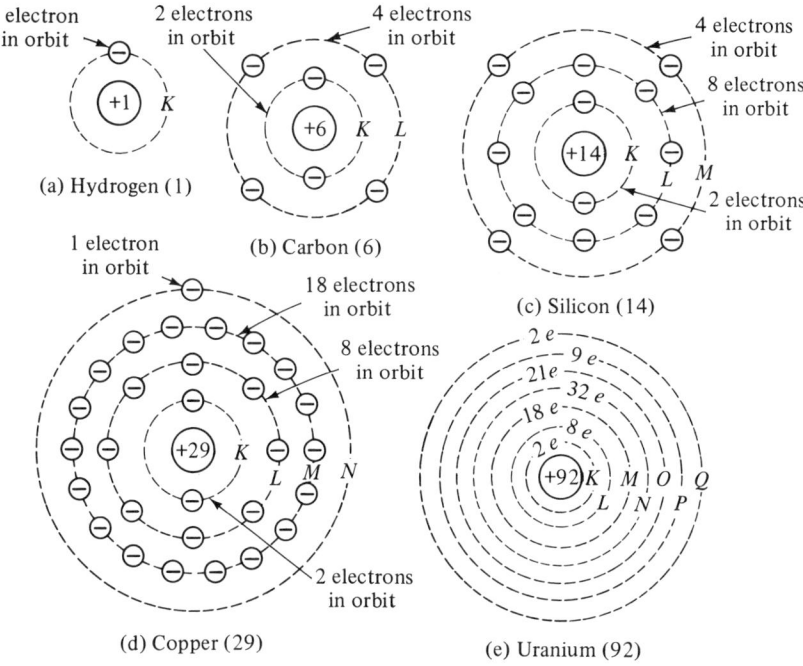

FIGURE 1-2 *Bohr Atom Diagrams*

Examples of the rule are as follows:

K is the first band, $n = 1$
$2n^2 = 2(1)^2 = 2$ electrons

L is the second band, $n = 2$
$2n^2 = 2(2)^2 = 8$ electrons

M is the third band, $n = 3$
$2n^2 = 2(3)^2 = 18$ electrons

A copper atom is shown in Figure 1-2(d) with a total of 29 protons and 29 electrons. The first three bands follow the rule of 2, 8, and 18 electrons. The twenty-ninth electron is in the fourth band, N. Uranium [Figure 1-2(e)] is the heaviest element and has 92 protons and electrons. There are other elements with more than 92 protons, but uranium has 146 neutrons where other elements have a closer balance between protons and neutrons. Since the atomic weight is approximately the number of protons plus the number of neutrons, the atomic weight of uranium is $92 + 146 \approx 238$.

The outermost band of an atom is also known as the **valence band**. This is the band that determines the ease with which an atom gains or loses electrons. If it only has one electron, it can lose that electron very easily. On the other hand, a valence of seven needs only one more for stability; hence, it can gain one easily. Notice that the copper atom, Figure 1-2(d), has one valence electron, indicating an unstable and mobile atom. Semiconductors such as silicon and carbon have four valence electrons; therefore, they are at a midway point of decision.

Table 1-1 lists several chemical elements. The atomic weights are not all twice the atomic number, indicating that the number of neutrons does not necessarily equal the number of protons.

TABLE 1-1 *Selected Chemical Elements*

Element	Symbol	At. No.	At. Wt.	Remarks
Copper	Cu	29	64	Conductors
Silver	Ag	47	108	
Gold	Au	79	179	
Carbon	C	6	12	Semiconductors
Silicon	Si	14	28	
Germanium	Ge	32	72	
Hydrogen	H	1	1	Gases
Oxygen	O	8	16	

1.2 Interaction of Atoms

An **element** is a substance that cannot be decomposed further by chemical action. Atoms are examples of elements. The fact that a particle of gold or silver the size of a needle point contains millions of atoms gives one an understanding of the size involved. A **molecule** is a particle with two or more elements (atoms) of the *same* kind. Several elements of gold are defined as a molecule of gold. A **compound** is two or more *different* elements. For example, mixing the elements of hydrogen and oxygen produces the compound of water (H_2O).

When atoms are grouped together, they tend to share their valence electrons, and a state called **covalent bonding** exists. For example, carbon has four electrons in its valence band; the maximum is eight. If two carbon atoms are sharing, a portion of the valence bands will merge so that each atom will act as though it had the additional electrons. When many carbon atoms are grouped together, multiple sharing will occur, and all valence bands will be filled to eight.

A two-dimensional crystal lattice is illustrated in Figure 1-3. Nine carbon atoms are shown, and the nucleus of each is labeled with a "6" to show its atomic number. Only the four valence electrons are shown around each nucleus. Due to the close proximity of other atoms, each valence atom forms an electron pair with its neighbor. Dotted lines encompass each electron. As far as each nucleus is concerned, it has four

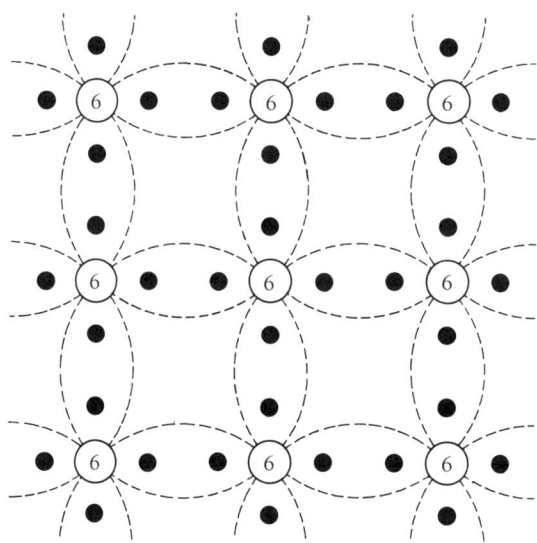

FIGURE 1-3 *Covalent Bonding of Carbon Atoms*

electron pairs or eight electrons. And because eight is the maximum number of electrons for a stable atom, the group forms a stable crystal lattice. Stability occurs when the valence bands are full, and the atoms are not trying to gain electrons.

To summarize, an atom that has a full valence band is considered inert and unable to combine with others. When the valence bands are not full, the atoms can combine with one another.

1.3 Excitation of Atoms

An orbiting electron has two kinds of energy. There is **kinetic** energy which is a result of the electron's motion around the nucleus, and there is **potential** energy due to the electrostatic attraction of the nucleus. If some form of external energy is applied, both the kinetic and potential energies will increase. Remember that the electrons in the inner bands are closer to the nucleus and have a greater electrostatic attraction. Also, the inner electrons tend to act as a shield which causes the outer electrons to be influenced more by external energy. An indication of energy is the mobility of the valence electrons. For example, copper has one electron in its valence band. No amount of bonding or combinations will fill the band. Therefore, the valence band is continually in motion, ready to accept other electrons. Copper, then, has high inherent energy, and its energy can easily be increased by external means.

A semiconductor, such as carbon, silicon, or germanium, requires more energy to mobilize the valence electrons since the band is full by means of covalent bonding.

Two common methods of increasing the energy of an atom by external means are heat and light. Heat may be applied by raising the ambient temperature around the atomic structure. Light may be ambient lighting or a concentrated beam of light. External energy causes electrons to seek a higher band. Valence electrons are affected first since they are already in the highest band. The energy of the electrons in the individual bands does not change—the change is the energy resulting from electrons jumping to a higher level band. When an electron escapes its atom, it is called a **free electron**. It is free to drift around until it recombines with another atom. If the external energy is sufficient, many electrons will move around from atom to atom.

Conductors are examples of the high mobility of electrons. Electrons can become so mobile that they lose their identity with any particular nucleus. When the electrons escape their nucleus, the atom becomes ionized (a positive ion). The characteristics of the particle are maintained by the bond that forms between positive ions and negative electrons.

1.4 Electrical Charges

As previously stated, the protons in the nucleus have a positive charge, and the electrons have a negative charge. Charges are measured in coulombs. One coulomb is equal to 6.28×10^{18} electrons or protons that are stored in a dielectric. Charge is abbreviated Q, and the formula for Q is

$$Q = \frac{\text{Number of electrons or protons}}{6.28 \times 10^{18}} \qquad (1\text{--}1)$$

where Q is in coulombs. Electrons are of primary interest because they carry electrical current. A charge can be positive $(+Q)$ or negative $(-Q)$.

Example 1-1 What is the charge in coulombs if a neutral dielectric is charged with 18.84×10^{18} protons?

Solution
$$+Q = \frac{18.84 \times 10^{18} \text{ protons}}{6.28 \times 10^{18}} = 3 \text{ coulombs}$$

Example 1-2 What is the net charge on a dielectric that is first charged with 6.28×10^{18} protons; then, a charge of 18.84×10^{18} electrons is added?

Solution
$$+Q = \frac{6.28 \times 10^{18} \text{ protons}}{6.28 \times 10^{18}} = 1 \text{ coulomb}$$

$$-Q = \frac{18.84 \times 10^{18} \text{ electrons}}{6.28 \times 10^{18}} = -3 \text{ coulombs}$$

$$\text{net charge} = +Q - Q = 1 \text{ coulomb} - 3 \text{ coulombs}$$
$$= -2 \text{ coulombs} = -Q$$
$$\text{or, electrons} = -Q \times 6.28 \times 10^{18} = 2 \times 6.28 \times 10^{18}$$
$$= 12.56 \times 10^{18}$$

When charges or electrons move in one direction, there is electron current flow. Electrons are much too small to measure, and coulombs are simply a quantity of electrons. Since current is charge movement, it can be defined in coulombs per second.

$$I = \frac{Q}{t} \tag{1-2}$$

where I is the current in amperes (amp); Q is the charge in coulombs, and t is the time in seconds (sec).

Example 1-3 If 20 coulombs flow past a given point for 2 sec, how much is the current flow?

Solution
$$I = \frac{Q}{t} = \frac{20 \text{ coulombs}}{2 \text{ sec}} = 10 \text{ amp}$$

Example 1-4 How many electrons are required for a current of 4 amp in 10 sec?

Solution
$$Q = It = 4 \text{ amp} \times 10 \text{ sec} = 40 \text{ coulombs}$$
$$\text{electrons} = Q \times 6.28 \times 10^{18} = 40 \times 6.28 \times 10^{18}$$
$$= 251.2 \times 10^{18} \text{ electrons}$$

Charges must flow if current is to flow. In addition, charges must flow in a continuous path if current flow is to be maintained at a specific level. As an example, consider a length of conducting material. If one end is charged positively, it will attract electrons. However, the electrons will soon bunch up and overload the end and repel further movement. The other end will simultaneously be drained of electrons. Figure 1-4 is an illustration of this phenomenon.

Batteries are one source of electrical charge. They convert chemical energy into electrical energy. Generators are another source of electrical charge, but they convert mechanical energy into electrical energy. When a conductor is connected across an electrical energy source such as the

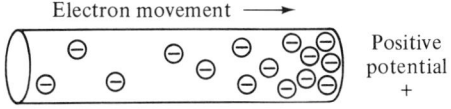

FIGURE 1-4 *Electrons in a Conductor*

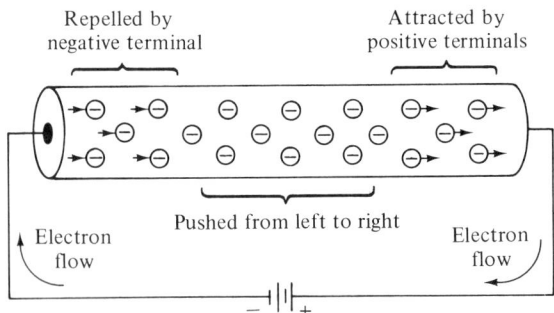

FIGURE 1-5 *Battery as Source of Electrons*

battery illustrated in Figure 1-5, electrons flow in a direction governed by the polarity of the terminals. For example, the positive terminal attracts the free electrons since electrons are negative charges (unlike charges attract). The negative terminal repels free electrons (like charges repel). Free electrons on the conductor are pushed along by those coming from behind. Thus, a battery or generator absorbs charges at one terminal while supplying charges from the other terminal.

Because a battery or generator is a driving force, it is often referred to as **electromotive force (emf)**. This driving force does not actually move electrons at high speed as one might expect. The number of moving charges is so large that their average velocity is something less than a millimeter per second.

1.5 Potential Difference

Potential difference is the electrical difference between charges. When a difference exists between charges, there is potential to do work. For example, a difference in charge potential has the ability to move electrons toward the most positive charge. In terms of work, it takes 0.7376 ft-lb to move 6.28×10^{18} electrons between any two points, or 0.7376 ft-lb to move one coulomb of charge.

Four examples of potential difference are shown in Figure 1-6. The charges are in coulombs (Q), and the electrons move upward in all cases. In (a), the lower plate is neutral, and the upper plate has a positive charge of 5 coulombs. The difference in charge is $5 - 0 = +5$ coulombs. In (b), the upper plate is neutral while the lower plate has a negative charge of 5 coulombs. The difference is $0 - (-5) = +5$ coulombs. If the upper plate were the reference, then the charge would be -5 coulombs. However, since electrons react to a difference in potential, they move in a positive direction. In the drawing, the neutral plate is more positive than the lower

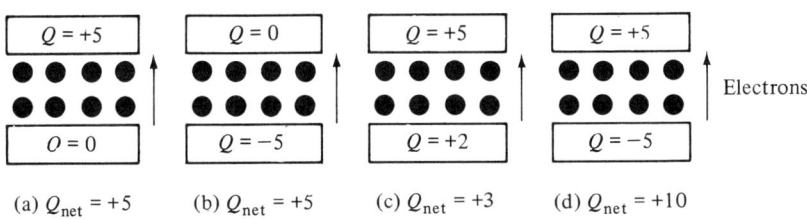

FIGURE 1-6 *Potential Difference in Coulombs*

plate. In (c), the upper plate is 3 coulombs more positive than the lower plate. Drawing (d) shows the upper plate having a positive charge of 5 coulombs while the lower plate has a negative charge of 5 coulombs. The net difference is $5 - (-5) = +10$ coulombs. The electrons travel in a positive direction from -5 to 0 (5 coulombs), then another 5 coulombs from 0 to $+5$.

The plates in Figure 1-6 could represent plates on each side of a dielectric, the opposite ends of a conductor, or the opposite ends of a semiconductor. The important point is that electrons move from negative to positive. Their movement through a material is erratic due to collisions with each other. The electrons drift much slower than they would in a vacuum where there are not any atoms to impede their movement. Therefore, electron flow is sometimes called **drift current**.

The unit of measure for potential difference is the **volt**, abbreviated v. One volt is the amount of work it takes to move one coulomb of charge. The symbol for potential difference is E, which represents electromotive force (emf). Potential difference, then, is the **voltage** between two points. A common flashlight battery has a potential difference of 1.5 volts (1.5 v), and an automobile battery has a potential difference of 12 v. Power outlets in the home have a potential difference of approximately 120 v between the two contact points.

If a potential is applied at each end of a conductor, the current flow is the same at all points along the conductor. If the potential is increased, more electrons flow so the current increases. Conversely, as the potential decreases, fewer electrons flow, and the current decreases. If the potential is zero, electrons cannot flow, and the current is zero.

1.6 Resistivity

Resistivity is a measure of the degree to which a material opposes or resists the flow of electric current. The resistivity of a material is determined by its atomic structure, the temperature of the material, and the density of free charge carriers that are available to move under external force such as a potential difference. Units of resistivity are ohms-centimeter (ohms-cm), which are independent of the geometry of the material. Table 1–2 lists the resistivity of several common materials. The materials are listed in the order of their resistivity. Notice the difference between conductors and insulators. Conductors are in the order of 10^{-6} ohms-cm, while insulators are 10^{12} to 10^{17} ohms-cm. Semiconductors fall into a group somewhere midway between conductors and insulators.

TABLE 1–2 *Resistivity of Common Materials*

Material	Resistivity (ohms-cm)	Remarks
Silver (Ag)	1.1×10^{-6}	Conductors
Copper (Cu)	1.7×10^{-6}	
Carbon (C)	4.0×10^{-3}	Semiconductors
Germanium (Ge)	4.6×10	
Silicon (Si)	2.3×10^3	
Glass	17×10^{12}	Insulators
Mica	1.0×10^{17}	

The resistivity of most metallic conductors increases with temperature, while the resistivity of non-metallic conductors decreases with temperature. The resistivity of tungsten lamp bulbs, for example, increases with temperature. The resistivity of carbon, however, decreases when its temperature increases.

1.7 Resistance

Resistance is similar to resistivity because it is also a measure of the degree to which a material opposes or resists the flow of electric current. Resistance is determined by the length and cross-sectional area, in addition to the atomic structure, the temperature of the material, and the density of free charge carriers. Units of resistance are in ohms, abbreviated Ω. One ohm is defined as the resistance offered to a steady electric current by a column of mercury 1 sq mm in cross-section and 106.3 centimeters (cm) long at 0°C. For example, an iron wire of those dimensions has a resistance of approximately 0.1 ohm. A formula for calculating the electrical resistance of a piece of wire is

$$R = \rho \frac{L}{A} \qquad (1\text{–}3)$$

where R is resistance in ohms; ρ is the specific resistance in ohms-cm; L is the length in cm and A is the cross-sectional area in cm².

Example 1-5 What is the resistance of a copper wire with a cross-sectional area of 0.01 cm² and 0.1×10^6 cm long?

Solution The specific resistance of copper from Table 1-2 is 1.7×10^{-6} ohms-cm.

$$R = \rho \frac{L}{A} = 1.7 \times 10^{-6} \frac{0.1 \times 10^6}{0.01}$$
$$= 1.7 \times 10^{-6} \times 10 \times 10^6 = 1.7 \times 10 = 17 \text{ ohms}$$

Example 1-6 How thick is a piece of mica having a cross-sectional area of 5 cm² if its resistance (through the thickness) is 2×10^{15} ohms?

Solution The specific resistance for mica (taken from Table 1-2) is 1.0×10^{17} ohms-cm.

$$L = \frac{RA}{\rho} = \frac{2 \times 10^{15} \times 5}{1 \times 10^{17}} = 2 \times 10^{-2} \times 5$$
$$= 10 \times 10^{-2} = 0.1 \text{ cm}$$

Copper has many free electrons, so its resistance is quite low and permits high current to flow (refer to Example 1-5). Semiconductors and insulators have fewer free electrons, so their resistance is high, and less current flows (refer to Example 1-6 for an insulator). As an example of work, one ohm develops 0.24 calories (cal) of heat when one amp flows for one sec.

Household light bulbs are examples of resistance and work done in terms of heat. A 60-watt (w) bulb has a resistance of 240 ohms and a current flow of 0.5 amp. A 150-watt bulb (which is much brighter) has a resistance of 96 ohms, but has a current flow of 1.25 amp. It would appear, and rightfully so, that as the resistance decreases, the current flow increases. The 150-watt bulb is brighter, has a lower resistance, and the current flow is higher.

1.8 Conductance

Conductance (abbreviated G) is the measure of the degree to which a material allows current to flow. It is the reciprocal of resistance.

$$G = \frac{1}{R} \tag{1-4}$$

Conductance units are mhos. A mho is ohm spelled backward, perhaps signifying the reciprocal.

Example 1-7 What is the conductance of a conductor if its resistance is 0.04 ohms?

Solution $$G = \frac{1}{R} = \frac{1}{0.04} = 25 \text{ mhos}$$

Example 1-8 What is the resistance of a semiconductor if its conductance is 2×10^{-3} mhos?

Solution $$R = \frac{1}{G} = \frac{1}{2 \times 10^{-3}} = 0.5 \times 10^3 = 500 \text{ ohms}$$

QUESTIONS

1-1 If an atom has 22 protons, how many electrons does it have?

1-2 How is the atomic number related to an atom?

1-3 How is the approximate atomic weight of an atom determined?

1-4 What is meant by an orbital band or shell?

1-5 What is meant by the term covalent bonding?

1-6 What are the two kinds of energy an orbiting electron has?

1-7 What are two common ways to increase the energy of an atom?

1-8 How many electrons are there in 50 coulombs?

1-9 How many coulombs do 15.7×10^{18} electrons represent?

1-10 What is the net charge on a dielectric that is first charged with 12.56×10^{18} electrons and then has a charge of 34.54×10^{18} protons added?

1-11 How much time is required for five coulombs to develop 25 amp?

1-12 How much current flows if 109.9×10^{18} electrons pass a given point in five seconds?

1-13 Refer to Figure 1-6(b). If the upper plate has a charge of -4 coulombs, and the lower plate has a charge of -13 coulombs, what is the net charge?

1-14 The text gives three examples of potential difference. What are at least two more?

1-15 Why does a potential difference cause current flow?

1-16 What is the difference between resistivity and resistance?

1-17 How long is a silver wire that measures 0.055 ohms with a cross-sectional area of 0.03 cm²?

1-18 What is the cross-sectional area of a 400-cm length of copper wire that has a resistance of 0.17 ohms?

1-19 What is the resistance of a material that has a conductance of 0.0004 mhos?

1-20 What is the conductance of a material that has a resistance of 12.5×10^6 ohms?

2 Magnetism

The principles of magnetism are practiced whenever mechanical energy is converted to electrical energy, or whenever electrical energy is converted to mechanical energy. A generator is an example of mechanical to electrical conversion. The generator is rotated by some mechanical means, and its output is a voltage potential. An automobile generator is operated from the engine. A P.G. & E. power generator is operated hydraulically from water pressure. Examples of electrical to mechanical conversion are solenoids, door bells, buzzers, and speaker systems.

2.1 Magnets

Three types of common magnets are shown in Figure 2–1. A bar magnet, (a), is a common magnet that might be found around the home; they come in various shapes and sizes. Figure 2–1 (b) illustrates a rod type magnet which is commonly used as core material for electrical coils and solenoids. The horseshoe magnet, (c), is considerably stronger, given the same material and magnetizing strength, because the north and south poles are closer together.

The lodestone is a natural magnet and was the first to be used by man. If a lodestone is suspended, as in Figure 2–2(a), it will swing around and align itself with the earth's magnetic north and south poles. A bar magnet, (b), is an artificial magnet that demonstrates the same characteristics. An artificial magnet can be made by stroking a magnetic material with a natural (lodestone) magnet or another artificial magnet. In drawing (c), a compass is shown for comparison. Notice that magnets have a north and a south pole.

The north pole of a magnet tends to seek the earth's north pole, so the three magnets of Figure 2-2 are pointing toward the upper-right, which would be the north pole of the earth. If two magnets are brought close together, and their poles are opposite (north-to-south), there will be an attraction between them. But if like poles are brought close to each other (north-to-north or south-to-south), there will be a repelling of one from the other. So it can be said that unlike poles attract, and like poles repel. Each pole (N and S) has equal strength. And the force of attraction or repulsion decreases inversely as the square of the distance between the poles.

The characteristic of magnets having separate poles is retained when a magnet is broken into smaller pieces. Each piece will be a separate magnet with its own north and south poles. Magnets have a power of attraction or repulsion across a distance. This power is unchanged when a non-magnetic material is placed in the magnetic field. Materials that are considered non-magnetic are paper, glass, wood, aluminum, copper, brass, gold,

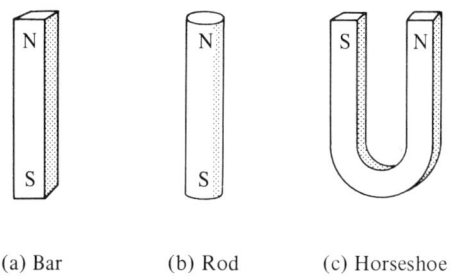

(a) Bar   (b) Rod   (c) Horseshoe

FIGURE 2-1 *Magnets*

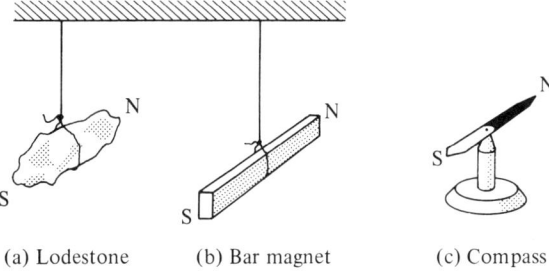

(a) Lodestone   (b) Bar magnet   (c) Compass

FIGURE 2-2

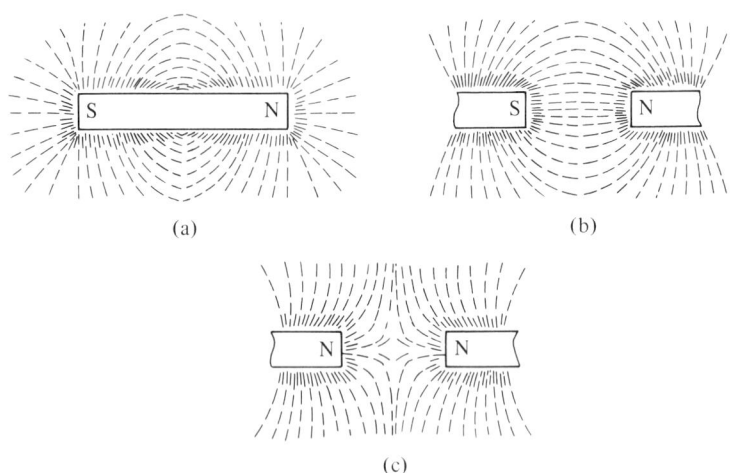

FIGURE 2-3 *Magnetic Lines of Force*

silver, and, of course, air or a vacuum. Steel and iron of various alloys are examples of strong magnetic materials.

Because they are polarized, magnets develop very distinct magnetic fields. Figure 2-3 (a) is a drawing of what iron filings would look like if sprinkled on a piece of paper placed over the top of a magnetic bar. Notice that the lines of force flare out at the ends, and the tendency is to complete a path between the poles. The power of attraction would be stronger close to the magnet, causing the concentration of "filings" adjacent to the bar. In (b), the lines of force are drawn for unlike poles. The lines are strongest directly between the north and south poles. However, they take on the character of (a) at the sides (above and below the poles). Like poles, north to north, are drawn in (c). Half-way between the poles, the lines of force bend and flare out. This phenomenon illustrates the act of repulsion because the lines do not connect in a continuous fashion as in drawing (b).

The force field around a magnet has a specific north to south direction. The field is strongest inside the magnet, and its direction is south to north. Figure 2-4 demonstrates the directional behavior of the lines. The external and internal force lines form a continuous loop. If compasses were placed as shown, their north poles would point toward the south pole of the magnet. Conversely, their south poles would point toward the north pole of the magnet. And if a compass were moved around the magnet, it would trace the directional characteristics of the force field.

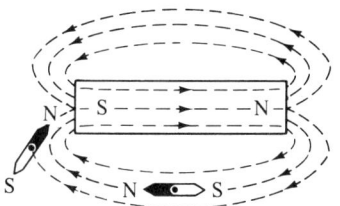

FIGURE 2-4 *Direction of Magnetic Lines of Force*

2.2 Magnetic Theory

The atomic structure of magnetic metals follows the same theory of electrons orbiting around a nucleus as explained in Chapter 1. In addition, each of the electrons generates a magnetic field. The most obvious field is the result of the orbit (see Figure 2-5). Every electron of every atom produces its own orbital fields, but the orbits are usually random because the electrons orbit in various directions. This causes a canceling effect so that the net field is quite weak.

Another field is created by the action of each electron spinning on its own axis. This field resembles a minute permanent magnet. Figure 2-5 has two electrons spinning in the same direction. This results in a good magnetic field because the net field is additive. And, when a material is magnetized, electrons are much more likely to spin in the same direction than to maintain similar orbital paths.

If each of the various electron fields were visualized as a small pin

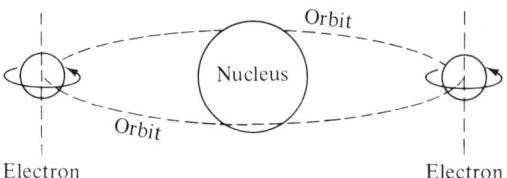

FIGURE 2-5 *Magnetic Effect of Electrons*

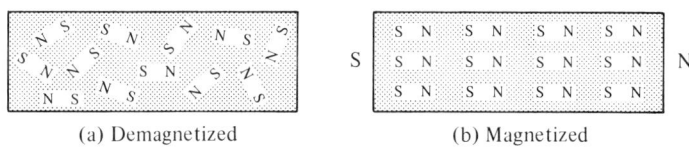

FIGURE 2-6 *Magnetic Particle Theory*

magnet, a piece of material might resemble the simplified drawing in Figure 2-6(a). The magnets are shown in random fashion to represent a nonmagnetic state. If the material is magnetized by external means, stroking with another magnet, for instance, the tiny magnets tend to align themselves in a single direction as shown in (b), establishing the north and south poles of the material.

2.3 Magnetic Flux

The strength of a magnet is measured in terms of magnetic flux lines. The total number of flux lines would then be the total strength, abbreviated ϕ. Figure 2-7 shows a total of six lines.

$$\phi = \text{total number of lines} \qquad (2\text{-}1)$$

The strength ϕ, then, is six lines. In the CGS system of measurement, lines are called **maxwells.**

$$1 \text{ line} = 1 \text{ maxwell} \qquad (2\text{-}2)$$

The figure has a strength of six lines or six maxwells. The MKS system of measurement uses a larger unit called the **weber**. One weber equals 1×10^8 lines or maxwells.

$$\phi \text{ (webers)} = 1 \times 10^8 \text{ lines or maxwells} \qquad (2\text{-}3)$$

Example 2-1 If a permanent magnet has 2500 magnetic flux lines, what is its strength in the CGS system and in the MKS system?

Solution In the CGS system one line equals one maxwell. Therefore, 2500 lines equals 2500 maxwells. In the MKS system 1×10^8 lines equals one weber.

$$\text{webers} = \frac{2500 \text{ lines}}{1 \times 10^8 \text{ lines}} = 2500 \times 10^{-8}$$
$$= 25 \times 10^{-6} = 25\mu \text{ webers}$$

Example 2-2 If a magnet has a strength of 3000μ webers, what is its strength in maxwells?

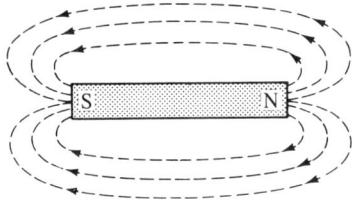

FIGURE 2-7 *Total Flux* ϕ

[2.3] Magnetism

Solution
$$\text{Maxwells} = \text{webers} \times 10^8 = 3000 \times 10^{-6} \times 10^8$$
$$= 0.3 \times 10^6 \text{ maxwells}$$

A more common method of measuring magnetic strength is in terms of **flux density**, abbreviated B. **Flux density** is the number of magnetic field lines per unit area. The area is taken as a cross-section perpendicular to the direction of the lines.

$$B = \frac{\phi \text{ (lines or maxwells)}}{A \text{ (cm}^2\text{)}} \qquad (2\text{--}4)$$

Figure 2–8 illustrates a cross-section of one cm² with three lines passing through it. Flux density for the figure would then be three lines per cm². The unit for flux density is called a **gauss**. A gauss is one line per cm². The flux density of Figure 2–8 in gauss is

$$B = \frac{3 \text{ lines}}{\text{cm}^2} = 3\frac{\text{lines}}{\text{cm}^2} = 3 \text{ gauss}$$

Example 2–3 If an area 5 cm by 10 cm contains 600 maxwells, what is the flux density in gauss?

Solution
$$B = \frac{600 \text{ maxwells}}{5 \text{ cm} \times 10 \text{ cm}} = \frac{600}{50} \times \frac{\text{maxwells}}{\text{cm}^2} = 12 \text{ gauss}$$

Example 2–4 An area of 40 cm² measures 50 gauss. How many lines of magnetic flux intersect the cross-section?

Solution
$$\text{lines} = B \times \text{cm}^2 = 50 \text{ gauss} \times 40 \text{ cm}^2 = 2000 \text{ lines}$$

Example 2–5 A measurement of 150 gauss covers an area that contains 30μ webers. What is the area?

Solution
$$\text{lines} = \text{webers} \times 10^8 = 30 \times 10^{-6} \times 10^8$$
$$= 30 \times 10^2 = 3000$$

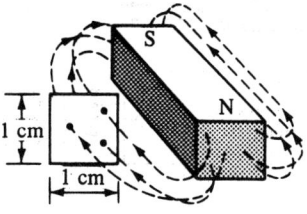

FIGURE 2–8 *Flux Density B*

$$A\,(cm^2) = \frac{\phi\,(lines)}{B} = \frac{3000\ lines}{150\ gauss} = 20\ cm^2$$

Example 2-6 How many gauss does an area of 1.75 cm² contain if there are 3.5×10^{-6} webers within that area?

Solution lines = webers × 10^8 = $3.5 \times 10^{-6} \times 10^8$ = 350 lines

$$B = \frac{\phi\,(lines)}{A\,(cm^2)} = \frac{350\ lines}{1.75\ cm^2} = 200\,\frac{lines}{cm^2} = 200\ gauss$$

2.4 Magnetic Induction

Magnetic induction occurs when a magnet is brought into the proximity of a piece of magnetic material such as iron or steel. The act of induction causes the magnetic material to become magnetized because the lines of force pass through it and align the internal "magnets." Figure 2-9 illustrates induction. The two pieces of iron are placed in the path of the force lines. One is at the end of the permanent magnet. Notice that the lines become concentrated when they contact the iron. The iron is then magnetized in the same direction as the permanent magnet. The force lines come out of the north pole of the iron and circle back through a second piece of iron. Again the iron concentrates the field and becomes magnetized. But the second piece of iron is magnetized south to north from right to left in the direction of the lines.

When a material is magnetized, the strength of the magnetized material is determined by the number of lines that pass through it. For example, if 100 lines enter the end, and the end is 2 cm², the flux density is calculated from the formula (2-4).

$$B = \frac{\phi}{A} = \frac{100\ lines}{2\ cm^2} = 50\,\frac{lines}{cm^2}$$
$$= 50\ gauss$$

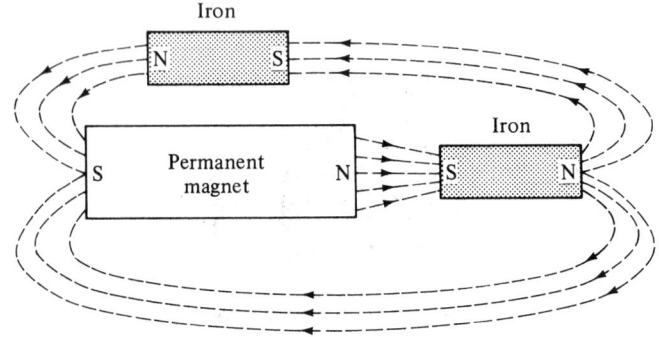

FIGURE 2-9 *Magnetic Induction*

The most effective way to magnetize a material by magnetic induction is to bring the material up against a permanent magnet. In this way, the strongest lines of force will act upon the material.

2.5 Permeability

Permeability is the ability of a material to concentrate magnetic flux; the symbol for permeability is μ. If a given material is easily magnetized, it has high permeability. High permeability allows a large amount of flux to be produced by a magnetic force.

Since air and vacuum do not change a magnetic field, they have a reference permeability of one. Materials like steel and iron have permeabilities in the order of 10^2 to 10^4. Magnetic flux follows the direction of least resistance. So if a material of higher permeability than air or vacuum is placed close to a magnet, the lines of force will actually swerve in order to pass through the high μ material. This factor can be used to a distinct advantage when attempting to induce magnetic lines in another material. For example, an air gap can be made small so that the field is more concentrated and thus stronger.

Three types of air gaps are illustrated in Figure 2-10. The first, (a), is the length of a bar magnet. If the bar magnet were bent into the shape of a horseshoe, as in (b), the air gap would be much smaller, and the field would be much stronger. Large air gaps allow the lines to spread out, which weakens the field. A narrow air gap is desired for maximum induction. The material is then placed in or adjacent to the gap so that the field is concentrated into the material. The drawing in Figure 2-10(c) illustrates the basic shape of magnetic recorder heads. The air gap is made quite small, and the magnet is shaped to help direct the lines downward. If a piece of magnetic tape, such as that used in a tape recorder, is passed across the gap (bottom of the magnet), the force field bends downward to pass through the magnetic coating.

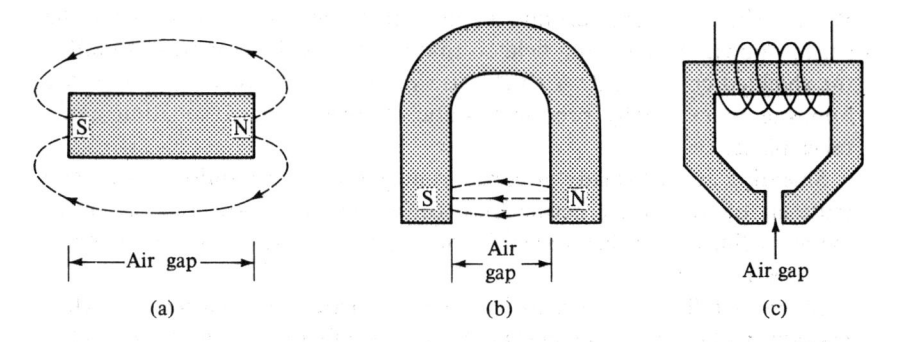

FIGURE 2-10 *Examples of Air Gaps*

A magnet such as the one in (c) is usually made of soft iron. Soft iron has the characteristic of becoming highly magnetized by induction, but removing the field causes the soft iron to lose its magnetism.

2.6 Magnetic Materials and Types of Magnets

Magnetic materials fall into several classifications. **Ferro-magnetic** materials are strongest in terms of magnetic strength. Examples are steel, iron, cobalt, nickel, and commercial alloys such as alnico and permaloy. These materials all magnetize in the same direction as the magnetizing field. They are also all high permeability materials in the order of 10^2 to 10^4. There are certain alloys that have a permeability as high as 10^5 although they tend to saturate easily at low values of flux density.

Ferrite materials are made of ceramics having the ferromagnetic properties of iron. Ferrites exhibit high permeability, so they are easily saturated at low magnetizing values of flux density. The chief advantage of ferrite is its insulating quality, whereas iron is a conductor. The losses due to "eddy currents" are minimal. Eddy currents occur when current is induced by a magnetic field and flows into the magnetic material. Eddy currents induce a field that opposes the principle field, and the result is some loss of overall field strength.

Paramagnetic materials are aluminum, chromium, platinum, and manganese. Their permeability is hardly more than one, and, as a result, their magnetic strength is very weak. As with ferromagnetic materials, paramagnetic materials magnetize in the same direction as the magnetizing field. Some of the **diamagnetic** materials are copper, zinc, mercury, gold, and silver. Their permeability is less than one, and they can be very weakly magnetized, but only in a direction opposite to the magnetizing field.

Magnetic shielding materials are used to isolate the effects of one material on another. A shield has to be a good conductor (metal) to prevent the induction of varying magnetic fields or static electric charges. When a shield is placed around a component, any induced currents in the shield oppose the inducing field so that the net field is too weak to affect the shielded component. A shield also provides induced charges that are opposite to externally produced static charges, and cancellation again takes place.

Shielding against steady or constant magnetic fields requires a magnetic material of very high permeability in which to concentrate as much magnetic flux as possible. Examples of steady fields are permanent magnets and electromagnets.

Magnets fall into two broad categories: **permanent magnets** and **electromagnets**. Permanent magnets (PM) are usually made of alnico (an alloy of aluminum, nickel, and cobalt) and cobalt steel. Magnets made of these materials will retain their strength indefinitely, unless upset by an external field, temperature, or shock, any one of which is sufficient to rearrange the molecular structure within the material. Typical permanent magnets have a flux density B of 10^2 to over 10^4 lines/cm^2 or gauss.

Permanent magnets in everyday use range from those used on automatic can openers to cupboard door latches to speakers in radio and television sets.

Electromagnets are normally made of soft iron which is easily magnetized, has high permeability, and loses its magnetism when the magnetizing force is removed. An electromagnet is made with a coil of wire wrapped around the iron core. When current flows through the wire, a magnetic field is set up which becomes concentrated in the iron. The iron, as a result, becomes highly magnetized as long as the current flows. When the current is stopped, the iron loses its magnetism. The magnetic strength can be increased by increasing the current and/or the number of turns around the core.

Electromagnets are used for door bells, door buzzers, and relays. Industrial applications include large lifting magnets capable of lifting more than a ton of magnetic material.

2.7 Electric Field

Several references have been made to the magnetic field created by the flow of current through a conductor. Such a field is concentric out from the conductor, with decreasing strength as the square of the distance away from the conductor. The effect is shown in Figure 2-11. The direction of the field is dependent upon the direction of the current. When electron current flows from minus to plus and upwards, the magnetic lines of force take a direction that is clockwise.

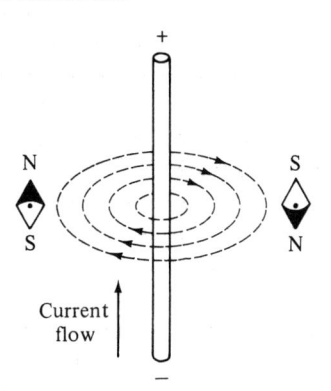

FIGURE 2-11 *Magnetic Field around a Conductor*

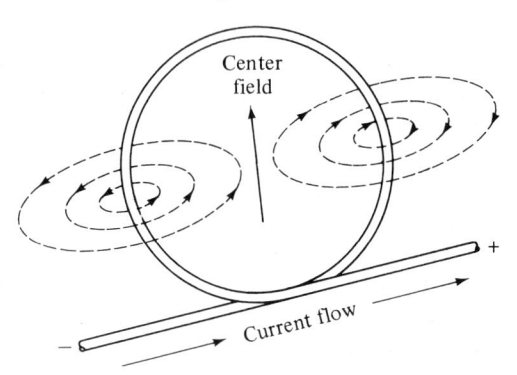

FIGURE 2-12 *Field around a Loop*

A method of determining the direction of the field or the current is called the **left-hand rule**. Grasp the conductor in the left hand with the thumb pointing in the direction of electron current flow (− to +). The fingers then point in the direction of the field. Conversely, if the field is known (through the use of a compass), grasp the conductor in the same way with the fingers pointing in the direction of the field. The thumb then points in the direction of electron current flow.

The same left-hand rule can be applied to the loop in Figure 2-12. Two concentric fields are shown; the one on the left side is counter-clockwise, and the one on the right side is clockwise. The same field effect is evident all the way around the loop. However, note the effect in the center. The fields merge with a common direction out through the loop in the drawing. If a magnetic material were placed inside the loop, the magnetic flux field would pass through it in a single direction, causing it to become magnetized. The various PM materials would become permanently magnetized. Soft iron, however, would be magnetized only as long as it was held inside the loop or as long as the current continued to flow.

If a soft iron bar or rod is magnetized in the loop, it has the power of attraction and induction in the same way a permanent magnet has. The same laws can be used that were applied to total flux, flux density B, and permeability μ. A stronger magnetic field can be created by increasing the number of loops so that the field is effective along the full length of the iron. The flux field is then continuous through the coil as shown in Figure 2-13. Also, the field loops back in a continuous loop in a pattern similar to a bar magnet.

A variation of the left-hand rule can be applied to the coil in Figure 2-13. The fingers are placed on the coil pointing in the direction of electron current flow. The thumb points in the direction of the field which is also the

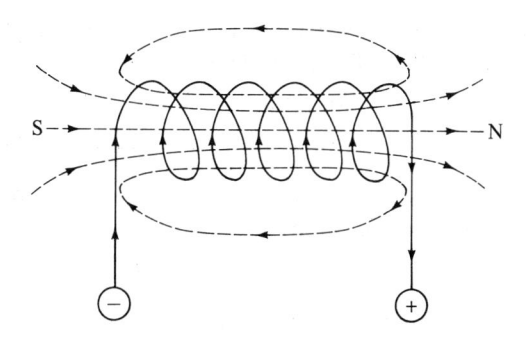

FIGURE 2-13 *Field through a Coil*

north pole. Conversely, if the direction of the field or pole is known, the left thumb can point in that direction while the fingers trace the current flow.

Suppose a coil were wrapped around a length of non-magnetic tubing, and suppose the tubing were longer than the coil. Then a soft iron rod could be slipped into the tubing so the tubing would act as a guide. The moment current flowed through the coil, the magnetic field of the coil would attract the rod into the coil and hold the rod there in its center where the field would be the strongest. A magnetic field produced by current flow through a conductor has the same properties as a magnetic field produced by a permanent magnet. Therefore, the principle of attractive force of the coil is the same as for a permanent magnet.

The two type of electric solenoids that have been introduced have magnetic properties. The sliding type was just discussed. The iron core type (Figure 2-14) operates by attracting another magnetic material. A relay operates in this manner. A metal bar pivots against a spring whenever the solenoid is energized. When deenergized, the bar pivots back. Electrical circuit contacts then open and close as a function of the solenoid.

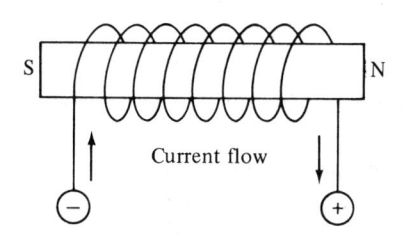

FIGURE 2-14 *Solenoid*

Ampere-turns is a term given for determining the strength of one coil in comparison to another. The formula is

$$\text{Amp-turns} = NI \tag{2-5}$$

where N is the number of turns, and I is the current in amperes.

Example 2-7 What is the amp-turns rating of a solenoid that has 600 turns and 0.1 amp of current?

Solution $\text{Amp-turns} = NI = 600 \times 0.1 = 60$

Example 2-8 If the amp-turns rating of a solenoid is 60, and the current is 10 amp, how many turns are there?

Solution $$N = \frac{\text{amp-turns}}{I} = \frac{60}{10 \text{ amp}} = 6 \text{ turns}$$

Examples 2-7 and 2-8 have the same amp-turns rating, which proves they must be of equal strength. The formula and comparison is perfectly valid, even though the solenoid in Example 2-8 has one hundred times as much current. Note that the solenoid in Example 2-7 has one hundred times as many turns. Thus, the strength of a solenoid is directly proportional to the number of turns and the current.

2.8 Field Intensity

Field intensity is the magnetic force per unit length of its path. Intensity is measured in oersteds which has a symbol of H. Since a permanent magnet and an electromagnet generate the same kind of field, the oersted is valid for both.

The field intensity around a single current-carrying wire is calculated

$$H = \frac{2I}{10r} \tag{2-6}$$

Where H is in oersteds, I is in amperes, and r is the radius in cm from the center of the wire [see Figure 2-15 (a)].

The intensity at the center of a current-carrying loop is calculated

$$H = \frac{2\pi I}{10r} \tag{2-7}$$

Again, H is in oersteds, I is in amperes, and r is the radius in centimeters [see Figure 2-15 (b)].

The intensity inside a multiple loop coil is

$$H = \frac{2\pi NI}{10r} \tag{2-8}$$

where N is the number of turns. The other symbols are the same as for formulas (2-6) and (2-7) [see Figure 2-15 (c)].

[2.8]   Magnetism   27

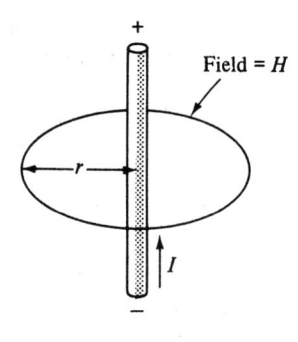

(a) $H = \frac{2I}{10r}$

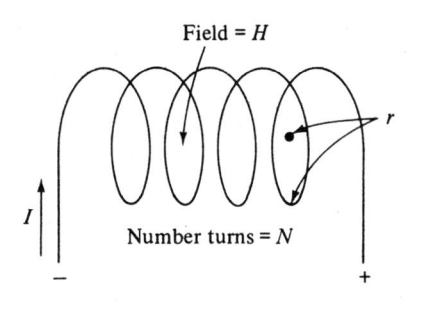

(c) $H = \frac{2\pi NI}{10r}$

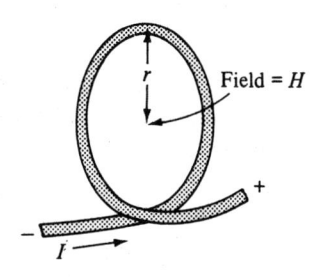

(b) $H = \frac{2\pi I}{10r}$

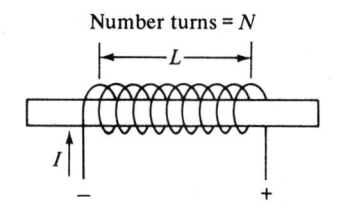

(d) $H = \frac{4\pi NI}{10L}$

FIGURE 2–15 *Field Intensity Formulas*

For the center of a long solenoid with a small radius in comparison to its length,

$$H = \frac{4\pi NI}{10L} \quad (2\text{-}9)$$

where L is the length of the wound portion of the solenoid in cm [see Figure 2-15 (d)].

Examples of formulas (2–6) through (2–9) are as follows.

Example 2–9   What is the field intensity at a radius of 1.8 cm from a straight wire with a current of 36 amp?

Solution   From formula (2–6),

$$H = \frac{2I}{10r} = \frac{2 \times 36}{10 \times 1.8} = \frac{72}{18} = 4 \text{ oersteds}$$

28 Magnetism [2.8]

Example 2–10 What is the radius of a single loop with a current flow of 30 amp and a field intensity of 6.28 oersteds at its center?

Solution Rearranging formula (2–7) gives

$$r = \frac{2\pi I}{10H} = \frac{2\pi \times 30}{10 \times 6.28} = \frac{30}{10} = 3 \text{ cm}$$

Example 2–11 What is the field intensity inside a 200-turn coil with a radius of 0.4 cm and a current of 2.5 amp?

Solution Using formula (2–8) yields

$$H = \frac{2\pi NI}{10r} = \frac{2\pi \times 200 \times 2.5}{10 \times 0.4}$$
$$= \frac{1000\pi}{4} = 250\pi = 784 \text{ oersteds}$$

Example 2–12 How many turns of wire are required to make a coil of π radius so that 50 milliamperes (ma) of current will produce an internal field of 15 oersteds?

Solution Rearranging formula (2–8) gives

$$N = \frac{10rH}{2\pi I} = \frac{10\pi \times 15}{2\pi \times 0.05} = \frac{150}{0.1} = 1500 \text{ turns}$$

Example 2–13 What is the field intensity of a solenoid with 180 turns over a length of 12 cm, a radius of 1.5 cm, and a current of 0.32 amp?

Solution From formula (2–9) the following is obtained:

$$H = \frac{4\pi NI}{10L} = \frac{4\pi \times 180 \times 0.32}{10 \times 12}$$
$$= \frac{230\pi}{120} = 6 \text{ oersteds}$$

Example 2–14 What is the required length of a solenoid if the center is to be 220 oersteds, the current is to be 1.1 amp, and the number of turns is to be 1000?

Solution Rearranging formula (2–9) yields

$$L = \frac{4\pi NI}{10H} = \frac{4\pi \times 1000 \times 1.1}{10 \times 220}$$
$$= \frac{4400\pi}{2200} = 2\pi = 6.28 \text{ cm}$$

Field intensity H is related to flux density B and permeability μ in the following way:

$$H = \frac{B}{\mu} \qquad (2\text{-}10)$$

where H is in oersteds, and B is in gauss or lines per cm², and μ is the permeability factor.

Example 2-15 What is the field intensity of an iron bar with a flux density of 300 lines per cm² and a permeability of 1200?

Solution $\qquad H = \dfrac{B}{\mu} = \dfrac{300}{1200} = 0.25$ oersteds

Example 2-16 What is the flux density of a material with a permeability of 20 and a field intensity of 4 oersteds?

Solution $\qquad B = \mu H = 20 \times 4 = 80$ gauss or 80 lines/cm²

2.9 Reluctance

Opposition to the production of flux is called **reluctance**, abreviated $\mathcal{R}$. Permeability is the ability to concentrate flux or the ease by which flux can be produced. Reluctance is therefore inversely proportional to permeability. For example, iron has high permeability but low reluctance; air has low permeability but high reluctance.

Because reluctance is the opposite of flux ϕ, and because flux is produced by ampere-turns, they are related as follows:

$$\mathcal{R} = \frac{mmf}{\phi} \qquad (2\text{-}11)$$

where ϕ is lines or maxwells, and mmf is *magnetomotive force* in ampere-turns. In the C.G.S. system of measurement, 1 amp-turn equals 1.256 gilberts. Reluctance is then

$$\mathcal{R} = \frac{\text{gilberts}}{\text{maxwells}} \qquad (2\text{-}12)$$

Example 2-17 What is the reluctance in amp-turns per line of an iron core with 1000 turns, 0.3 amp, and a total flux of 150 lines?

Solution $\qquad \mathcal{R} = \dfrac{NI\,(\text{amp-turns})}{\phi\,(\text{lines})} = \dfrac{1000 \times 0.3}{150} = \dfrac{300}{150}$
$\qquad\qquad\quad = 2$ amp-turns per line

Example 2-18 What is the reluctance of Example 2-17 in terms of gilberts per maxwell?

Solution

$$\phi = 150 \text{ lines} = 150 \text{ maxwells}$$
$$NI = 300 \text{ gilberts} = 1.256 \times 300 = 377$$
$$\mathscr{R} = \frac{\text{gilberts}}{\text{maxwells}} = \frac{377}{150} = 2.51 \text{ gilberts/maxwell}$$

Example 2-19 If the reluctance of a 25-turn coil is 12.56 gilberts/maxwell, and the total flux is 10 maxwells, what is the total current?

Solution

$$\text{mmf} = \mathscr{R}\phi = 12.56 \frac{\text{gilberts}}{\text{maxwell}} \times 10 \text{ maxwells}$$
$$= 125.6 \text{ gilberts}$$
$$NI = \frac{\text{gilberts}}{1.256} = \frac{125.6}{1.256} = 100 \text{ amp-turns}$$
$$I = \frac{\text{amp-turns}}{N} = \frac{100}{25} = 4 \text{ amp}$$

2.10 B and H Curve

Flux density B and field intensity H are often displayed in graphical form to visually analyze the effects of an increasing field. A given material can only be magnetized to a certain point and increasing the field causes little change in flux density. This point is known as **saturation**. The effect is illustrated in Figure 2-16 where saturation begins at about 40 oersteds.

Table 2-1 summarizes the data used to plot the B and H curve. The data are for an iron core solenoid 12.56 cm long with 1000 turns. Per-

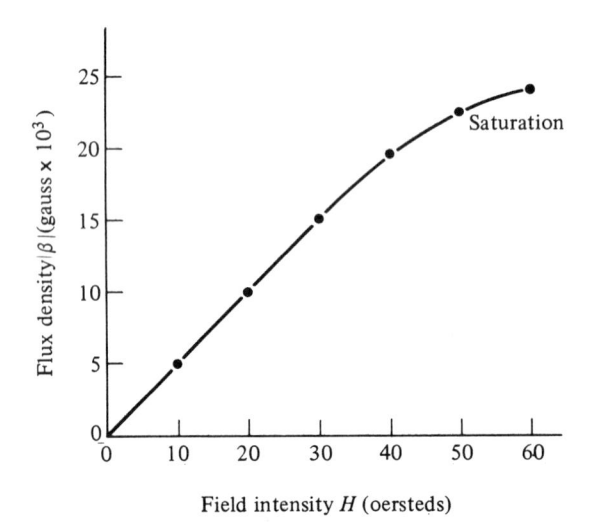

FIGURE 2-16 *B and H Curve*

TABLE 2-1 *Data For Figure 2-16, B-H Curve*

L length-cm	N turns	I amp	NI amp-turns	μ permeability	H oersteds	B gauss
12.56	1000	0.1	100	500	10	5,000
12.56	1000	0.2	200	500	20	10,000
12.56	1000	0.3	300	500	30	15,000
12.56	1000	0.4	400	490	40	19.600
12.56	1000	0.5	500	450	50	22,500
12.56	1000	0.6	600	400	60	24,000

meability of the iron is 500, but this will decrease when it begins to saturate. Formula (2-9) is used to calculate the field intensity H.

$$H = \frac{4\pi NI}{10L}$$

Since 4π is 12.56, and L is 12.56, the formula can be simplified for the table,

$$H = \frac{NI}{10}$$

so that values for the H column are one-tenth of those in the NI column. Field intensity is calculated by rearranging formula (2-10).

$$B = \mu H$$

Notice that at 400 amp-turns, the permeability decreased by 10. It decreases at a faster rate as the current I increases. If the current in the example continued to increase beyond 0.6 amp, the flux density would stop increasing and remain at a value somewhere just under 25,000.

In summary, the field intensity of the solenoid was produced as a result of its length, number of turns, and the amount of current. And as the length and turns remained constant, the intensity was increased in direct proportion to the increasing current. The flux density was therefore the product of permeability and field intensity. As long as the permeability remained constant at 500, the density B increased in direct proportion to the intensity H. However, as the iron became saturated, the permeability of the iron began to fall off, which reduced the rate at which the flux density increased.

2.11 Hysteresis

The term **hysteresis**, as applied to magnetism, means a retarding or lagging in magnetic strength due to a changing magnetic force. For instance, if the current in a solenoid made an abrupt change, the field intensity H

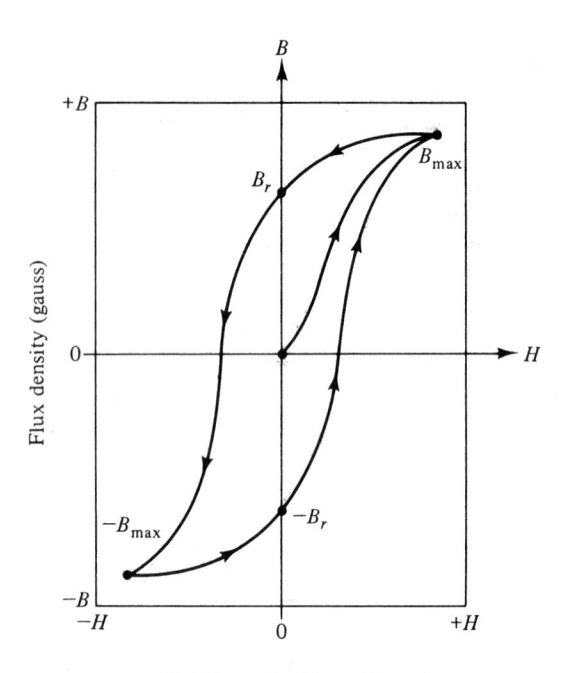

FIGURE 2-17 *B-H Hysteresis Loop*

would follow the change. But the flux density B through the iron core would experience a delay so that B would lag H by some amount. The delay or lag occurs because the molecular structure of the magnetic material cannot respond as rapidly as a current change.

Hysteresis is illustrated graphically in Figure 2-17. The curve is similar to the one in Figure 2-16. Field intensity H is shown horizontally, and flux density is shown vertically. The curve starts at the center when both H and B are at zero. As the H increases, the B increases until the maximum field is generated, resulting in a maximum flux density B_{max}. As long as the field is maintained, the flux will remain at B_{max}. As soon as H decreases to zero, the flux will lag to a residual point B_r. If the magnetic material is the permanent type, the flux will fall to B_r and remain there when the field is removed, hence, the term residual.

Assuming the magnetic material is soft iron and the field is reversed, the curve will continue from B_r, across the horizontal zero, and to a maximum negative flux density of $-B_{max}$. As H returns to zero, the flux will decrease to a residual point $-B_r$. And if the material is a PM type, the flux will remain at $-B_r$. Iron will trace a path in a direction toward B_{max}, since the flux in iron will tend to follow the field.

The curve, B_{max} to B_r to $-B_{max}$ to $-B_r$ to B_{max}, is the actual representation of what the flux does when the field is alternated between negative and positive values. Since the lag occurs as a result of the work to align and reverse the molecular structure of the material, a certain amount of heat is produced. The heat shows up as an energy loss known as **hysteresis loss**. Slow reversals of the field H have less hysteresis loss than fast reversals. And soft iron has a lower loss than steel because it is easier to change polarity in iron. In fact, for a slow reversal of H around soft iron, the hysteresis loss can be considered negligible.

Recall that a permanent magnet will remain magnetized at a point B_r or $-B_r$ when the field is removed. De-magnetizing a PM magnet is achieved by applying an alternating field with a coil so the flux will be forced to follow a pattern similar to that in Figure 2-17. As the field coil is slowly withdrawn, the amplitudes of H and B will slowly decrease. When the coil is far enough away, the field will no longer affect the magnet, and the flux will have been reduced to zero. This technique is often called **degaussing** a magnet.

2.12 Field Force on a Conductor

When a current-carrying conductor is placed in a magnetic field, the field produced by the conductor must interact with the magnetic field in which it was placed. The principle is similar to magnetics where like poles repel and unlike poles attract. That is, magnetic fields that are in the same direction add and become stronger. Fields in opposite directions cancel and become weaker. Motion, then, takes the path of least resistance from the stronger to the weaker field.

To achieve maximum interaction, a conductor must be placed at a right angle (90°) to the external field. In that way, the circular field of the conductor will, in varying degrees, be either with the external field or against it. A conductor is shown perpendicular to the page in Figure 2-18(a). The conductor has a dot in the center to indicate current coming out of the page. Applying the left-hand rule, the thumb points out from the page, and the circular field is clockwise. The external magnetic field is from north to south. The lines tend to cancel above the conductor while they add below the conductor. Hence, the stronger field is below and forces the conductor upward. If the conductor were stationary, and the magnet were free to move, the magnet would be forced downward.

Figure 2-18(b) illustrates a loop of wire fashioned so it can rotate. The loop is placed between two poles of a PM magnet. The left-hand conductor

Conductor forced upward

(a) Single conductor

Forced up   Forced down

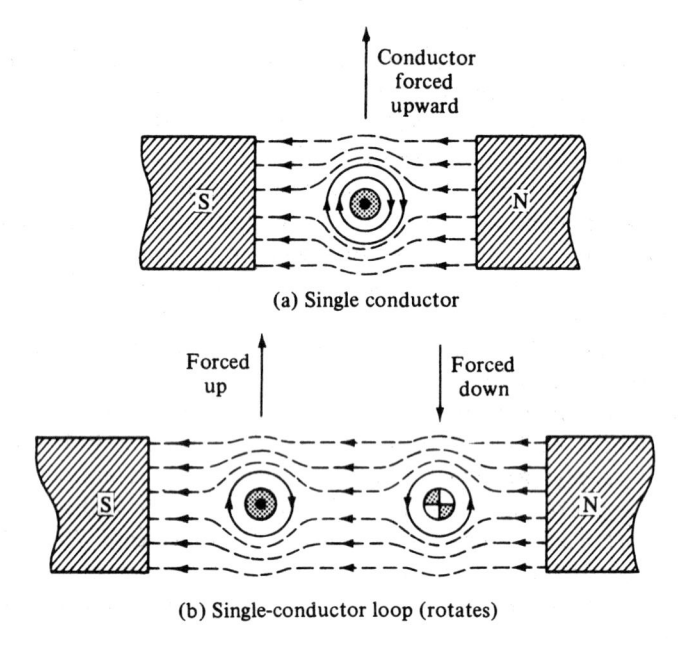

(b) Single-conductor loop (rotates)

FIGURE 2–18 *Field Force on Current-Carrying Conductors*

shows current coming out and a clockwise field. The right-hand conductor shows current going in and a counterclockwise field. The left side will be forced up as it was in drawing (a). The opposite is true on the right side. The field above the conductor adds and is stronger while the field below it cancels. The right-hand conductor is forced down. The result of the motion is a circular movement in a clockwise direction.

The illustrations with a single conductor and a single loop are also valid for multiple loops, or coils. When a coil is used, the individual effects are additive, which creates more force to move the coil (or magnet). But only the portions of the wires that are at right angles to the external field interact. Portions of wires that parallel the external field have no effect.

The force of magnetic fields that causes motion is called **torque**, which is multiplied by the number of turns. Torque is also increased by increasing the current and/or the external field strength H. Another way to increase torque is to increase the area of the loop (coil) which in turn causes an increased leverage.

2.13 Induced Current

If a straight conductor is placed 90° in an external magnetic field, current can be induced by moving the field or the conductor. The movement of magnetic flux across the conductor forces free electrons to move in a direction governed by the direction of the external field and the direction of movement. In Figure 2-19, the conductor is moved downward, or the magnet is moved upward. Either way, the external field resists movement, so the stronger (additive) field is under the conductor. The forced field around the conductor is somewhat circular and induces current flow in the conductor. The field around the conductor is clockwise in the drawing. With a clockwise field, the left-hand rule shows electron current coming out of the page.

Another rule, known as **Flemming's left-hand rule**, can be used to determine current flow. The index finger points to the south pole of the external field magnet. The thumb points in the direction the conductor moves, and the middle finger is held at a right angle and points in the direction of electron current flow.

Lenz's law helps explain the action of induction. It says in effect that electromagnetically induced currents have a direction such that the field created by the induced currents will tend to oppose the motion that produces them. Notice that in Figure 2-19 the field around the conductor opposes (additive effect) the external field underneath the conductor.

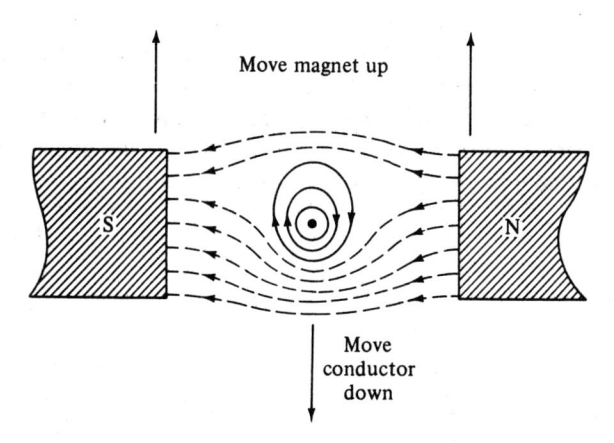

FIGURE 2-19 *Current Flow Due to Movement Against Field*

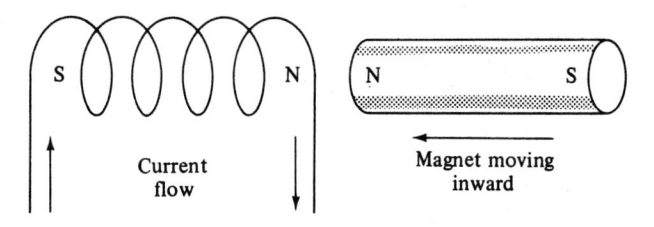

FIGURE 2-20 *Induced Current*

Figure 2-20 is an example of induced current as the result of moving a magnet into a coil. The polarity of the magnetized bar is north at the end going into the coil. According to Lenz's law, the coil must set up a reaction opposing the motion. Therefore the right end of the coil will also have a polarity of north. Induced current is achieved by the same principles used for a single conductor. The moving flux forces free electrons to move, and the direction of the electrons is determined by one of the earlier left-hand rules. The thumb points to the north pole, and the fingers indicate current flow.

2.14 Induced Voltage

The motion between a conductor and an external field also produces a voltage potential. In Figure 2-21 the magnet is shown moving upward. Notice the congestion of lines beneath the conductor. Flemming's left-hand rule still applies. Even though the magnet moves, the effect is the same as though the conductor moves. The index finger points to the south pole; the thumb points down, and the middle finger, held at a right angle, points in the direction of electron flow.

If the conductor is open, the electrons will "pile up" on the end and produce a negative charge. The other end will be deficient in electrons and will be charged positively. The charges would neutralize, of course, if the motion causing flux changes were to stop. If a coil were used in the same way, each turn would produce its own emf. The individual emf's are additive, which results in a larger overall voltage potential.

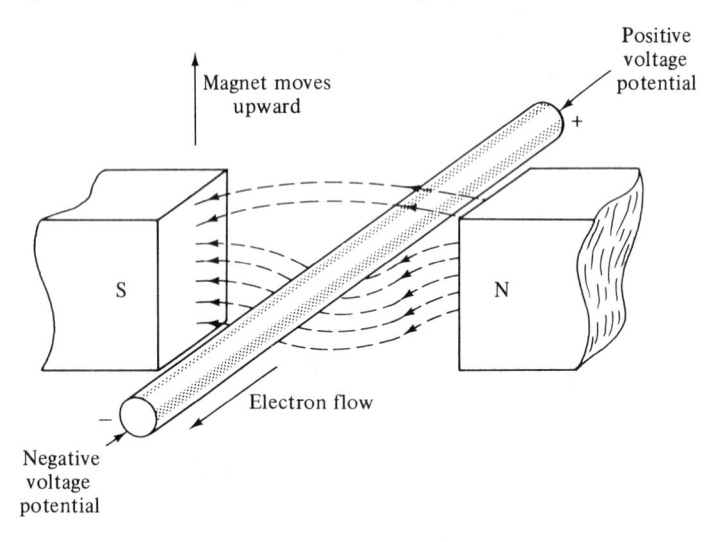

FIGURE 2-21 *Voltage Polarity*

In Figure 2-20, a charge potential is developed across the open ends of the coil. Electron flow is already indicated as a result of the moving magnet. Again, the electrons will "pile up" on the end of the wire (north end). That end is then charged with a negative polarity while the other end is positive.

The amount of induced voltage from magnetic flux lines cutting across a coil depends upon the amount of flux, the number of turns in the coil, and the speed at which the flux cuts through the coil. Increasing any of the three will increase the induced voltage. A formula for calculating induced voltage is

$$e = N \frac{\Delta \phi}{\Delta t \times 10^8} \qquad (2\text{-}13)$$

where e is in volts, N is the number of turns, $\Delta \phi$ is the change in the field in maxwells, and Δt is the amount of time in seconds for $\Delta \phi$.

Table 2-2 illustrates the use of the formula. The table is made up for a total time of 10 seconds, in 2-second intervals. The first column shows increasing field strength up to 500×10^6 maxwells. The change, $\Delta \phi$, for each 2-second interval is 100×10^6 maxwells. The number of turns is arbitrarily selected to be 100. Voltage e was calculated from formula (2-13).

$$e = N \frac{\Delta \phi}{\Delta t \times 10^8} = 100 \times \frac{100 \times 10^6}{2 \times 10^8} = 100 \times \tfrac{1}{2} = 50 \text{ v}$$

Figure 2-22 is a plot of the field strength ϕ versus time. The graph shows a linear increase in the field as time increases. At some point along the graph, a change in ϕ can be related to a change in time. The increments could be larger or smaller with the same results because the relationship is proportional.

Example 2-20 What is the induced voltage for Table 2-2 if a point in Figure 2-22 is selected between 8 and 9 sec?

Solution The time interval between 8 and 9 sec is a change in ϕ of $450 \times 10^6 - 400 \times 10^6 = 50 \times 10^6$ maxwells. The

TABLE 2-2 *Induced Voltage as Result of Flux Change*

ϕ maxwells $\times 10^6$	$\Delta \phi$ maxwells $\times 10^6$	t sec	Δt sec	N turns	e v
100	100	2	2	100	50
200	100	4	2	100	50
300	100	6	2	100	50
400	100	8	2	100	50
500	100	10	2	100	50

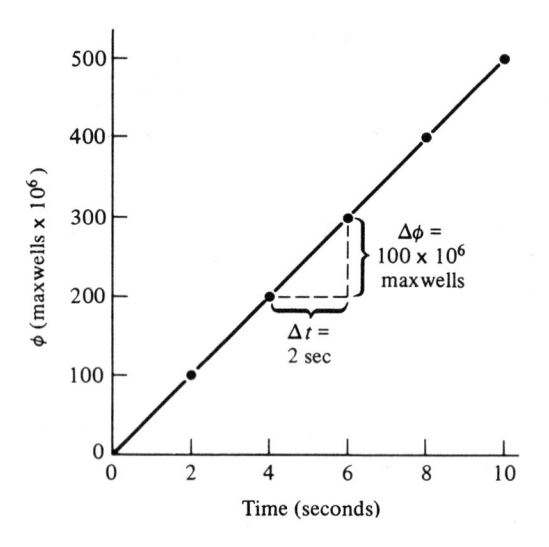

FIGURE 2-22 *Field Strength Versus Time*

time interval is 1 sec. According to formula (2-13) the result would be the same even if it had been calculated with the higher ratio of $\Delta\phi/\Delta t$.

$$e = N\frac{\Delta\phi}{\Delta t \times 10^8} = 100 \times \frac{50 \times 10^6}{1 \times 10^8}$$
$$= 100 \times 0.5 = 50 \text{ v}$$

Example 2-21 What is the flux change in maxwells during a change of 20 msec if a 40-turn coil produces 0.02 v?

Solution Rearranging formula (2-13) produces

$$\Delta\phi = \frac{e \times \Delta t \times 10^8}{N} = \frac{0.02 \times 20 \times 10^{-3} \times 10^8}{40}$$
$$= \frac{40 \times 10^3}{40} = 1 \times 10^3 = 1000 \text{ maxwells}$$

Example 2-22 How long does a coil have to be in motion to produce 80 v if the number of turns is 2400 and the field changes from 4.5×10^6 to 6.5×10^6 maxwells?

Solution $\Delta\phi = 6.5 \times 10^6 - 4.5 \times 10^6 = 2 \times 10^6$ maxwells. Rearranging formula (2-13) results in

$$\Delta t = \frac{N \times \Delta\phi}{e \times 10^8} = \frac{2400 \times 2 \times 10^6}{80 \times 10^8}$$
$$= \frac{48 \times 10^8}{80 \times 10^8} = 0.6 \text{ sec}$$

Magnetism

QUESTIONS

2-1 What are two methods of magnetizing a material?

2-2 What are the "laws" of attraction and repulsion?

2-3 Is a magnetic field strongest inside or outside a magnet? In what direction are the lines of force inside and outside of a magnet?

2-4 Which end of a compass would point to the north pole of a permanent magnet?

2-5 What atomic action creates the strongest magnetic field? Why?

2-6 If a magnet has a strength of 40,000 μ webers, what is its strength in maxwells?

2-7 Convert 95 μ webers to maxwells.

2-8 What is the flux density in gauss if an area 2.2 cm by 5 cm contains 880 lines?

2-9 What is the area of a magnetic pole that has a strength of 180 maxwells and a flux density of 60 gauss?

2-10 If an area of 2 cm² has a flux density of 30 gauss, what is the field strength in webers?

2-11 What is the strength of an iron bar that measures 1 $\times$ 1.3 cm at the end if it is placed in a magnetic field so that 260 maxwells pass through it from end-to-end?

2-12 What is an approximation of the permeability of iron?

2-13 What is the advantage of a small air gap between magnetic poles?

2-14 What is a ferromagnetic material?

2-15 What are ferrite materials?

2-16 When a conductor carries current, what is a good way to determine the direction of the magnetic field around the conductor?

2-17 If the north pole of a solenoid is known (via a compass), what is a good way to determine the direction of electron current flow?

2-18 Coil A has 180 turns with 0.15 amp. Coil B has 75 turns with 0.3 amp. Which coil has the strongest field and by how much?

2-19 How much current is required in a single conductor to create a 7-oersted field at a radius of 0.2 cm?

2-20 What is the field intensity at the center of a single loop with a radius of 0.5 cm and carrying 10 amp?

2-21 What is the radius of a 2000-turn coil that produces a field of 120 oersteds with 60 ma of current?

2-22 What is the field intensity of a long solenoid with 400 turns of wire over a distance of 12.56 cm, which carries a current of 1.5 amp?

2-23 What is the permeability of an iron bar that has a flux density of 90 lines per cm^2 and a field intensity of 0.3 oersteds?

2-24 If an iron core of a solenoid has a reluctance of 30 amp-turns per line and a total flux field of 200 lines, what is the ampere-turns ratio?

2-25 What is the reluctance in gilberts per maxwell of a 50-maxwell solenoid with 1000 turns and 0.25 amp?

2-26 In Table 2-1, if the permeability for 40 oersteds were 500 instead of 490, what would be the flux density?

2-27 Define saturation in terms of a *B* and *H* curve.

2-28 What is hysteresis?

2-29 What is residual magnetism?

2-30 Why does a magnetic field force a current-carrying conductor in a given direction?

2-31 When current is induced in a conductor, what determines the direction of the current?

2-32 What causes a voltage potential across the open wires of a coil when passed through a magnetic field?

2-33 How much voltage is induced across a coil of 500 turns if the field changes at a rate of 40 maxwells in 4 sec?

2-34 How many turns must a solenoid have to produce 800 v for a flux change of 30×10^6 maxwells per 0.3 sec?

3 Resistance

When current flows through an electrical circuit, it meets various kinds of resistance. Resistance is any phenomenon that restricts the flow of current. It may be wiring, or it may be special components called resistors and inductors which are designed to resist current flow in varying amounts. Whenever current is restricted, a difference in voltage potential exists at each end of the resistive path.

3.1 Definitions

Electric current is measured in amperes, and one **ampere** is the flow of one coulomb of charge per second. One **coulomb** equals 3×10^9 electrostatic units of charge and also equals 6.28×10^{18} electrons. Therefore, a current flow of one ampere is a flow of 6.28×10^{18} electrons per second past any specific point.

Resistance is measured in ohms. One **ohm** is the resistance offered to a steady flow of current by a column of mercury 106.3 cm long, having a cross section of 1 sq mm, at 0°C.

One **volt** is the potential difference required to generate one ampere of current through one ohm of resistance.

Basic conversions for electrical work are listed in Table 3–1. The use of prefixes and symbols tends to simplify notations used in electrical work.

TABLE 3-1

Prefix	Symbol	Math Unit	Example
mega	M	1×10^6	megohm
kilo	k	1×10^3	kilohm, kilovolt
milli	m	1×10^{-3}	milliampere, millivolt
micro	μ	1×10^{-6}	microampere, microvolt

For example, ten megohms written 10 MΩ is much simpler than ten million or 10,000,000 ohms, and ten microvolts written 10 μv is simpler than 10 one-millionth or 0.000,010 volts.

3.2 Basic Ohm's Law

Since resistance is the opposite of conductance, or actually the reciprocal, ohms and conductance can be expressed as follows:

$$\text{resistance} = \frac{1}{\text{conductance}} \quad \text{conductance} = \frac{1}{\text{resistance}}$$

Resistance, current, and voltage are interdependent. Their relationship is called Ohm's law

$$R = \frac{E}{I} \tag{3-1}$$

where R is resistance in ohms, E is voltage in volts, and I is current in amperes.

Ohm's law states that for a constant voltage source, current is inversely proportional to resistance. As resistance increases, current flow is restricted and therefore decreases. If the resistance is reduced, the current increases. If the resistance is reduced, the current increases. Figure 3–1 (a) shows a 1-ohm resistor connected across a voltage source. A current meter is shown in series for the purpose of measuring the current flow through the circuit. The table at (b) shows a constant resistance of 1 ohm and the resulting current as the voltage E is increased from 0 to 6 v in 1-v increments. Notice that the current increases linearly in 1-amp steps. The table was tabulated using Ohm's law [formula (3–1)]. When $E = 0$, no current is flowing, and $I = 0$. When $E = 1$ v,

$$I = \frac{E}{R} = \frac{1 \text{ v}}{1 \text{ ohm}} = 1 \text{ amp}$$

and when $E = 2$ v,

$$I = \frac{E}{R} = \frac{2 \text{ v}}{1 \text{ ohm}} = 2 \text{ amp}$$

[3.2] Resistance 43

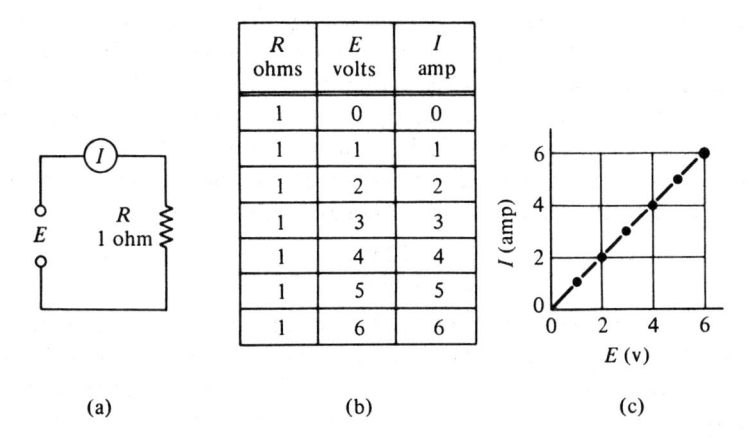

R ohms	E volts	I amp
1	0	0
1	1	1
1	2	2
1	3	3
1	4	4
1	5	5
1	6	6

(a) (b) (c)

FIGURE 3–1 *With a Constant Value of R, an Increasing Voltage Causes a Linear Increase in Current*

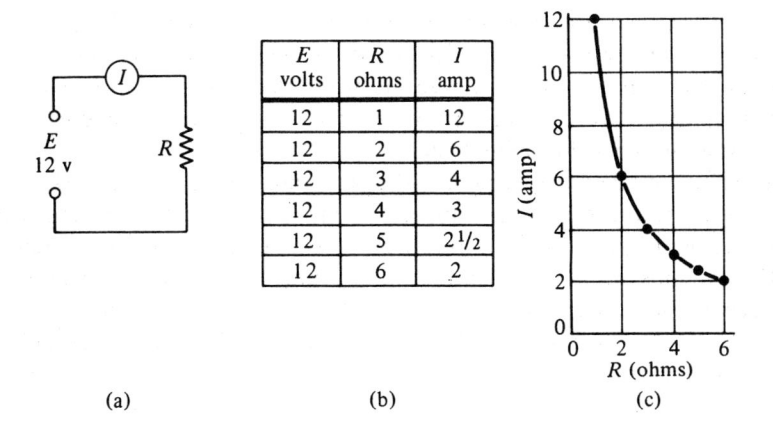

E volts	R ohms	I amp
12	1	12
12	2	6
12	3	4
12	4	3
12	5	2½
12	6	2

(a) (b) (c)

FIGURE 3–2 *With a Constant Value of E, an Increasing Resistance Causes a Non-Linear Decrease in Current*

and so on. The linear relationship holds for any value of E as long as R remains constant.

The graph in Figure 3–1 (c) is a plot of the table in (b). The horizontal (X) axis shows voltage from 0 to 6 v. The vertical (Y) axis shows current from 0 to 6 amp. A straight line that intersects each point on the graph illustrates the linear relationship.

Figure 3–2 illustrates a non-linear function when the voltage is held at some constant value, in this case, 12 v. Drawing (a) shows the current meter in series with the resistance to measure current flow in the circuit. The table in (b) has $E = 12$ v and shows changes in current as a result of increasing the resistance. Notice that a value of zero resistance would short out the voltage source. In the case of the circuit in (a), $R =$ zero ohms would place the current meter directly across the voltage source. A current meter is a very low resistance device, much lower than the normal circuit resistance. As a result, the meter would overheat and probably burn out.

Returning to the table in Figure 3-2 (b), when $R = 1$ ohm,

$$I = \frac{E}{R} = \frac{12 \text{ v}}{1 \text{ ohm}} = 12 \text{ amp}$$

and so on for the remaining values of R. In Figure 3-2 (c), the current is plotted on the Y axis as a function of increasing R on the X axis. Notice that the current decreases rapidly at first, and then begins to change at a lower rate. The graph illustrates a non-linear relationship when the voltage is held at a constant value.

3.3 Finding the Current

The analysis of current is the analysis of electron flow. The positive terminal of a battery attracts electrons into the battery. The electrons leaving the wire are replaced by free electrons that are mobilized by the potential difference of the battery terminals. This activity causes free electron movement throughout the circuit.

The electrons begin movement at the same time, move at the same speed, and move at the same magnitude. Therefore, the current, in terms of amperes, is the same amount in any part of a serially connected circuit.

Finding the current in a circuit is nothing more than dividing the applied voltage by the resistance of the circuit. In Figure 3-3, a battery is used as a voltage source. Assume $E = 20$ volts, and $R = 5$ ohms. By rearranging formula (3-1), the current is found

$$I = \frac{E}{R} = \frac{20 \text{ v}}{5 \text{ ohms}} = 4 \text{ amp}$$

As another example, assume $E = 10$ v, and $R = 0.5$ M ohms.

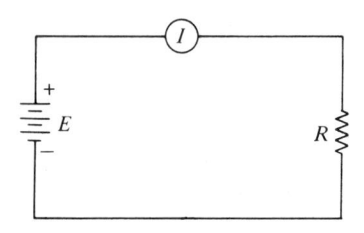

FIGURE 3-3

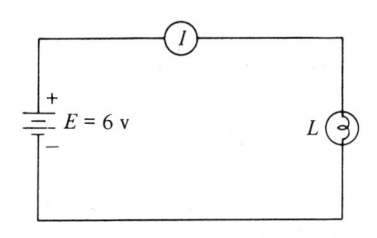

$$I = \frac{E}{R} = \frac{10 \text{ v}}{0.5 \times 10^6 \text{ ohms}} = 20 \times 10^{-6} \text{ amp} = 20 \text{ μa}$$

In Figure 3-4, a lamp (L) is shown connected across a battery of 6 v. If the resistance of the lamp is 12 ohms, the current is

$$I = \frac{E}{R} = \frac{6 \text{ v}}{12 \text{ ohms}} = 0.5 \text{ amp}$$

3.4 Finding the Resistance

The resistance of a circuit is determined in much the same manner as the current. Refer back to Figure 3-3. Assume that $E = 15$ v, and that the current meter reads 3 amp. The resistance is

$$R = \frac{E}{I} = \frac{15 \text{ v}}{3 \text{ amp}} = 5 \text{ ohms}$$

As a second example, assume $E = 3$ v, and $I = 1.5$ μa. R is found

$$R = \frac{E}{I} = \frac{3 \text{ v}}{1.5 \times 10^{-6} \text{ amp}} = 2 \times 10^6 \text{ ohms} = 2 \text{ M ohms}$$

Suppose the current meter in Figure 3-4 reads 50 ma. What is the resistance of the lamp L?

$$R = \frac{E}{I} = \frac{6 \text{ v}}{50 \times 10^{-3} \text{ amp}} = 120 \text{ ohms}$$

3.5 Finding the Voltage

If current and resistance are known, voltage can be determined by using Ohm's law.

$$E = IR$$

Referring to Figure 3-4, assume that the resistance of the lamp is 10 ohms and that 2.5 amp are required to illuminate the lamp. The required voltage is

$$E = IR = 2.5 \text{ amp} \times 10 \text{ ohms} = 25 \text{ v}$$

FIGURE 3-5

Suppose a 50-ohm lamp is being used for an indicator lamp in an electrical control panel. Fifty ma is required to raise the intensity of the lamp for adequate viewing. The voltage requirement is

$$E = IR = 50 \times 10^{-3} \text{ amp} \times 50 \text{ ohms} = 2.5 \text{ v}$$

The voltage across a resistor can be determined once the current and resistance are known. For example, refer to Figure 3-5. If $I = 1$ ma, and $R = 100$ ohms,

$$E = IR = 1 \times 10^{-3} \times 100 = 0.1 \text{ v or 100 mv}$$

That means that if the voltage on the positive end is 1.0 v the voltage on the other end is $1.0 - 0.1 = 0.9$ v. If the *voltage* on the negative end were 0.5 v, the positive end would be $0.5 + 0.1 = 0.6$ v.

3.6 Finding the Power

The unit for electrical power is called the **watt** (w). By definition, the work done to move a charge of one coulomb in one second by one volt is one watt. Power in watts equals coulombs per second times volts. One coulomb per second equals one ampere, thus power equals volts times amperes:

$$P = EI \tag{3-2}$$

where P is power in watts; E is voltage in volts, and I is current in amperes. In Figure 3-3, assume $E = 10$ v, and $I = 2$ amp. The 10 v is also the potential across the resistor since the current meter is of insignificant resistance—essentially a short circuit. The 2 amp is the current through the resistor. Therefore, the power dissipated in the resistor is

$$P = EI = 10 \text{ v} \times 2 \text{ amp} = 20 \text{ w}$$

Suppose the resistance is 50 ohms, and the voltage is 5 v. One way to solve the problem is

$$I = \frac{E}{R} = \frac{5 \text{ v}}{50 \text{ ohms}} = 0.1 \text{ amp}$$
$$P = EI = 5 \text{ v} \times 0.1 \text{ amp} = 0.5 \text{ w}$$

Another way to solve the problem involves a single formula. Basic Ohm's law [formula (3-1)] has already been manipulated to determine current, which can be substituted into the power equation (3-2).

$$I = \frac{E}{R}$$
$$P = EI = E\frac{E}{R} = \frac{E^2}{R}$$

The previous example is solved:
$$P = \frac{E^2}{R} = \frac{5 \text{ v}^2}{50 \text{ ohms}} = \frac{25 \text{ v}^2}{50 \text{ ohms}} = 0.5 \text{ w}$$

A further substitution can be accomplished as well:
$$E = IR$$
$$P = EI = (IR)I = I^2 R$$

where IR is substituted for E in the power formula.

As another example, the current meter in Figure 3–5 reads 3 amp for $R = 20$ ohms.
$$P = I^2 R = 3 \text{ amp}^2 \times 20 \text{ ohms}$$
$$= 9 \text{ amp}^2 \times 20 \text{ ohms} = 180 \text{ w}$$

3.7 Examples of Problems

Abbreviations and symbols for some of the most common electrical quantities are summarized in Table 3–2.

TABLE 3–2

Electrical Quantity	Abbreviation	Electrical Value	Symbol
current	I	amperes	amp, A, a
resistance	R	ohms	Ω
voltage	E	volts	v
power	P	watts	w

Current is symbolized with a capital A in terms of amperes, i.e., 0.5 A, 1 A, 10 A, etc. In terms of milliamperes or microamperes, a lower case is used, i.e., ma, μa, 100 μa, etc.

Resistance is symbolized Ω for ohms. Examples are 0.5Ω, 10Ω, 2 kilohms = 2 kΩ, 15 megohms = 15 MΩ. Voltage is simply v for volts (1 v), 10 millivolts = 10 mv, and 50 microvolts = 50 μv. Power in watts is symbolized w, i.e., 10 w, 30 milliwatts = 30 mw, etc.

When any two of the four functions (current, resistance, voltage, and power) are known, the remaining two can be determined. Formulas (3–1) and (3–2) have been manipulated in terms of each one of the four quantities to produce Table 3–3.

TABLE 3-3

Current	$I = \dfrac{E}{R}$ (a)	$I = \dfrac{P}{E}$ (b)	$I = \sqrt{\dfrac{P}{R}}$ (c)
Resistance	$R = \dfrac{E}{I}$ (d)	$R = \dfrac{P}{I^2}$ (e)	$R = \dfrac{E^2}{P}$ (f)
Voltage	$E = IR$ (g)	$E = \dfrac{P}{I}$ (h)	$E = \sqrt{RP}$ (i)
Power	$P = EI$ (j)	$P = I^2 R$ (k)	$P = \dfrac{E^2}{R}$ (l)

It is not necessary to memorize all twelve formulas. Only the basic formulas must be memorized, and their manipulation in terms of I, R, E, and P should be practiced until they become reasonably familiar.

The following examples illustrate the use of the formulas from Table 3-3; notice each formula is labeled (a), (b), etc., for ease of reference.

Example 3-1 In Figure 3-6 when the resistor dissipates 4.5 w, what are the values of the resistor and the current through the resistor?

Solution The two known quantities are voltage (15 v) and power (4.5 w). Formula (f) provides R, and formula (b) provides I.

(f) $\quad R = \dfrac{E^2}{P} = \dfrac{(15 \text{ v})^2}{4.5 \text{ w}} = \dfrac{225 \text{ v}^2}{4.5 \text{ w}} = 50 \Omega$

(b) $\quad I = \dfrac{P}{E} = \dfrac{4.5 \text{ w}}{15 \text{ v}} = 0.3 \text{ A}$

The answers are cross-checked:

$E = IR = 0.3 \text{ A} \times 50 \, \Omega = 15 \text{ v}$ (the specified voltage)
$P = I^2 R = 0.3^2 \text{ A}^2 \times 50 \, \Omega = 4.5 \text{ w}$ (the specified power)

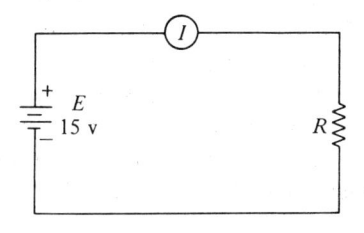

FIGURE 3-6

Example 3-2 In Figure 3-6 when the resistor equals 1 kilohm, what is the power dissipated by the resistor, and what is the reading on the current meter?

Solution The two known quantities are voltage (15 v) and resistance (1 kΩ). Formula (1) provides P, and formula (a) provides I.

(1) $$P = \frac{E^2}{R} = \frac{(15 \text{ v})^2}{1 \text{ k}\Omega} = \frac{225 \text{ v}^2}{1 \text{ k}\Omega} = 225 \text{ mw}$$

Notice that dividing volts² by kΩ yields power in milliwatts

$$\frac{225 \text{ v}^2}{1 \times 10^3 \Omega} = \frac{225 \text{ v}^2 \times 10^{-3}}{1 \Omega} = 0.225 \text{ w} = 225 \text{ mw}$$

(a) $$I = \frac{E}{R} = \frac{15 \text{ v}}{1 \text{ k}\Omega} = 15 \text{ ma}$$

Notice that dividing volts by kΩ yields current in milliamperes.

The answers are cross-checked:

$$E = \frac{P}{I} = \frac{225 \text{ mw}}{15 \text{ ma}} = 15 \text{ v (the specified voltage)}$$

$$R = \frac{P}{I^2} = \frac{225 \text{ mw}}{(15 \text{ ma})^2} = \frac{225 \times 10^{-3} \text{ w}}{(15 \times 10^{-3} \text{ A})^2}$$

$$= \frac{225 \times 10^{-3} \text{ w}}{225 \times 10^{-6} \text{ A}^2} = 1\Omega \times 10^3 = 1 \text{ k}\Omega$$

Notice that 1 kΩ was specified at the beginning of Example 3-2.

Example 3-3 In Figure 3-7 when the resistor dissipates 5 w, what is the current and voltage source?

Solution The known quantities are power (5 w) and resistance (20 Ω). Formula (c) provides I, and formula (i) provides E.

(c) $$I = \sqrt{\frac{P}{R}} = \sqrt{\frac{5 \text{ w}}{20 \Omega}} = \sqrt{0.25 \text{ A}^2} = 0.5 \text{ A}$$

(i) $$E = \sqrt{RP} = \sqrt{20 \Omega \times 5 \text{ w}} = \sqrt{100 \text{ v}^2} = 10 \text{ v}$$

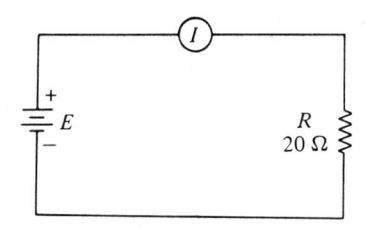

FIGURE 3-7

The answers are cross-checked:

$$R = \frac{E}{I} = \frac{10 \text{ v}}{0.5 \text{ A}} = 20 \text{ }\Omega \text{ (the specified resistance)}$$
$$P = EI = 10 \text{ v} \times 0.5 \text{ A} = 5 \text{ w (the specified power)}$$

Example 3-4 In Figure 3-7 when the current meter shows a reading of 60 ma, what are the voltage across the resistor and the power dissipation of the resistor?

Solution The known quantities are current and resistance. Formula (g) provides E, and formula (k) provides P.

(g) $E = IR = 60 \times 10^{-3} \text{ A} \times 20 \text{ }\Omega = 1.2 \text{ v}$

(k) $P = I^2 R = (60 \times 10^{-3} \text{ A})^2 \times 20 \text{ }\Omega$
$ = 3600 \times 10^{-6} \text{ A}^2 \times 20 \text{ }\Omega = 72 \times 10^{-3} \text{ w}$
$ = 72 \text{ mw}$

The answers are cross-checked:

$$I = \frac{P}{E} = \frac{72 \text{ mw}}{1.2 \text{ v}} = 60 \text{ ma (the specified current)}$$
$$R = \frac{E^2}{P} = \frac{(1.2 \text{ v})^2}{72 \text{ mw}} = \frac{1.44 \text{ v}^2}{72 \text{ mw}} = 20 \text{ }\Omega \text{ (the specified resistance)}$$

3.8 Ohm's Law with Two Power Sources

The use of more than one power supply provides the flexibility in design necessary for the operation of many electronic devices. Complex computers for instance, often use half a dozen power supplies of different voltages. The latter portion of this text is devoted to circuits involving vacuum tubes and solid state devices, and practical examples will demonstrate the need and advantage of using more than one power source.

Figure 3-8 (a) illustrates a resistor that is connected to two batteries. A ground connection is shown for purposes of a common reference point. Current flow is from the negative terminal of the larger battery. The current flows through the small battery and from right to left through the resistor

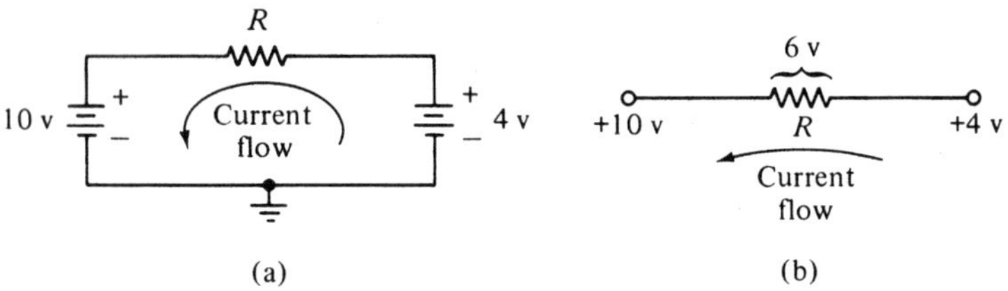

FIGURE 3-8

(R). When two batteries are connected so that their polarities oppose, the resulting voltage potential is the difference (6 v for the example). Conversely, if the batteries are connected plus to minus in the same direction, the voltages add.

The circuit in Figure 3-8 (a) is redrawn in (b). It shows a potential of +10 v on the left end of the resistor. In circuit work, the ground is assumed to be the return path for the other side of a voltage source. The right end of the resistor has a potential of +4 v. Careful study will show that the two circuits are identical. However, it is easier to see the voltage difference and the direction of current flow in the second drawing.

Figure 3-9 (a) illustrates two batteries turned around, directly opposite of the ones in Figure 3-8. Current flows from the negative terminal of the larger supply and left to right through the resistor. Then it travels through the 4-v supply, returning to the positive terminal of the 10-v supply. The circuit is redrawn in (b); it shows -10 v on the left and -4 v on the right. The return to common ground through the supplies is assumed. A voltmeter connected between ground and either end of the resistor would measure a negative quantity of the amount shown, -10 v or -4 v. Since the batteries oppose each other, the difference in potential is 6 v, as indicated across the resistor.

The last two figures (3-8 and 3-9) demonstrated opposing voltage supplies. Figure 3-10 (a) shows aiding supplies. Current flows from the negative side of the larger supply (10 v) into the positive terminal of the 4-v supply. But in leaving the 4-v supply, current leaves through the negative

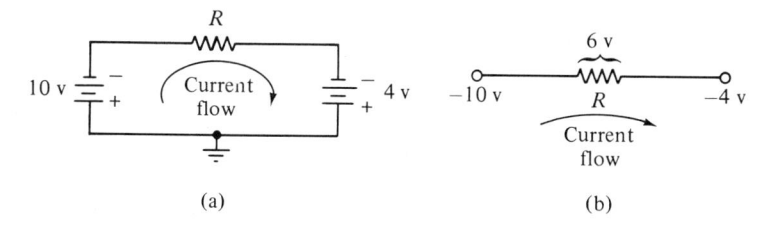

FIGURE 3-9

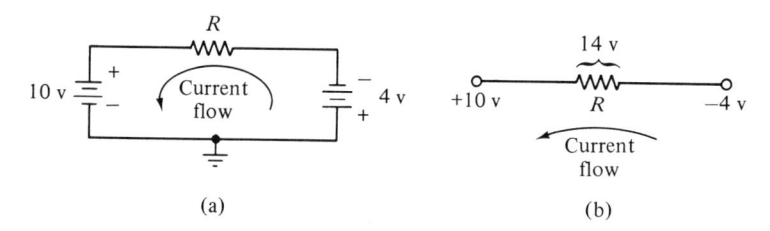

FIGURE 3-10

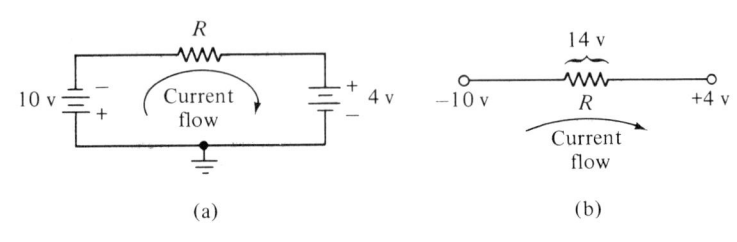

```
                    R
                  ─MW─
          ┌───────────────┐                            14 v
          │ -   ┌─────┐  +│                           ─MW─
    10 v ═╡    │Current│  ╞═ 4 v      o──────────────────────────o
          │ +   │flow  │  -│         -10 v      R              +4 v
          └───────┬───────┘                     Current
                  ═                              flow
                  
                (a)                               (b)

                        FIGURE 3-11
```

terminal. The two batteries are in series plus to minus, so the potentials add. The circuit in (b) shows the relationship of -4 v and $+10$ v. One is -4 v to ground, and the other is ground to $+10$ v for a total of 14 v across R.

The circuit of Figure 3-10 is shown in Figure 3-11 (a) with the batteries reversed. The current leaves the negative terminal of the 10-v battery, passes through the resistor left to right, through the 4-v battery plus to minus, and returns to the positive terminal of the 10-v battery. As the batteries are aiding (polarities in the same direction), the potentials add to a total of 14 v. Circuit (b) shows -10 v to ground and ground to $+4$ v.

A close study of each circuit in Figures 3-8 to 3-11 demonstrates that the voltage across the resistor is the difference between potentials. If both potentials are plus (plus on each end of the resistor), the net potential is a direct subtraction. If both potentials are minus, the net potential is again subtraction (remember that one sign changes in subtraction; one of the minus signs becomes plus). The difference in a plus and a minus potential is really the total of the two quantities since one sign has to change to obey the law of subtraction.

Figures 3-12 through 3-16 are examples of the application of Ohm's law to circuits with two power supply voltages.

Example 3-5   In Figure 3-12 determine the current and power dissipation of the resistor.

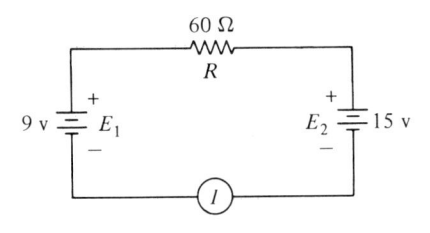

```
                          60 Ω
                          ─MW─
                            R
          ┌─────────────────────────────────┐
          │     +                    +      │
         9 v ═  E₁                   E₂  ═ 15 v
          │     -                    -      │
          └────────────────┬────────────────┘
                          (I)

                      FIGURE 3-12
```

[3.8]   Resistance   53

Solution E_R (voltage across R) = 15 v − 9 v = 6 v.

$$I = \frac{E}{R} = \frac{6\text{ v}}{60\ \Omega} = 0.1\text{ A}$$

$$P = IE = 0.1\text{ A} \times 6\text{ v} = 0.6\text{ w}$$

Example 3–6 In Figure 3–13 determine the value of R and its power dissipation.

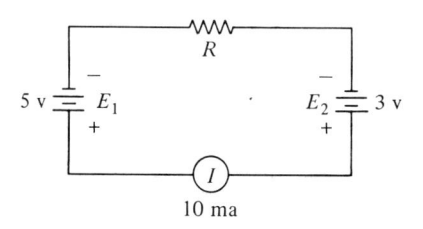

FIGURE 3–13

Solution $E_R = (-5\text{ v}) - (-3\text{ v}) = -5\ V + 3\text{ v} = -2\text{ v}$

$$R = \frac{E}{I} = \frac{2\text{ v}}{10\text{ ma}} = 0.2\text{ k}\Omega = 200\ \Omega$$

$$P = I^2 R = (10\text{ ma})^2 \times 0.2\text{ k}\Omega = 20\text{ mw}$$

Example 3–7 In Figure 3–14 determine the current and value of R if the power dissipated by R is 120 mw.

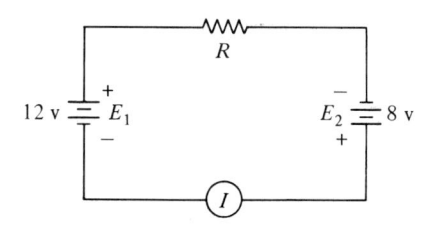

FIGURE 3–14

Solution $E_R = (12\text{ v}) - (-8\text{ v}) = 12\text{ v} + 8\text{ v} = 20\text{ v}$

$$I = \frac{P}{E} = \frac{120\text{ mw}}{20\text{ v}} = 6\text{ ma}$$

$$R = \frac{E}{I} = \frac{20\text{ v}}{6\text{ ma}} = 3.33\text{ k}\Omega$$

Also, from Table 3–3,

$$R = \frac{E^2}{P} = \frac{(20\text{ v})^2}{120\text{ mw}} = \frac{400\text{ v}^2}{120\text{ mw}} = 3.33\text{ k}\Omega$$

Example 3-8 In Figure 3-15 determine the values of E_2, the voltage drop across $R(E_R)$, and the power dissipated by R.

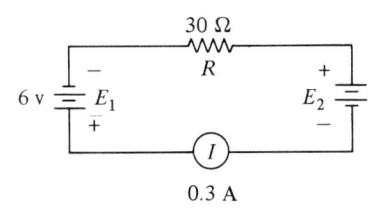

FIGURE 3-15

Solution   Determine E_R first since I and R are given.

$$E_R = IR = 0.3 \text{ A} \times 30 \text{ }\Omega = 9 \text{ v}$$
$$E_1 + E_2 = E_R \therefore E_2 = E_R - E_1 = 9 \text{ v} - 6 \text{ v} = 3 \text{ v}$$
$$P = IE_R = 0.3 \text{ A} \times 9 \text{ v} = 2.7 \text{ w}$$

Example 3-9 In Figure 3-16 determine the values of E_1 and the voltage drop across R.

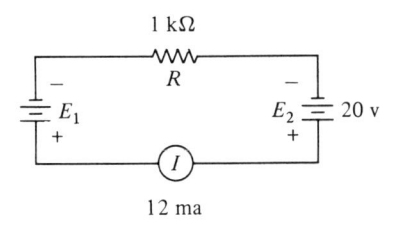

FIGURE 3-16

Solution   Since I and R are given, the first step is to find E_R. Notice that the negative terminals of both E_1 and E_2 are connected to the resistor.

$$E_R = IR = 12 \text{ ma} \times 1 \text{ k}\Omega = 12 \text{ v}$$
$$E_1 = E_2 - E_R = -20 \text{ v} - 12 \text{ v} = -32 \text{ v}$$

or

$$E_1 = E_R - E_2 = 12 \text{ v} - 20 \text{ v} = -8 \text{ v}$$

Thus, E_1 can be -32 v or -8 v to satisfy the stated conditions.

QUESTIONS

3-1   If $R = 11$ ohms, and $I = 3$ A, what is E?

3-2   If $E = 4$ v, and $I = 8$ ma, what is R?

3-3   If $E = 0.5$ v, and $R = 10$ kΩ, what is I?

3-4   If $R = 0.2$ ohms, and $I = 25$ μa, what is E?

3-5   If $R = 3$ k ohms, and $I = 70$ ma, what is P?

3-6   What is E for Question 3-5?

3-7   If $P = 75$ w, and $R = 15$ ohms, what is E?

3-8   What is I for Question 3-7?

3-9   If $R = 120$ ohms, and $I = 30$ ma, what is P?

3-10 What is E for Question 3-9?

3-11 If $E = 18$ v, and $I = 3.14 \times 10^{18}$ electrons/sec, what is P?

3-12 What is R for Question 3-11?

3-13 Refer to Figure 3-12. If E_1 and E_2 are 14 v and 2 v respectively, what is the power dissipated by R?

3-14 Refer to Figure 3-13. If E_1 and E_2 are 6 v and 20 v respectively, what is the value of R?

3-15 Refer to Figure 3-14. If E_1 and E_2 are 18 v and 3 v respectively, and the power dissipation in R is 3 w, what are the current and value of R?

3-16 Refer to Figure 3-16. If E_2 is 5 v, what is the voltage for E_1?

4 Series Circuits

In the last chapter, various combinations of Ohm's law were studied and practiced. The study was limited to single resistors. This chapter and the next two are concerned with more than one resistor in a circuit. Resistance, current, voltage, and power will be studied for resistors connected in series, in parallel, and in both series and parallel together.

4.1 Kirchhoff's Law

The circuits studied so far have been series circuits with a single resistor and one or two power sources. Because the resistor and battery or batteries were connected serially, the current was the same in all parts of the circuit. For example, one ampere through the resistor meant that there was one ampere through the battery, or one ampere through both batteries when two were connected serially.

The first of Kirchhoff's laws states that in any closed circuit, the algebraic sum of the voltage sources and IR drops is zero. If a resistor is connected across a 10-v battery, the voltage source is 10 v, and the IR drop across the resistor is 10 v, providing the algebraic sum of zero. If two batteries, E_1 and E_2, are used to bias each end of a resistor, the algebraic sum of the batteries is first determined. Depending upon how they are connected, the algebraic sum may be the difference, or it may be the sum of them

[4.2] Series Circuits 57

(straight addition). The algebraic sum of the two batteries must then equal the potential across the resistor.

Suppose two resistors are connected across a battery. As already stated, the same amount of current flows through each resistor and the battery. According to Ohm's law, the current flow and resistance yield an IR drop (E) across each resistor; the amounts are dependent upon the resistance. According to Kirchhoff's law, these IR drops must be equal to the applied potential of the battery. The same law holds when more than one battery is used. The algebraic sum of the batteries will be the same as the sum of the IR drops across the serially connected resistors, and then the algebraic sum of the voltage sources and IR drops will be zero.

4.2 DC Current

Figure 4–1 (a) illustrates two resistors in series with a voltage source of 10 v. In the last chapter it was found that the current meter resistance was low enough to be considered a short circuit or part of the wiring, which means

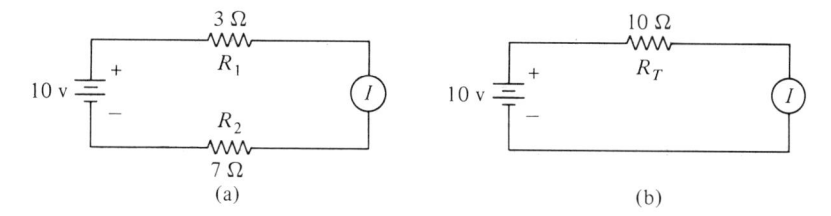

FIGURE 4–1 *Two Resistors in Series*

that the circuit resistance is represented by the two resistors. When resistors are serially connected, the resistance adds so that adding resistance in a line increases the total resistance (R_T). Circuit (b) shows the equivalent circuit with the two resistors replaced by a single resistor R_T. Note that the 3-ohm and 7-ohm resistors total 10 ohms.

Regardless of the number of resistors, as long as they are connected in series, the total resistance is the sum of the individual resistors.

$$R_T = R_1 + R_2 + R_3 + \cdots + R_n \qquad (4\text{--}1)$$

In Figure 4–1 (a), $R_T = R_1 + R_2 = 3\,\Omega + 7\,\Omega = 10\,\Omega$, as shown in (b).

The dc current through the circuit is determined by Ohm's law from the equivalent circuit in (b):

$$I = \frac{E}{R_T} = \frac{10\text{ v}}{10\,\Omega} = 1\text{ A}$$

Figure 4–2 (a) is an example of three resistors. Formula (4–1) gives

$$R_T = R_1 + R_2 + R_3 = 1\text{ k}\Omega + 2\text{ k}\Omega + 5\text{ k}\Omega = 8\text{ k}\Omega$$

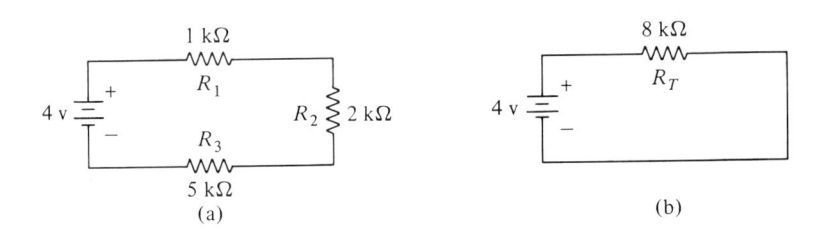

FIGURE 4-2 Three Resistors in Series

The current is

$$I = \frac{E}{R_T} = \frac{4 \text{ v}}{8 \text{ k}\Omega} = 0.5 \text{ ma}$$

Figure 4-3(a) shows two resistors in series with two batteries. The circuit is the same even if the resistors are both shown between the batteries, and

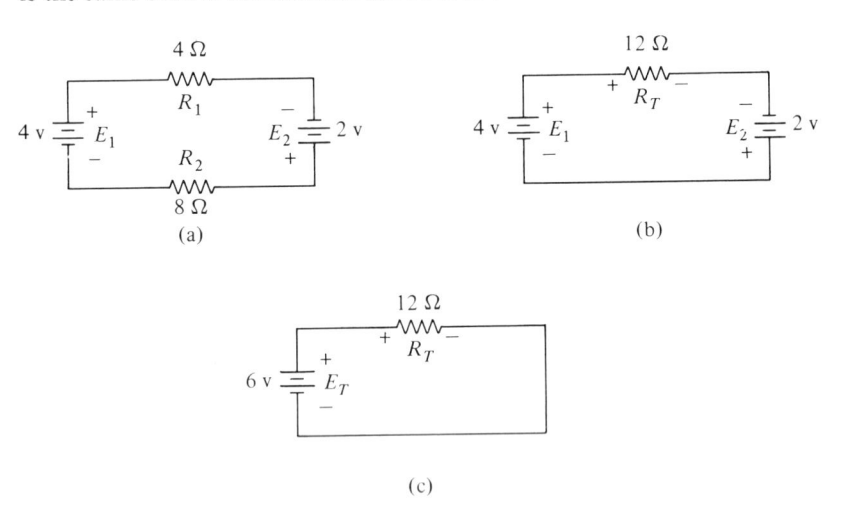

FIGURE 4-3 Two Resistors and Two Batteries in Series

the resistors can be in the upper line or lower line. Since $R_T = R_1 + R_2$, $R_T = 4 + 8 = 12$ ohms as shown in circuit (b). The resistor could just as well have been in the lower line. The voltage across the resistor is $E_1 - E_2 = (4 \text{ v}) - (-2 \text{ v}) = 6 \text{ v}$, which becomes $E_T = 6$ v as shown in circuit (c). Notice the polarity on each side of R_T is unchanged from (b) to (c). The current is simply

$$I = \frac{E_T}{R_T} = \frac{6 \text{ v}}{12 \text{ }\Omega} = 0.5 \text{ A}$$

[4.3] Series Circuits 59

Figure 4-4(a) is a 4-resistor, 2-battery circuit. Circuit (b) shows
$$R_T = R_1 + R_2 + R_3 + R_4 = 3\,\Omega + 1\,\Omega + 2\,\Omega + 6\,\Omega = 12\,\Omega$$
Circuit (c) shows
$$E_T = E_1 - E_2 = (-30\text{ v}) - (-12\text{ v}) = -18\text{ v}$$
and the current is
$$I = \frac{E_T}{R_T} = \frac{18\text{ v}}{12\,\Omega} = 1.5\text{ A}$$

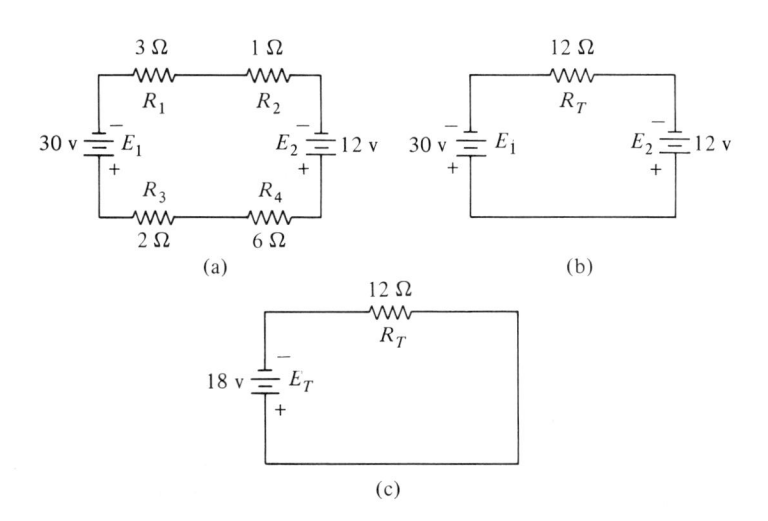

FIGURE 4-4 Four Resistors and Two Batteries in Series

4.3 Voltage Drops

The voltage drops around a series circuit must be equal to the applied voltage (Kirchhoff's law). Each resistor in a circuit has a voltage or IR drop. If the current and resistance are known, the IR drop for that resistor is $E = IR$ (Ohm's law). Once the IR drop is determined for each resistor, their sum will be equal to the source voltage. When more than one source is used, the difference is the effective potential.

In Figure 4-1(a), the current was found to be 1 A. The IR drop for R_1 is
$$E_{R1} = IR_1 = 1\text{ A} \times 3\,\Omega = 3\text{ v}$$
and for R_2
$$R_{R2} = IR_2 = 1\text{ A} \times 7\,\Omega = 7\text{ v}$$
The sum of the IR drops is the applied potential.
$$E_{R1} + E_{R2} = 3\text{ v} + 7\text{ v} = 10\text{ v}$$

In Figure 4–2(a), the current was 0.5 ma. The *IR* drops are

$$E_{R1} = 0.5 \text{ ma} \times 1 \text{ k}\Omega = 0.5 \text{ v}$$
$$E_{R2} = 0.5 \text{ ma} \times 2 \text{ k}\Omega = 1.0 \text{ v}$$
$$E_{R3} = 0.5 \text{ ma} \times 5 \text{ k}\Omega = 2.5 \text{ v}$$
$$E_T = E_{R1} + E_{R2} + E_{R3} = 0.5 \text{ v} + 1.0 \text{ v} + 2.5 \text{ v} = 4 \text{ v}$$

Figure 4–3(a) shows two source voltages. The current is 0.5 A. The *IR* drops are

$$E_{R1} = 0.5 \text{ A} \times 4 \text{ }\Omega = 2 \text{ v}$$
$$E_{R2} = 0.5 \text{ A} \times 8 \text{ }\Omega = 4 \text{ v}$$
$$E_T = E_{R1} + E_{R2} = 2 \text{ v} + 4 \text{ v} = 6 \text{ v}$$

The circuit in Figure 4–3(c) shows that E_T is 6 v, which agrees with the sum of the *IR* drops.

Figure 4–4(a) is a 4-resistor, 2-source circuit. Since the current is 1.5 A, the *IR* drops are

$$E_{R1} = 1.5 \text{ A} \times 3 \text{ }\Omega = 4.5 \text{ v}$$
$$E_{R2} = 1.5 \text{ A} \times 1 \text{ }\Omega = 1.5 \text{ v}$$
$$E_{R3} = 1.5 \text{ A} \times 2 \text{ }\Omega = 3.0 \text{ v}$$
$$E_{R4} = 1.5 \text{ A} \times 6 \text{ }\Omega = 9.0 \text{ v}$$
$$E_T = 4.5 \text{ v} + 1.5 \text{ v} + 3.0 \text{ v} + 9.0 \text{ v} = 18 \text{ v}$$

Circuit (c) agrees with $E_T = 18$ v.

This knowledge of *IR* drops can be applied to determining junction voltages. Consider the circuit in Figure 4–5. The current is

$$I = \frac{E}{R_1 + R_2} = \frac{10 \text{ v}}{5 \text{ }\Omega + 5 \text{ }\Omega} = 1 \text{ A}$$

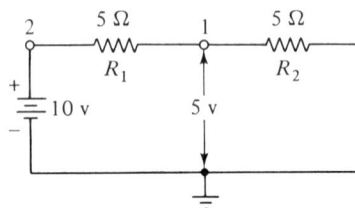

FIGURE 4–5 *Junction Voltage for Two Equal Resistors*

The *IR* drop for each resistor is $1 \text{ A} \times 5 \text{ }\Omega = 5$ v. The junction potential, between point 1 and ground, is actually the *IR* drop of R_2, which is 5 v. The junction potential between point 2 and ground is the sum of the *IR* drops for R_1 and R_2, which is $5 \text{ v} + 5 \text{ v} = 10$ v. Point 2 in this case is the battery potential, 10 v. Figure 4–6 is another example.

[4.3] Series Circuits 61

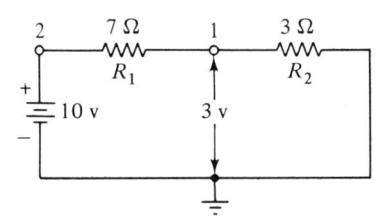

FIGURE 4-6 *Junction Voltage for Two Different Resistors*

$$I = \frac{E}{R_1 + R_2} = \frac{10 \text{ v}}{7 \, \Omega + 3 \, \Omega} = 1 \text{ A}$$
$$E_{R1} = 1 \text{ A} \times 7 \, \Omega = 7 \text{ v}$$
$$E_{R2} = 1 \text{ A} \times 3 \, \Omega = 3 \text{ v}$$

The junction potential between point 1 and ground is 3 v, the *IR* drop of R_2. Point 2 is the sum of the *IR* drops of R_1 and R_2, 7 v + 3 v = 10 v, the battery potential.

Figure 4–7 is a more complicated example. The total resistance R_T is $R_1 + R_2 + R_3 + R_4$.

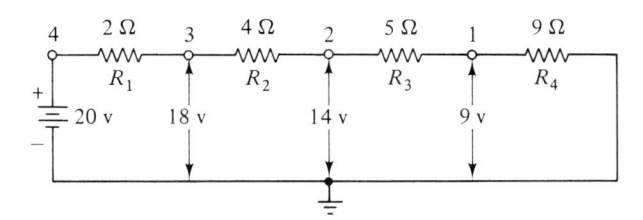

FIGURE 4-7 *Junction Voltages for Four Resistors*

$$I = \frac{E}{R_T} = \frac{E}{R_1 + R_2 + R_3 + R_4}$$
$$= \frac{20 \text{ v}}{2 \, \Omega + 4 \, \Omega + 5 \, \Omega + 9 \, \Omega} = 1 \text{ A}$$
$$E_{R1} = 1 \text{ A} \times 2 \, \Omega = 2 \text{ v}$$
$$E_{R2} = 1 \text{ A} \times 4 \, \Omega = 4 \text{ v}$$
$$E_{R3} = 1 \text{ A} \times 5 \, \Omega = 5 \text{ v}$$
$$E_{R4} = 1 \text{ A} \times 9 \, \Omega = 9 \text{ v}$$
$$E_T = E_{R1} + E_{R2} + E_{R3} + E_{R4} = 2 \text{ v} + 4 \text{ v} + 5 \text{ v} + 9 \text{ v} = 20 \text{ v}$$

Beginning at point 1, the potential is $E_{R4} = 9$ v.

Point 2 = $E_{R4} + E_{R3} = 9 \text{ v} + 5 \text{ v} = 14 \text{ v}$
Point 3 = $E_{R4} + E_{R3} + E_{R2} = 9 \text{ v} + 5 \text{ v} + 4 \text{ v} = 18 \text{ v}$

Point 4 = $E_{R4} + E_{R3} + E_{R2} + E_{R1}$
= 9 v + 5 v + 4 v + 2 v = 20 v (the battery potential)

Notice also that junction potentials can be determined by starting at the battery. Since point 4 is $E_T = 20$ v,

Point 3 = $E_T - E_{R1}$ = 20 v − 2 v = 18 v
Point 2 = $E_T - (E_{R1} + E_{R2})$ = 20 v − (2 v + 4 v)
= 20 v − 6 v = 14 v
Point 1 = $E_T - (E_{R1} + E_{R2} + E_{R3})$ = 20 v − (2 v + 4 v + 5 v)
= 20 v − 11 v = 9 v

Another method is to determine junction potentials from some point in-between. For example, in Figure 4–7, point 2 is 14 v.

Point 3 = point 2 + E_{R2} = 14 v + 4 v = 18 v
Point 4 = point 2 + $E_{R2} + E_{R1}$ = 14 v + 4 v + 2 v = 20 v
Point 1 = point 2 − E_{R3} = 14 v − 5 v = 9 v

Figure 4–8 is an example with two source voltages. The effective voltage (E_T) across all three resistors is

$$E_T = E_1 - E_2 = -12 \text{ v} - (-3 \text{ v}) = -12 \text{ v} + 3 \text{ v} = -9 \text{ v}$$

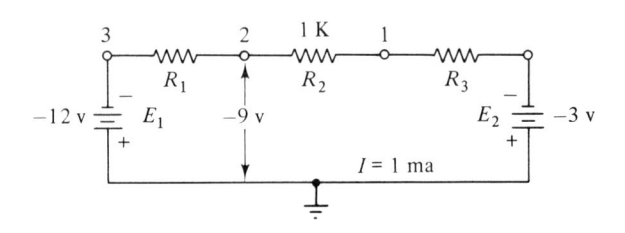

FIGURE 4–8 *Junction Voltages for Three Resistors and Two Batteries*

The current is given as 1ma. Since point 2 is specified as −9v, point 1 is point 2 − E_{R2}.

$E_{R2} = IR_2$ = 1 ma × 1 kΩ = 1 v
Point 1 = −9 v − (−1 v) = −9 v + 1 v = −8 v

Notice that point 2 is −9 v, and the total IR drop of all three resistors is 9 v. But point 2 is a point between −3 v and −12 v. The IR drop of R_1 is

$$E_1 - \text{point 2} = -12 \text{ v} - (-9 \text{ v}) = -3 \text{ v}$$

The resistance of R_1 is

$$R_1 = \frac{E_{R1}}{I} = \frac{3 \text{ v}}{1 \text{ ma}} = 3 \text{ k}\Omega$$

Series Circuits

The *IR* drop of R_3 is
$$E_{R3} = \text{point } 1 - E_2 = -8 \text{ v} - (-3 \text{ v}) = -5 \text{ v}$$
The resistance of R_3 is
$$R_3 = \frac{E_{R3}}{I} = \frac{5 \text{ v}}{1 \text{ ma}} = 5 \text{ k}\Omega$$

4.4 Power Dissipation

The power dissipated when there is more than one resistor in a circuit is first determined for each resistor individually using the formula, $P = EI$. The *IR* drop for a resistor is calculated as described in the last section. The product of the current and *IR* drop is the power dissipated by the resistor. Power can be calculated directly by using the formula, $P = I^2R$. The total power dissipated in the circuit is the sum of the power dissipated by each resistor:

$$P_T = P_1 + P_2 + P_3 + \cdots + P_n \tag{4-2}$$

For Figure 4–1(a),
$$I = \frac{E}{R_1 + R_2} = \frac{10 \text{ v}}{3\,\Omega + 7\,\Omega} = 1 \text{ A}$$
$$P_{R1} = I^2R = (1 \text{ A})^2 \times 3\,\Omega = 3 \text{ w}$$
$$P_{R2} = I^2R = (1 \text{ A})^2 \times 7\,\Omega = 7 \text{ w}$$

Figure 4–2 is
$$I = \frac{E}{R_1 + R_2 + R_3} = \frac{4 \text{ v}}{1 \text{ k}\Omega + 2 \text{ k}\Omega + 5 \text{ k}\Omega} = 0.5 \text{ ma}$$
$$P_{R1} = I^2R = (0.5 \text{ ma})^2 \times 1 \text{ k}\Omega = 0.25 \text{ ma}^2 \times 1 \text{ k}\Omega = 0.25 \text{ mw}$$
$$P_{R2} = (0.5 \text{ ma})^2 \times 2 \text{ k}\Omega = 0.25 \text{ ma}^2 \times 2 \text{ k}\Omega = 0.5 \text{ mw}$$
$$P_{R3} = (0.5 \text{ ma})^2 \times 5 \text{ k}\Omega = 0.25 \text{ ma}^2 \times 5 \text{ k}\Omega = 1.25 \text{ mw}$$

The total power dissipated in the circuit is the sum of the power dissipated by each resistor. Using formula (4–2) yields

$$P_T = P_{R1} + P_{R2} + P_{R3} = 0.25 \text{ mw} + 0.5 \text{ mw} + 1.25 \text{ mw} = 2 \text{ mw}$$

Figure 4–4(c) shows $I = 18 \text{ v}/12\,\Omega = 1.5 \text{ A}$. In circuit (a), the individual power dissipated is

$$P_{R1} = I^2R_1 = (1.5 \text{ A})^2 \times 3\,\Omega = 2.25 \text{ A}^2 \times 3\,\Omega = 6.75 \text{ w}$$
$$P_{R2} = (1.5 \text{ A})^2 \times 1\,\Omega = 2.25 \text{ A}^2 \times 1\,\Omega = 2.25 \text{ w}$$
$$P_{R3} = (1.5 \text{ A})^2 \times 2\,\Omega = 2.25 \text{ A}^2 \times 2\,\Omega = 4.5 \text{ w}$$
$$P_{R4} = (1.5 \text{ A})^2 \times 6\,\Omega = 2.25 \text{ A}^2 \times 6\,\Omega = 13.5 \text{ w}$$
$$P_T = 6.75 \text{ w} + 2.25 \text{ w} + 4.5 \text{ w} + 13.5 \text{ w} = 27 \text{ w}$$

The result can be double-checked by using E_T and R_T of circuit (c). $E_T = 18$ v, which is the IR drop across R_T.

$$P = \frac{E^2}{R} = \frac{(18 \text{ v})^2}{12 \text{ }\Omega} = \frac{324 \text{ v}^2}{12 \text{ }\Omega} = 27 \text{ w}$$

For one more example, refer to Figure 4–8.

$$P_{R1} = (E_1 - \text{point 2})(I) = (-12 \text{ v} + 9 \text{ v})(1 \text{ ma}) = 3 \text{ mw}$$
$$P_{R2} = I^2 R_2 = (1 \text{ ma})^2 \times 1 \text{ k}\Omega = 1 \text{ mw}$$
$$\text{Point 1} = \text{point 2} - IR_2 = -9 \text{ v} - (-1 \text{ ma} \times 1 \text{ k}\Omega)$$
$$= -9 \text{ v} + 1 \text{ v} = -8 \text{ v}$$
$$P_{R3} = (\text{point 1} - E_2)(I) = (-8 \text{ v} + 3 \text{ v})(1 \text{ ma}) = 5 \text{ mw}$$
$$P_T = P_{R1} + P_{R2} + P_{R3} = 3 \text{ mw} + 1 \text{ mw} + 5 \text{ mw} = 9 \text{ mw}$$

or

$$P_T = (E_1 - E_2)(I) = (-12 \text{ v} + 3 \text{ v})(1 \text{ ma}) = 9 \text{ mw}$$

4.5 Determining an Unknown Quantity

The first step in determining an unknown quantity is to examine the circuit. Collect all known values and mark the drawing if it helps in showing relative values. Then use the various forms of Ohm's law to identify unknowns until the desired quantity is found. Often it is quicker to jump in and calculate randomly rather than take an excessive amount of time to study the circuit.

Example 4–1 Figure 4–9 is an example in which $V_3 = 14$ v. The problem is to determine the values of R_3 and V_2 and the power dissipated by each resistor.

Solution Since $V_3 = 14$ v, and V_1 is the source voltage (30 v), the current in the circuit is

$$I = \frac{V_1 - V_3}{R_1 + R_2} = \frac{30 \text{ v} - 14 \text{ v}}{3 \text{ }\Omega + 5 \text{ }\Omega} = 2 \text{ A}$$

then

$$R_3 = \frac{V_3}{I} = \frac{14 \text{ v}}{2 \text{ A}} = 7 \text{ }\Omega$$
$$V_2 = V_1 - IR_1 = 30 \text{ v} - (2 \text{ A} \times 3 \text{ }\Omega)$$
$$= 30 \text{ v} - 6 \text{ v} = 24 \text{ v}$$

Power is

$$P_{R1} = I(V_1 - V_2) = 2 \text{ A}(30 \text{ v} - 24 \text{ v})$$
$$= 2 \text{ A} \times 6 \text{ v} = 12 \text{ w}$$
$$P_{R2} = I(V_2 - V_3) = 2 \text{ A}(24 \text{ v} - 14 \text{ v})$$
$$= 2 \text{ A} \times 10 \text{ v} = 20 \text{ w}$$

[4.5] Series Circuits 65

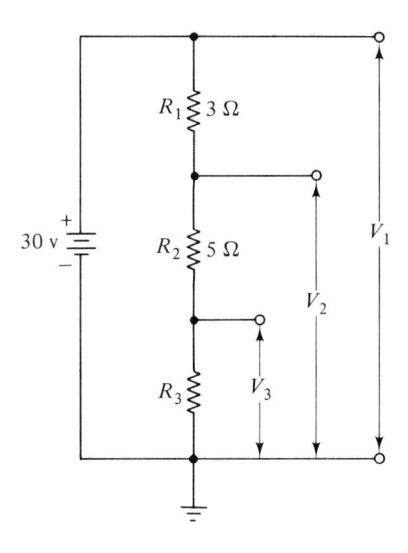

FIGURE 4-9 *Determining Unknown Values*

$$P_{R3} = IV_3 = 2 \text{ A} \times 14 \text{ v} = 28 \text{ w}$$
$$P_T = P_{R1} + P_{R2} + P_{R3} = 12 \text{ w} + 20 \text{ w} + 28 \text{ w} = 60 \text{ w}$$

An alternate solution is

$$P_T = I^2 R_T = (2 \text{ A})^2(3 \text{ }\Omega + 5 \text{ }\Omega + 7 \text{ }\Omega)$$
$$= 4 \text{ A}^2 \times 15 \text{ }\Omega = 60 \text{ w}$$

Example 4-2 What is the value of E_1 in Figure 4-10?

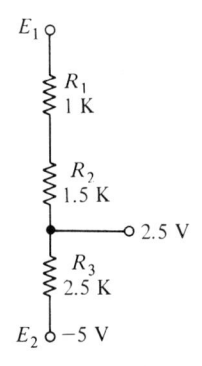

FIGURE 4-10 *Determining an Unknown Supply Voltage E*

Solution The only values given that can be used are R_3 and its IR drop. The IR drop of R_3 is

$$E_{R3} = 2.5 \text{ v} - (-5 \text{ v}) = 7.5 \text{ v}$$

The current through R_3 is the same current that flows through the rest of the resistors, so

$$I_3 = \frac{E_{R3}}{R_3} = \frac{7.5 \text{ v}}{2.5 \text{ k}\Omega} = 3 \text{ ma}$$

The IR drops of R_2 and R_1 are

$$E_{R2} = IR_2 = 3 \text{ ma} \times 1.5 \text{ k}\Omega = 4.5 \text{ v}$$
$$E_{R1} = IR_1 = 3 \text{ ma} \times 1 \text{ k}\Omega = 3 \text{ v}$$
$$E_1 = E_2 + E_{R3} + E_{R2} + E_{R1}$$
$$= -5\text{v} + 7.5\text{v} + 4.5 \text{ v} + 3 \text{ v} = 10 \text{ v}$$

Example 4-3 What is the value of R_2 in Figure 4-11?

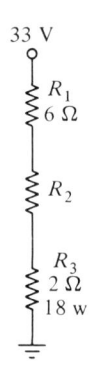

FIGURE 4-11 *Determining an Unknown Resistance R_2*

Solution R_3 is 2 ohms, and P_{R3} is 18 w. The current, which flows through all resistors, is determined from formula (c) in Table 3-3,

$$I = \sqrt{\frac{P}{R}} = \sqrt{\frac{18 \text{ w}}{2 \Omega}} = \sqrt{9 \text{ A}^2} = 3 \text{ A}$$

$$R_2 = \frac{E_{R2}}{I}$$

$$E_{R2} = 33 \text{ v} - (E_{R1} + E_{R3})$$
$$E_{R1} = IR_1 = 3 \text{ A} \times 6 \Omega = 18 \text{ v}$$
$$E_{R3} = IR_3 = 3 \text{ A} \times 2 \Omega = 6 \text{ v}$$
$$E_{R2} = 33 \text{ v} - (18 \text{ v} + 6 \text{ v}) = 9 \text{ v}$$
$$R_2 = \frac{E_{R2}}{I} = \frac{9 \text{ v}}{3 \text{ A}} = 3 \Omega$$

Series Circuits

The circuit is double-checked:

$$I = \frac{E_T}{R_1 + R_2 + R_3} = \frac{33 \text{ v}}{6\,\Omega + 3\,\Omega + 2\,\Omega}$$
$$= \frac{33 \text{ v}}{11\,\Omega} = 3 \text{ A}$$

QUESTIONS

4-1 In Figure 4-1(a), what is the total resistance if $R_1 = 70$ ohms, and $R_2 = 30$ ohms?

4-2 In Figure 4-1(a), what is the total resistance if $R_1 = 4$ k ohms, and $R_2 = 16$ k ohms?

4-3 How much current flows for Question 4-1?

4-4 How much current flows for Question 4-2?

4-5 In Figure 4-2(a), what is the total current if the battery voltage is $E = 20$ v?

4-6 Refer to Figure 4-3(a). Assume the battery polarities are reversed; E_1 is doubled; E_2 is halved, and the resistors are tripled. What are the total resistance and current in the circuit?

4-7 Refer to Figure 4-4(a). Assume $R_2 = 5$ ohms, and E_1 is unknown. Determine E_1 if $I_T = 3$ A.

4-8 In Figure 4-2(a), what are the voltage drops across the resistors if $E = 32$ v?

4-9 For Figure 4-12, what are the values of R_2 and R_4, and what are the voltage drops across R_1, R_3, R_4, and R_5?

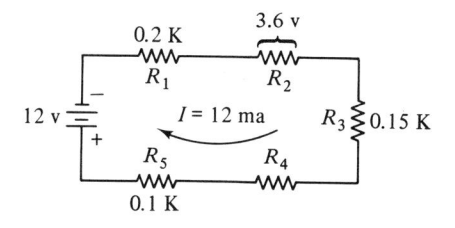

FIGURE 4-12

4-10 In Figure 4-6, what is the junction voltage at point 1 (E_{P1}) if $R_2 = 18$ ohms?

4-11 What are the junction voltages in Figure 4-7 if R_1 and R_3 are exchanged, and R_2 and R_4 are exchanged?

4-12 In Figure 4-8, if $E_1 = -15$ v, and $E_{P2} = -12$ v, what are the values of R_1 and R_3?

4-13 In Figure 4-8, if $E_1 = -15$ v, $E_2 = +3$ v, and $I = 2$ ma, what are the values of the resistors, and what are the junction voltages?

4-14 What is the power dissipated by a 40-ohm resistor with an IR drop of 10 v?

4-15 What power is dissipated by a resistor if the IR drop is 12 v, and the current is 20 ma? What value does the resistor have?

4-16 If the power dissipated by a 2.2-k ohm resistor is 19.8 mw, what are the current and voltage across the resistor?

4-17 What is the total power dissipated by the two resistors in Figure 4-6?

4-18 For Figure 4-13, what are the values of R_1, R_2, and V_3? What are the total current in the circuit and the total power dissipated by the resistors?

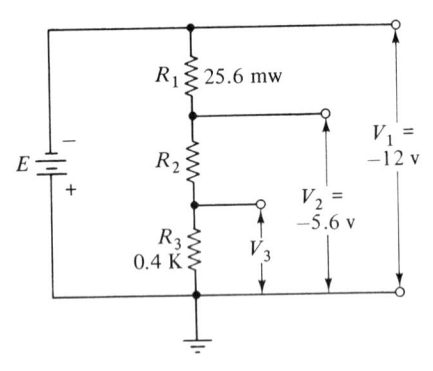

FIGURE 4-13

5 Parallel Circuits

This chapter discusses various combinations of paralleled-resistors and one or two source voltages.

5.1 Kirchhoff's Law

Series circuits were concerned with IR drops that could vary from resistor to resistor, and with current that was the same through all the resistors in the circuit. Every resistor passed an equal amount of current. When resistors are connected in parallel, the IR drop across each one is the same, but the current may be different from one resistor to the next.

The second part of Kirchhoff's law states that the algebraic sum of currents into a junction is equal to the current out of the junction. The concept is shown in Figure 5-1. The total current I_T is shown in the series wires to and from the battery terminals. Following its path from the negative terminal, I_T enters junction 1. At junction 1, the current separates, and some goes left (I_1), and the rest goes right (I_2). At junction 2, the two currents, I_1 and I_2, rejoin to make up I_T which enters the positive terminal of the battery. Thus, the current that enters junction (1) is equal to the current that leaves junction (2).

Kirchhoff's law holds for any number of parallel branches. Figure 5-1 could just as well have had a half dozen resistors in parallel. The current

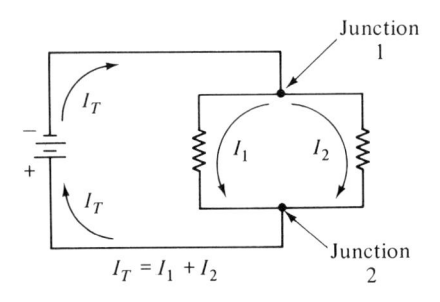

FIGURE 5–1 *Junction Current*

entering the junction divides between the resistors and rejoins when leaving the parallel net. Therefore, for a parallel circuit,

$$I_T = I_1 + I_2 + I_3 + \cdots + I_n \tag{5-1}$$

where I_T is also the current passing through the source voltage or the battery.

5.2 DC Currents

Figure 5–2 shows a total current of 3 A. $I_1 = 1$ A, and $I_2 = 2$ A for $I_T = 3$ A. This agrees with Equation (5-1). Consider Figure 5–3; the total current is the sum of the individual currents I_1, I_2, and I_3:

$$I_T = 2\,\text{A} + 3\,\text{A} + 5\,\text{A} = 10\,\text{A}$$

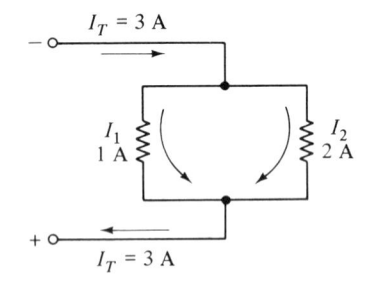

FIGURE 5–2 *Parallel Current*

Suppose, for Figure 5–3, that $I_T = 15$ A, and I_1 is an unknown. Find the current for I_1.

$$I_T = I_1 + I_2 + I_3$$
$$I_1 = I_T - I_2 - I_3 = 15\,\text{A} - 3\,\text{A} - 5\,\text{A} = 7\,\text{A}$$

The important thing to remember is that I_T is the same going into a net as it is coming out of a net.

[5.2]   Parallel Circuits   71

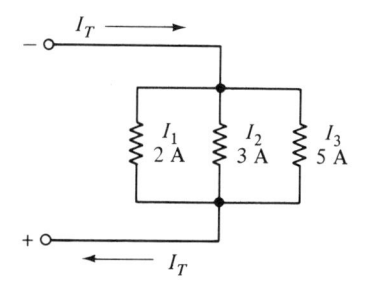

FIGURE 5-3 *Parallel Current for Three Resistors*

Figure 5-4(a) gives the source voltage and values of the two parallel resistors. The problem here is to determine individual currents for the resistors and the total current through the battery. The 12-v source is applied across both resistors, so that each one has an *IR* drop of 12 v. Ohm's law provides the current for each resistor.

$$I_1 = \frac{E}{R_1} = \frac{12 \text{ v}}{3 \text{ }\Omega} = 4 \text{ A}$$

$$I_2 = \frac{E}{R_2} = \frac{12 \text{ v}}{6 \text{ }\Omega} = 2 \text{ A}$$

The total current is the sum of I_1 and I_2:

$$I_T = I_1 + I_2 = 4 \text{ A} + 2 \text{ A} = 6 \text{ A}$$

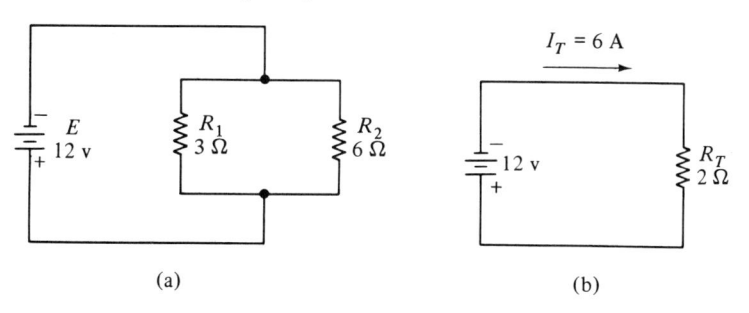

FIGURE 5-4 *Parallel Resistance*

Figure 5-5 is a 3-resistor example in which E, I_T, and two resistors are given. The problem in this example is to determine the current through each resistor. The source voltage is $E = 15$ v, and is applied across each of the resistors. By Ohm's law,

$$I_1 = \frac{E}{R_1} = \frac{15 \text{ v}}{1.5 \text{ k}\Omega} = 10 \text{ ma}$$

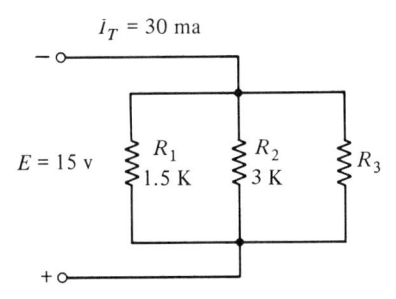

FIGURE 5-5 *Determining an Unknown Parallel Resistance*

$$I_2 = \frac{E}{R_2} = \frac{15 \text{ v}}{3 \text{ k}\Omega} = 5 \text{ ma}$$
$$I_T = I_1 + I_2 + I_3$$
$$I_3 = I_T - I_1 - I_2 = 30 \text{ ma} - 10 \text{ ma} - 5 \text{ ma} = 15 \text{ ma}$$

In addition, it is interesting to note that the value of R_3 is

$$R_3 = \frac{E}{I_3} = \frac{15 \text{ v}}{15 \text{ ma}} = 1 \text{ k}\Omega$$

5.3 Circuit Resistance

In the last section, various methods of analyzing branch currents were discussed. Individual resistances were also discussed, but only from the standpoint of their individual currents. In terms of the source voltage and total current, the total resistance is

$$R_T = \frac{E}{I_T}$$

Since each resistor has the same applied voltage from the source, the total current is

$$I_T = \frac{E}{R_T}$$

From formula (5-1), the total current is also the sum of each branch:

$$\frac{E}{R_T} = \frac{E}{R_1} + \frac{E}{R_2} + \frac{E}{R_3} + \cdots + \frac{E}{R_n}$$

Dividing both sides by E gives

$$\frac{1}{R_T} = \frac{1}{R_1} + \frac{1}{R_2} + \frac{1}{R_3} + \cdots + \frac{1}{R_n} \qquad (5\text{-}2)$$

Parallel Circuits

Further algebraic manipulation gives

$$R_T = \frac{1}{\frac{1}{R_1} + \frac{1}{R_2} + \frac{1}{R_3} + \cdots + \frac{1}{R_n}} \tag{5-3}$$

Either formula, (5-2) or (5-3), may be used depending upon personal preference. Formula (5-2) is most often used because it is the easiest to remember and apply.

For example, refer to Figure 5-4(a). Total current has already been determined as

$$I_T = \frac{E}{R_1} + \frac{E}{R_2} = \frac{12 \text{ v}}{3 \, \Omega} + \frac{12 \text{ v}}{6 \, \Omega} = 4 \text{ A} + 2 \text{ A} = 6 \text{ A}$$

The total resistance is

$$R_T = \frac{E}{I_T} = \frac{12 \text{ v}}{6 \text{ A}} = 2 \, \Omega$$

The circuit is redrawn in Figure 5-4(b) with a single resistor R_T of 2 ohms. The circuit obeys Ohm's law $E = IR = 6 \text{ A} \times 2 \, \Omega = 12 \text{ v}$, and it agrees with its counterpart in (a).

Notice that R_T is lower than the smaller resistance in the circuit. While studying series circuits, it made sense to increase resistance when adding resistors. In parallel circuits, additional paths are provided for passing current. The resistance of each individual resistor does not change, but providing additional paths for current flow does lower the overall resistance. The overall or total resistance, then, must be less than the lowest one. For example, if a small resistor is connected across a source voltage, and then another is added in parallel, the total resistance of the circuit is less than the first resistor since the second resistor provided another current path in the circuit. The additional path provides additional current flow, and Ohm's law shows resistance to be inversely proportional to current.

By applying formula (5-2), the results can be cross-checked for Figure 5-4:

$$\frac{1}{R_T} = \frac{1}{3 \, \Omega} + \frac{1}{6 \, \Omega} = \frac{1}{2 \, \Omega}$$
$$R_T = 2 \, \Omega$$

Formula (5-3) gives

$$R_T = \frac{1}{\frac{1}{R_1} + \frac{1}{R_2}} = \frac{1}{\frac{1}{3 \, \Omega} + \frac{1}{6 \, \Omega}} = \frac{1}{\frac{1}{2 \, \Omega}} = 2 \, \Omega$$

A simpler formula is available when only two resistors are to be equated to a total resistance.

$$R_T = \frac{R_1 R_2}{R_1 + R_2} \tag{5-4}$$

For Figure 5–4(a),

$$R_T = \frac{R_1 R_2}{R_1 + R_2} = \frac{3\,\Omega \times 6\,\Omega}{3\,\Omega + 6\,\Omega} = \frac{18\,\Omega^2}{9\,\Omega} = 2\,\Omega$$

If two resistors in parallel have the same value, the total resistance is half the value of one of them:

$$R_1 = R_2 \qquad R_T = \frac{R_1}{2} \tag{5-5}$$

Suppose two resistors are 10 Ω each. According to formula (5–5),

$$R_T = \frac{R_1}{2} = \frac{10\,\Omega}{2} = 5\,\Omega$$

and by formula (5–4),

$$R_T = \frac{R_1 R_2}{R_1 + R_2} = \frac{10\,\Omega \times 10\,\Omega}{10\,\Omega + 10\,\Omega} = \frac{100\,\Omega^2}{20\,\Omega} = 5\,\Omega$$

These calculations verify the method.

An example of three resistors in parallel is shown in Figure 5–6(a).

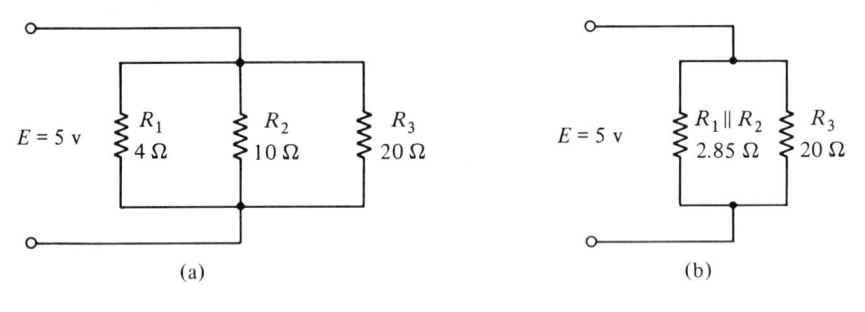

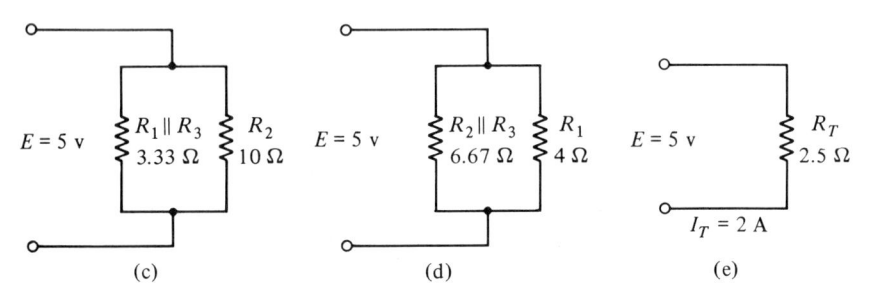

FIGURE 5–6 *Calculating Resistance in Parallel*

[5.3] Parallel Circuits

$$\frac{1}{R_T} = \frac{1}{4\,\Omega} + \frac{1}{10\,\Omega} + \frac{1}{20\,\Omega} = \frac{5}{20} + \frac{2}{20} + \frac{1}{20} = \frac{8}{20} = \frac{2}{5}$$

$$R_T = \frac{5}{2} = 2.5\,\Omega$$

$$I_T = \frac{E}{R_T} = \frac{5\,\text{v}}{2.5\,\Omega} = 2\,\text{A}$$

Individual currents are

$$I_1 = \frac{E}{R_1} = \frac{5\,\text{v}}{4\,\Omega} = 1.25\,\text{A}$$

$$I_2 = \frac{E}{R_2} = \frac{5\,\text{v}}{10\,\Omega} = 0.5\,\text{A}$$

$$I_3 = \frac{E}{R_3} = \frac{5\,\text{v}}{20\,\Omega} = 0.25\,\text{A}$$

$$I_T = I_1 + I_2 + I_3 = 1.25\,\text{A} + 0.5\,\text{A} + 0.25\,\text{A} = 2\,\text{A}$$

The circuit of Figure 5–6(a) is redrawn in (b), (c), and (d). Each of the revisions shows the equivalent resistance of two resistors in parallel along with the third resistor. The examples demonstrate that resistance can be equated from any number of parallel resistors to combinations of a lesser number. The total resistance ($R_T = 2.5$ ohms) is unchanged, and R_T is shown in (e). By Ohm's law, it agrees with the circuit in (a).

$$E = I_T R_T = 2\,\text{A} \times 2.5\,\Omega = 5\,\text{v}$$

The parallel values for circuits (b), (c), and (d) are calculated:

(b) $\quad R_1 \| R_2 = \dfrac{R_1 R_2}{R_1 + R_2} = \dfrac{4\,\Omega \times 10\,\Omega}{4\,\Omega + 10\,\Omega} = 2.85\,\Omega$

(c) $\quad R_1 \| R_3 = \dfrac{R_1 R_3}{R_1 + R_3} = \dfrac{4\,\Omega \times 20\,\Omega}{4\,\Omega + 20\,\Omega} = 3.33\,\Omega$

(d) $\quad R_2 \| R_3 = \dfrac{R_2 R_3}{R_2 + R_3} = \dfrac{10\,\Omega \times 20\,\Omega}{10\,\Omega + 20\,\Omega} = 6.67\,\Omega$

The following examples show $R_T = 2.5$ ohms for each of the three equivalent circuits:

(b) $\quad R_T = \dfrac{(R_1 \| R_2)R_3}{(R_1 \| R_2) + R_3} = \dfrac{2.85\,\Omega + 20\,\Omega}{2.85\,\Omega + 20\,\Omega} = 2.5\,\Omega$

(c) $\quad R_T = \dfrac{(R_1 \| R_3)R_2}{(R_1 \| R_3) + R_2} = \dfrac{3.33\,\Omega \times 10\,\Omega}{3.33\,\Omega + 10\,\Omega} = 2.5\,\Omega$

(d) $\quad R_T = \dfrac{(R_2 \| R_3)R_1}{(R_2 \| R_3) + R_1} = \dfrac{6.67\,\Omega \times 4\,\Omega}{6.67\,\Omega + 4\,\Omega} = 2.5\,\Omega$

For an example of determining an unknown resistance, refer back to Figure 5–5. Through algebraic manipulation, formula (5–2) becomes

$$\frac{1}{R_3} = \frac{1}{R_T} - \frac{1}{R_1} - \frac{1}{R_2}$$

R_T must be determined first:

$$R_T = \frac{E}{I_T} = \frac{15 \text{ v}}{30 \text{ ma}} = 0.5 \text{ k}\Omega$$

Now, R_3 can be determined:

$$\frac{1}{R_3} = \frac{1}{0.5 \text{ k}\Omega} - \frac{1}{1.5 \text{ k}\Omega} - \frac{1}{3 \text{ k}\Omega} - \frac{1}{1 \text{ k}\Omega}$$

$$R_3 = \frac{1 \text{ k}\Omega}{1} = 1 \text{ k}\Omega$$

Formula (5–5) stated that for two resistors of equal value the total resistance is half the value of either one of them. The formula can be expanded to state that for any number of equal resistors, the total resistance is the number of resistors divided into the value of any one of them.

$$R_1 = R_2 = \cdots = R_n$$
$$R_T = \frac{R}{n} \tag{5-6}$$

where n is the number of equal resistors.

Consider Figure 5–7. The total resistance by formula (5–6) is

$$R_T = \frac{R}{n} = \frac{20 \text{ }\Omega}{4} = 5 \text{ }\Omega$$

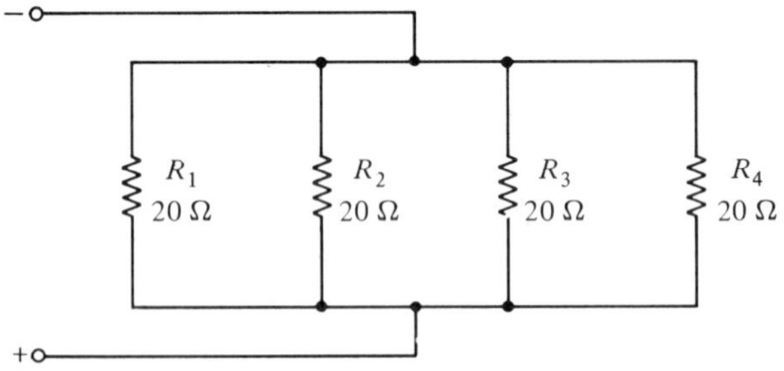

FIGURE 5–7 *Four Equal Resistances in Parallel*

This answer can be verified by using formula (5–2):

$$\frac{1}{R_T} = \frac{1}{R_1} + \frac{1}{R_2} + \frac{1}{R_3} + \frac{1}{R_4}$$
$$= \frac{1}{20 \text{ }\Omega} + \frac{1}{20 \text{ }\Omega} + \frac{1}{20 \text{ }\Omega} + \frac{1}{20 \text{ }\Omega} = \frac{4}{20 \text{ }\Omega}$$
$$R_T = \frac{20 \text{ }\Omega}{4} = 5 \text{ }\Omega$$

[5.3] Parallel Circuits 77

The application of formula (5–3) also verifies the answers:

$$R_T = \frac{1}{\frac{1}{R_1} + \frac{1}{R_2} + \frac{1}{R_3} + \frac{1}{R_4}}$$

$$= \frac{1}{\frac{1}{20\,\Omega} + \frac{1}{20\,\Omega} + \frac{1}{20\,\Omega} + \frac{1}{20\,\Omega}}$$

$$= \frac{1}{\frac{4}{20\,\Omega}} = \frac{20\,\Omega}{4} = 5$$

The following is a summary of resistance formulas.

$$\frac{1}{R_T} = \frac{1}{R_1} + \frac{1}{R_2} + \frac{1}{R_3} + \cdots + \frac{1}{R_n} \tag{5-2}$$

$$\frac{1}{R_1} = \frac{1}{R_T} - \frac{1}{R_2} - \frac{1}{R_3} - \cdots - \frac{1}{R_n}$$

$$R_T = \frac{1}{\frac{1}{R_1} + \frac{1}{R_2} + \frac{1}{R_3} + \cdots + \frac{1}{R_n}} \tag{5-3}$$

$$R_1 = \frac{1}{\frac{1}{R_T} - \frac{1}{R_1} - \frac{1}{R_2} - \frac{1}{R_3} - \cdots - \frac{1}{R_n}}$$

$$R_T = \frac{R_1 R_2}{R_1 + R_2} \tag{5-4}$$

$$R_1 = \frac{R_2 R_T}{R_2 - R_T} \qquad R_2 = \frac{R_1 R_T}{R_1 - R_T}$$

$$R_T = \frac{R}{n} \qquad R = nR_T \qquad n = \frac{R}{R_T} \tag{5-6}$$

(where n is the number of equal resistances).

Formula (5–2) above is also shown in terms of $1/R_1$. This term can be exchanged for any of the other single resistance terms ($1/R_2$, $1/R_3$, etc.).

Formula (5–3) is shown in terms of R_1 which can be exchanged for R_2, R_3, etc.

Formula (5–4) is shown in terms of R_1 and in terms of R_2 for convenience. Formula (5–6) is also shown for each of its individual terms. Several examples of these formulas follow.

Example 5-1 Figure 5–8 gives the values for R_T and all but one resistor, R_3. What is the value of R_3?

Solution From formula (5–2),

$$\frac{1}{R_3} = \frac{1}{R_T} - \frac{1}{R_1} - \frac{1}{R_2} - \frac{1}{R_4}$$

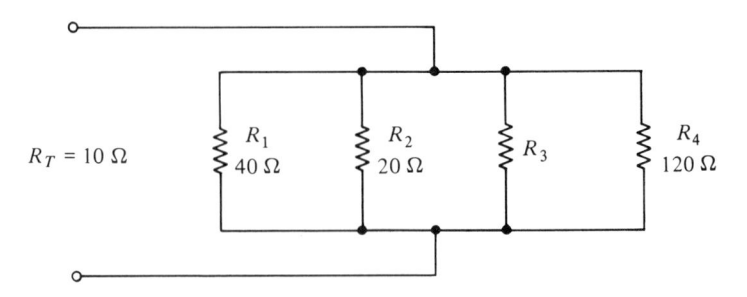

FIGURE 5-8 *Determining Unknown Resistance for Four in Parallel*

$$= \frac{1}{10\,\Omega} - \frac{1}{40\,\Omega} - \frac{1}{20\,\Omega} - \frac{1}{120\,\Omega}$$

$$= \frac{12}{120\,\Omega} - \frac{3}{120\,\Omega} - \frac{6}{120\,\Omega} - \frac{1}{120\,\Omega} = \frac{2}{120\,\Omega}$$

$$R_3 = \frac{120\,\Omega}{2} = 60\,\Omega$$

Example 5-2 Repeat Example 5-1 using formula (5-3).

Solution

$$R_3 = \frac{1}{\frac{1}{R_T} - \frac{1}{R_1} - \frac{1}{R_2} - \frac{1}{R_4}}$$

$$= \frac{1}{\frac{1}{10\,\Omega} - \frac{1}{40\,\Omega} - \frac{1}{20\,\Omega} - \frac{1}{120\,\Omega}}$$

$$= \frac{1}{\frac{2}{120\,\Omega}} = \frac{120\,\Omega}{2} = 60\,\Omega$$

Example 5-3 A total resistance of 4 ohms is required for a particular circuit. How much resistance must be placed in parallel with a 6-ohm resistor?

Solution Using formula (5-4) gives

$$R_1 = \frac{R_2 R_T}{R_2 - R_T} = \frac{6\,\Omega \times 4\,\Omega}{6\,\Omega - 4\,\Omega} = \frac{24\,\Omega^2}{2\,\Omega} = 12\,\Omega$$

Notice that R_1 was arbitrarily labeled as the unknown; it could just as easily have been R_2.

Example 5-4 Eight identical resistors are connected in parallel for a total resistance of 3 kΩ. What is the value of each of the resistors?

Solution   Applying formula (5–6) yields

$$R = nR_T = 8 \times 3 \text{ k}\Omega = 24 \text{ k}\Omega$$

5.4 DC Voltage

The dc voltage across parallel resistors is the same for each one. Therefore, if the current and resistance are known for any resistor in a parallel circuit, the *IR* drop will be the same for all other resistors in the circuit.

The source voltage E and total current I_T are determined in Figure 5–9.

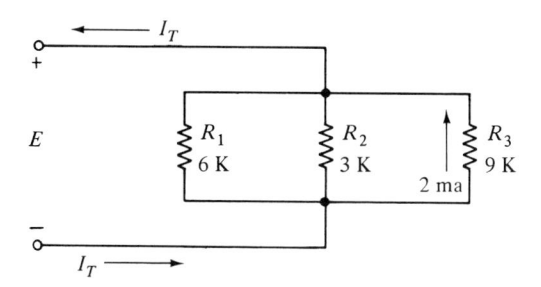

FIGURE 5–9 *Determining Source Voltage and Total Current*

The values of the three resistors and the current through R_3 are given. The first step is to find the *IR* drop across the resistors, which will also be the source voltage in the example:

$$E = I_3 R_3 = 2 \text{ ma} \times 9 \text{ k}\Omega = 18 \text{ v}$$

There are two approaches to calculating I_T. One way is to calculate the current for each of the other two resistors and then sum the three currents. The other approach is to calculate the parallel resistance, R_T, then divide E by R_T. The first approach is

$$I_1 = \frac{E}{R_1} = \frac{18 \text{ v}}{6 \text{ k}\Omega} = 3 \text{ ma}$$

$$I_2 = \frac{E}{R_2} = \frac{18 \text{ v}}{3 \text{ k}\Omega} = 6 \text{ ma}$$

$$I_T = I_1 + I_2 + I_3 = 3 \text{ ma} + 6 \text{ ma} + 2 \text{ ma} = 11 \text{ ma}$$

The other method is

$$\frac{1}{R_T} = \frac{1}{R_1} + \frac{1}{R_2} + \frac{1}{R_3} = \frac{1}{6 \text{ k}\Omega} + \frac{1}{3 \text{ k}\Omega} + \frac{1}{9 \text{ k}\Omega} = \frac{11}{18 \text{ k}\Omega}$$

$$R_T = \frac{18 \text{ k}\Omega}{11} = 1.636 \text{ k}\Omega$$

$$I_T = \frac{E}{R_T} = \frac{18 \text{ v}}{1.636 \text{ k}\Omega} = 11 \text{ ma}$$

As demonstrated, either approach yields the same answer, 11 ma. Note that the polarity of the source voltage is reversed from the previous examples. In Figure 5-1, for instance, I_T is shown entering the junction at the top, leaving the bottom junction, and returning to the positive terminal of the battery. In Figure 5-9, the same principles apply. I_T leaves the negative terminal, enters the lower junction, leaves the upper junction, and returns to the positive terminal. Regardless of polarity, the positive terminal of a voltage source attracts the electrons—current flows to the positive terminal and from the negative terminal.

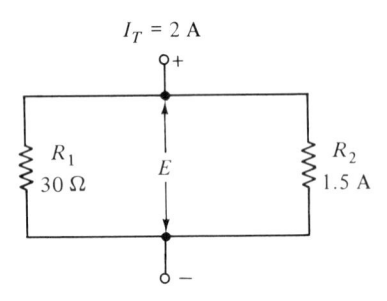

FIGURE 5-10 *Net Voltage Across Parallel Resistors*

There is a slight variation in circuitry in Figure 5-10. The positive and negative terminals are shown above and below the resistor net. E is still the potential between the terminals, and current still flows toward the positive terminal (up through the resistors in the diagram). The problem is to determine the net voltage E that provides the currents and resistance given in the circuit. Since I_T and I_2 are given,

$$I_1 = I_T - I_2 = 2 \text{ A} - 1.5 \text{ A} = 0.5 \text{ A}$$

The IR drop for R_1 is the source voltage:

$$E = I_1 R_1 = 0.5 \text{ A} \times 30 \text{ }\Omega = 15 \text{ v}$$

Additional analysis gives

$$R_2 = \frac{E}{I_2} = \frac{15 \text{ v}}{1.5 \text{ A}} = 10 \text{ }\Omega$$

$$R_T = \frac{R_1 R_2}{R_1 + R_2} = \frac{30 \text{ }\Omega \times 10 \text{ }\Omega}{30 \text{ }\Omega + 10 \text{ }\Omega} = \frac{300 \text{ }\Omega^2}{40 \text{ }\Omega} = 7.5 \text{ }\Omega$$

And to double-check the first solution for E,

$$E = I_T R_T = 2 \text{ A} \times 7.5 \text{ }\Omega = 15 \text{ v}$$

Figure 5-11(a) is a more complex circuit containing four parallel resistors. The resistance of one resistor is not given, but I_T and I_2 (current through the unknown resistor R_2) are given. The IR drop, $E_1 - E_2$, is

Parallel Circuits

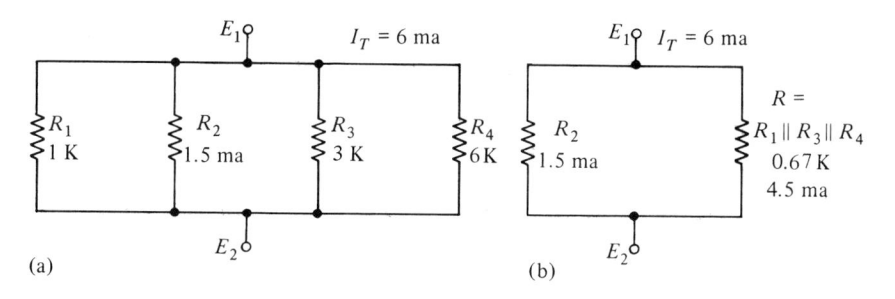

FIGURE 5-11 *Determining Source Voltage for Four Parallel Resistors*

to be determined. Since the current through R_2 is 1.5 ma, the remaining current (I_R) is distributed through the other resistors.

$$I_R = I_T - I_2 = 6 \text{ ma} - 1.5 \text{ ma} = 4.5 \text{ ma}$$

The total resistance (R_T) of the three given resistances is determined:

$$\frac{1}{R_T} = \frac{1}{R_1} + \frac{1}{R_3} + \frac{1}{R_4} = \frac{1}{1 \text{ k}\Omega} + \frac{1}{3 \text{ k}\Omega} + \frac{1}{6 \text{ k}\Omega}$$

$$= \frac{9}{6 \text{ k}\Omega}$$

$$R_T = \frac{6 \text{ k}\Omega}{9} = 0.67 \text{ k}\Omega$$

The circuit for Figure 5-11(a) is redrawn in (b). R_2 is shown with 1.5 ma, and the equivalent of the three resistors (0.67 kΩ) is shown with 4.5 ma. $E_1 - E_2$ is obtained using Ohm's law:

$$E_1 - E_2 = I_R R = 4.5 \text{ ma} \times 0.67 \text{ k}\Omega = 3 \text{ v}$$

To cross-check the answer $E_1 - E_2 = 3$ v, the current for each branch in circuit (a) is calculated:

$$I_1 = \frac{E_1 - E_2}{R_1} = \frac{3 \text{ v}}{1 \text{ k}\Omega} = 3 \text{ ma}$$

$$I_2 = 1.5 \text{ ma (given)}$$

$$I_3 = \frac{E_1 - E_2}{R_3} = \frac{3 \text{ v}}{3 \text{ k}\Omega} = 1 \text{ ma}$$

$$I_4 = \frac{E_1 - E_2}{R_4} = \frac{3 \text{ v}}{6 \text{ k}\Omega} = 0.5 \text{ ma}$$

And the total current should compare to the given amount of 6 ma:

$$I_T = I_1 + I_2 + I_3 + I_4$$
$$= 3 \text{ ma} + 1.5 \text{ ma} + 1 \text{ ma} + 0.5 \text{ ma}$$
$$= 6 \text{ ma}$$

Another check is to calculate R_2, derive the parallel resistance of all four resistors (R_T), and compare it with $(E_1 - E_2)/6$ ma.

$$R_2 = \frac{E_1 - E_2}{I_2} = \frac{3 \text{ v}}{1.5 \text{ ma}} = 2 \text{ k}\Omega$$

$$\frac{1}{R_T} = \frac{1}{R_1} + \frac{1}{R_2} + \frac{1}{R_3} + \frac{1}{R_4}$$

$$= \frac{1}{1 \text{ k}\Omega} + \frac{1}{2 \text{ k}\Omega} + \frac{1}{3 \text{ k}\Omega} + \frac{1}{6 \text{ k}\Omega} = \frac{12}{6 \text{ k}\Omega}$$

$$R_T = \frac{6 \text{ k}}{12} = 0.5 \text{ k}\Omega$$

In comparison,

$$R_T = \frac{E_1 - E_2}{I_T} = \frac{3 \text{ v}}{6 \text{ ma}} = 0.5 \text{ k}\Omega$$

The past example illustrates the usefulness of combining resistors and redrawing circuits. Recall from Section 5.3 that resistors can be combined in any sequence; two can be shown as one; three can be shown as two and one, or one alone, etc.

5.5 Power Dissipation

The power dissipation for individual resistors is determined by Ohm's law, $P = EI$. When resistors are connected in parallel, the voltage is the same across each of the resistors. Remembering that the current divides when it enters a junction, the current through each resistor may be different, or the current may be the same through each if the resistors are all of equal value.

To calculate the total power dissipated in a series circuit, the power dissipated by each resistor is summed. The power dissipation for parallel resistors is also summed for total power:

$$P_T = P_1 + P_2 + P_3 + \cdots + P_n \tag{5-7}$$

Consider the circuit in Figure 5-4. Total resistance and current have already been determined where

$$R_T = 2 \text{ }\Omega$$

and

$$I_T = 2 \text{ A}$$

The power dissipated by each resistor is

$$I_1 = \frac{E}{R_1} = \frac{12 \text{ v}}{3 \text{ }\Omega} = 4 \text{ A} \quad P_1 = EI = 12 \text{ v} \times 4 \text{ A} = 48 \text{ w}$$

$$I_2 = \frac{E}{R_2} = \frac{12 \text{ v}}{6 \text{ }\Omega} = 2 \text{ A} \quad P_2 = EI = 12 \text{ v} \times 2 \text{ A} = 24 \text{ w}$$

[5.5] Parallel Circuits 83

Total power is the sum of the individual powers:
$$P_T = P_1 + P_2 = 48 \text{ w} + 24 \text{ w} = 72 \text{ w}$$

Another way to calculate the total power for the circuit is with the source voltage and total current:
$$P_T = EI_T = 12 \text{ v} \times 6 \text{ A} = 72 \text{ w}$$

or
$$R_T = 2 \, \Omega$$

therefore,
$$P_T = \frac{E^2}{R_T} = \frac{(12 \text{ v})^2}{2 \, \Omega} = 72 \text{ w}$$

Figure 5–6(a) is a circuit that specifies the IR drop and resistance for each of the three resistors. Power dissipation is

$$P_1 = \frac{E^2}{R_1} = \frac{(5 \text{ v})^2}{4 \, \Omega} = 6.25 \text{ w}$$

$$P_2 = \frac{E^2}{R_2} = \frac{(5 \text{ v})^2}{10 \, \Omega} = 2.5 \text{ w}$$

$$P_3 = \frac{E^2}{R_3} = \frac{(5 \text{ v})^2}{20 \, \Omega} = 1.25 \text{ w}$$

$$P_T = P_1 + P_2 + P_3 = 6.25 \text{ w} + 2.5 \text{ w} + 1.25 \text{ w} = 10 \text{ w}$$

Referring to circuit (e) gives
$$P_T = \frac{E^2}{R_T} = \frac{(5 \text{ v})^2}{2.5 \, \Omega} = 10 \text{ w}$$

or
$$P_T = EI = 5 \text{ v} \times 2 \text{ A} = 10 \text{ w}$$

Example 5-5   The circuit in Figure 5–11(a) dissipates a total power of 18 mw. Determine the power dissipated by each resistor.

Solution   By inspection, if $E_1 - E_2$ is determined, the power dissipated by each resistor can be determined.

$$E_1 - E_2 = E = \frac{P_T}{I_T} = \frac{18 \text{ mw}}{6 \text{ ma}} = 3 \text{ v}$$

$$P_1 = \frac{E^2}{R_1} = \frac{(3 \text{ v})^2}{1 \text{ k}\Omega} = 9 \text{ mw}$$

$$P_2 = EI_2 = 3 \text{ v} \times 1.5 \text{ ma} = 4.5 \text{ mw}$$

$$P_3 = \frac{E^2}{R_3} = \frac{(3 \text{ v})^2}{3 \text{ k}\Omega} = 3 \text{ mw}$$

$$P_4 = \frac{E^2}{R_4} = \frac{(3 \text{ v})^2}{6 \text{ k}\Omega} = 1.5 \text{ mw}$$

The sum of the powers $P_1, P_2, P_3,$ and P_4 can be cross-checked against the specified total of 18 mw in the following manner:

$$P_T = 9 \text{ mw} + 4.5 \text{ mw} + 3 \text{ mw} + 1.5 \text{ mw} = 18 \text{ mw}$$

5.6 Determining an Unknown Quantity

To determine an unknown quantity, Ohm's law must be applied at a point where two or more quantities are already known. If more than one point is specified by two (or more) quantities, and a choice of where to start is difficult to make, it often pays to randomly select one of them and proceed.

Example 5-6 Figure 5–12 is an example of two resistors in parallel; two quantities are given for each resistor. Find the total current and wattage.

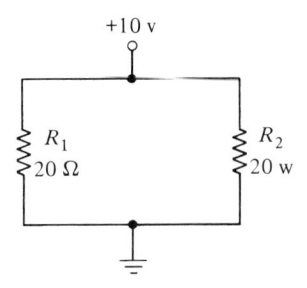

FIGURE 5-12 *Analysis of Power Dissipation*

Solution The circuit in this figure is drawn differently than in the previous circuits. The voltage source is shown as +10 v at the upper junction and ground is shown at the lower junction. The circuit implies a power source between the upper and lower terminals, such as a battery with its positive terminal connected to the upper junction.

Each resistor has the same IR drop of 10 v. The total current is the sum of I_1 and I_2. The individual currents can be calculated directly from Ohm's law.

$$I_1 = \frac{E}{R_1} = \frac{10 \text{ v}}{20 \text{ }\Omega} = 0.5 \text{ A}$$

$$I_2 = \frac{P_2}{E} = \frac{20 \text{ w}}{10 \text{ v}} = 2 \text{ A}$$

$$I_T = I_1 + I_2 = 0.5 \text{ A} + 2 \text{ A} = 2.5 \text{ A}$$

[5.6] Parallel Circuits

Another way to determine the total current is

$$R_2 = \frac{E^2}{P_2} = \frac{(10 \text{ v})^2}{20 \text{ w}} = \frac{100 \text{ v}^2}{20 \text{ w}} = 5 \text{ }\Omega$$

$$R_T = \frac{R_1 R_2}{R_1 + R_2} = \frac{20 \text{ }\Omega \times 5 \text{ }\Omega}{20 \text{ }\Omega + 5 \text{ }\Omega} = \frac{100 \text{ }\Omega^2}{25 \text{ }\Omega} = 4 \text{ }\Omega$$

$$I_T = \frac{E}{R_T} = \frac{10 \text{ v}}{4 \text{ }\Omega} = 2.5 \text{ A}$$

The total wattage can also be calculated two ways.

(1) $P_T = EI_T = 10 \text{ v} \times 2.5 \text{ A} = 25 \text{ w}$
(2) $P_T = P_1 + P_2$

$$P_1 = \frac{E^2}{R_1} = \frac{(10 \text{ v})^2}{20 \text{ }\Omega} = 5 \text{ w}$$

P_2 is given as 20 w

$P_T = 5 \text{ w} + 20 \text{ w} = 25 \text{ w}$

Example 5-7 Figure 5-13 is another example of two parallel resistors. What are the values of E_2, I_T, P_T, and R_2?

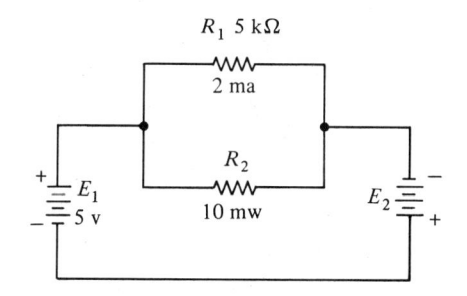

FIGURE 5-13 *Example for Two Voltage Sources*

Solution The unique thing about this circuit is the way the batteries are connected. Since the resistor circuit is of prime concern, a common ground connection may be assumed at the bottom of the circuit at the wire connecting the negative terminal of E_1 to the positive terminal of E_2. The voltage at the left side of the resistors is $E_1 = +5$ v with reference to common (ground). The voltage at the right side is $E_2 = -\text{v}$, as yet unknown.

Examination indicates that R_1 is the only part of the circuit that has two given quantities. The IR drop can be found from those quantities, I_1 and R_1:

$$E_1 - E_2 = I_1 R_1 = 2 \text{ ma} \times 5 \text{ k}\Omega = 10 \text{ v}$$

A 10-v drop across R_1 is also a 10-v drop across R_2. E_2 is specified as negative, so the IR drop must be subtracted from E_1:

$$E_2 = E_1 - 10 \text{ v} = 5 \text{ v} - 10 \text{ v} = -5 \text{ v}$$

Total current, the next unknown, is the sum of I_1 and I_2. I_1 is 2 ma, and I_2 is

$$I_2 = \frac{P_2}{E_1 - E_2} = \frac{10 \text{ mw}}{10 \text{ v}} = 1 \text{ ma}$$
$$I_T = I_1 + I_2 = 2 \text{ ma} + 1 \text{ ma} = 3 \text{ ma}$$

The total power is:

$$P_T = (E_1 - E_2)I_T = 5 \text{ v} - (-5 \text{ v}) \times 3 \text{ ma}$$
$$= 10 \text{ v} \times 3 \text{ ma} = 30 \text{ mw}$$

And the final unknown is

$$R_2 = \frac{E_1 - E_2}{I_2} = \frac{10 \text{ v}}{1 \text{ ma}} = 10 \text{ k}\Omega$$

Several ways of cross-checking the results are given:

$$R_T = \frac{R_1 R_2}{R_1 + R_2} = \frac{5 \text{ k}\Omega \times 10 \text{ k}\Omega}{5 \text{ k}\Omega + 10 \text{ k}\Omega} = 3.33 \text{ k}\Omega$$
$$I_T = \frac{E_1 - E_2}{R_T} = \frac{10 \text{ v}}{3.33 \text{ k}\Omega} = 3 \text{ ma}$$
$$P_1 = I_1^2 R_1 = (2 \text{ ma})^2 \times 5 \text{ k}\Omega = 20 \text{ mw}$$
$$P_T = P_1 + P_2 = 20 \text{ mw} + 10 \text{ mw} = 30 \text{ mw}$$

A comparison of the last four calculations with the problem solution shows perfect consistency in the methods.

Example 5-8 A final example is shown in Figure 5-14 for three resistors in parallel. Total power dissipated by the circuit is 360 mw.

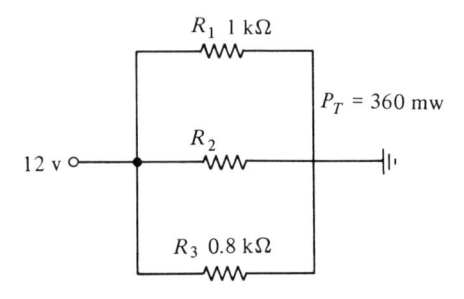

FIGURE 5-14 *Analysis for Total Power Dissipation*

Parallel Circuits 87

Find the value of R_2 and the power dissipated by each resistor.

Solution

$$I_T = \frac{P_T}{E} = \frac{360 \text{ mw}}{12 \text{ v}} = 30 \text{ ma}$$

$$I_1 = \frac{E}{R_1} = \frac{12 \text{ v}}{1 \text{ k}\Omega} = 12 \text{ ma}$$

$$I_3 = \frac{E}{R_3} = \frac{12 \text{ v}}{0.8 \text{ k}\Omega} = 15 \text{ ma}$$

$$I_2 = I_T - (I_1 + I_3)$$
$$= 30 \text{ ma} - (12 \text{ ma} + 15 \text{ ma}) = 3 \text{ ma}$$

$$R_2 = \frac{E}{I_2} = \frac{12 \text{ v}}{3 \text{ ma}} = 4 \text{ k}\Omega$$

$$P_1 = EI_1 = 12 \text{ v} \times 12 \text{ ma} = 144 \text{ mw}$$
$$P_2 = EI_2 = 12 \text{ v} \times 3 \text{ ma} = 36 \text{ mw}$$
$$P_3 = EI_3 = 12 \text{ v} \times 15 \text{ ma} = 180 \text{ mw}$$

The first four calculations were done to set up quantities that allowed a direct approach to answering the problem. The last four answered the problem in terms of R_2 and power.

QUESTIONS

5-1 The source voltage is 20 v in a closed circuit. What is the algebraic sum of the IR drops in the circuit?

5-2 If the algebraic sum of the currents into a junction is 100 ma, how much is the current out of the junction?

5-3 In Figure 5-3, if $I_T = 21$ A; $I_1 = 4$ A, and $I_2 = 7$ A, how much is I_3?

5-4 In Figure 5-4, if $E = 8$ v, $R_1 = 4$ k ohms, and $R_2 = 6$ k ohms, how much is I_T?

5-5 Refer to Problem 5-4. What value is R_T if $R_2 = 10$ k ohms?

5-6 If five resistors are connected in parallel, and each resistor is 10 ohms, what is the equivalent resistance?

5-7 If a 20-k ohm resistor and a 40-k ohm resistor are paralleled, what is their total resistance?

5-8 Refer to Figure 5-5. If $R_1 = 1$ k ohm, what is the value of R_3?

5-9 Four resistors in parallel are $R_1 = 15$ ohms; $R_2 = 25$ ohms, and $R_3 = 30$ ohms. If $R_T = 4.17$ ohms, what is the value of R_4?

5-10 Refer to Figure 5-9. If $I_3 = 3.5$ ma, what are the source voltage and total current?

5-11 Refer to Figure 5-10. If the total current is 3.5 A, what is the value of E?

5-12 Refer to Figure 5-11. If the total current is 15 ma, what is $E_1 - E_2$?

5-13 Refer to Figure 5-3. If the voltage between + and − is 7 v, what is the total power dissipated in the circuit?

5-14 Refer to Figure 5-5. What is the power dissipated by each resistor?

5-15 Refer to Figure 5-7. If $I_3 = 1$ ma, how much power is dissipated in the total circuit?

5-16 Refer to Figure 5-12. If R_2 dissipates 40 w, what are I_1, I_2, and R_2?

5-17 Refer to Figure 5-13. If I_1 is 10 ma, what are E_2, I_2 and P_T?

5-18 Refer to Figure 5-14. If $R_3 = 4$ k ohms, what are the values of R_2 and I_T?

6 Series Parallel Circuits

This chapter combines and extends the studies of Chapters 4 and 5. The various forms of Ohm's law will be applied to solve combinations of series and parallel circuits.

6.1 Kirchhoff's Law

The first part of Kirchhoff's law applies to Figure 6–1(a). It states that the algebraic sum of the *IR* drops and source voltages around a closed circuit equals zero, or the sum of the *IR* drops is equal to the sum of the source voltages. In Figure 6–1(a) there is only one source voltage, E_T. The *IR* drops are E_1 across the series resistor R_1, and E_2 across the parallel resistors R_2 and R_3. Applying Kirchhoff's law gives

$$E_1 + E_2 - E_T = 0$$

or

$$E_1 + E_2 = E_T$$

The circuit illustrates a series resistor (R_1) that is connected in series with parallel resistors $(R_2$ and $R_3)$. The circuit is redrawn in Figure 6–1(b). R_1 is the same, but R_2 and R_3 are shown as a single resistor. The illustration shows that series/parallel circuits can be equated to series circuits for ease of analysis.

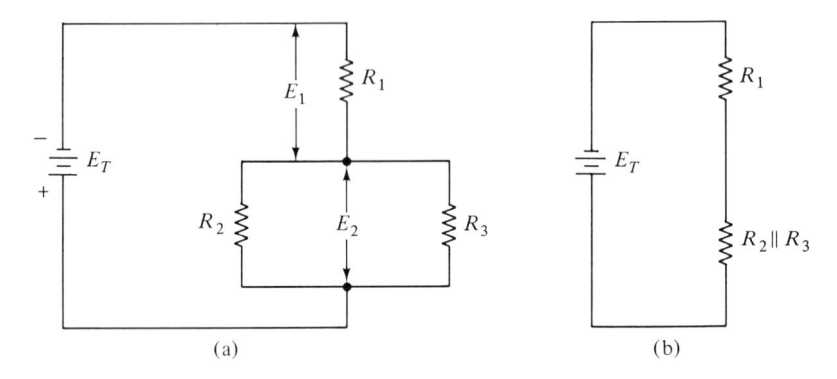

FIGURE 6-1 Basic Series/Parallel Circuit

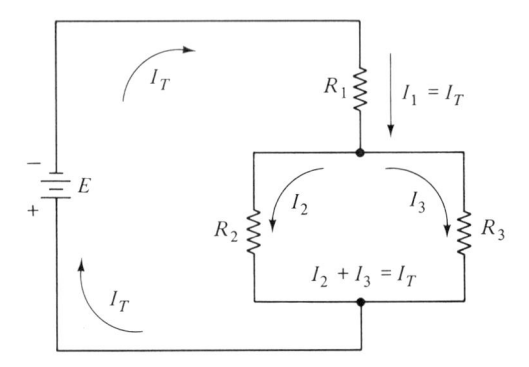

FIGURE 6-2 Series/Parallel Circuit for Current Flow

The second part of Kirchhoff's law applies to Figure 6-2. It states that the algebraic sum of the currents entering a junction is equal to the current leaving the junction. Notice that current I_T leaves the negative terminal of the battery. The current through R_1 is I_1 and is actually the same current as I_T, so $I_1 = I_T$. $I_1 = I_T$ enters the top of the junction of R_2 and R_3. The same current leaves the bottom of the junction. In parallel circuits, the sum of the individual currents equals the total. Therefore, $I_2 + I_3 = I_T = I_1$.

The laws apply for any combination. More than one resistor can be used for the series string. Any number can be used in parallel, and parallel branches can be placed in series strings. Regardless of complexity, the entire combination can be reduced to a single resistor.

6.2 Series/Parallel

Figures 6-1 and 6-2 are considered series/parallel circuits. In Figure 6-2, suppose $I_2 = 3$ A, and $I_3 = 7$ A. Then,

[6.2] Series / Parallel Circuits 91

$$I_T = I_2 + I_3 = 3\text{ A} + 7\text{ A} = 10\text{ A}$$
$$I_1 = I_T = 10\text{ A}$$

A series/parallel circuit is presented in a slightly different way in Figure 6-3. Actually, it is electrically identical to Figures 6-1 and 6-2. To analyze the circuit, the equivalent resistance (R_{eq}) of R_2 and R_3 is found:

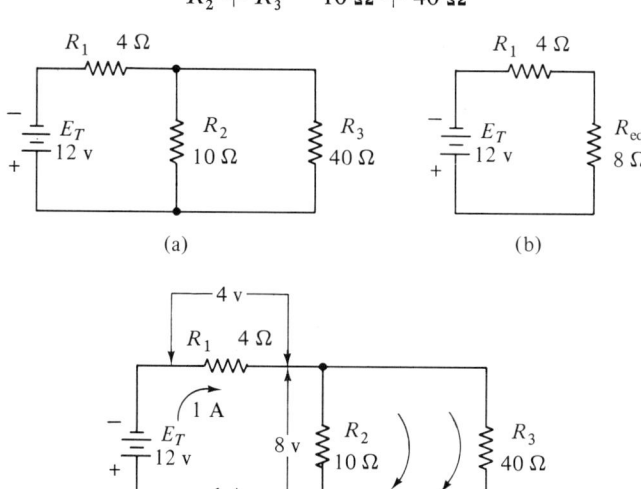

$$R_{eq} = \frac{R_2 R_3}{R_2 + R_3} = \frac{10\text{ Ω} \times 40\text{ Ω}}{10\text{ Ω} + 40\text{ Ω}} = 8\text{ Ω}$$

FIGURE 6-3 *Series/Parallel Analysis*

The equivalent circuit is shown in (b) as a series circuit. The current is

$$I_T = \frac{E_T}{R_1 + R_{eq}} = \frac{12\text{ v}}{4\text{ Ω} + 8\text{ Ω}} = 1\text{ A}$$

Since I_T is 1 A, the IR drop of R_1 is

$$E_{R1} = I_T R_1 = 1\text{ A} \times 4\text{ Ω} = 4\text{ v}$$

The IR drop across the parallel resistors is

$$E_{Req} = E_T - E_{R1} = 12\text{ v} - 4\text{ v} = 8\text{ v}$$

Notice that the IR drops agree with circuit (b) for an $I_T = 1$ A.

$$E_{R1} = I_T R_1 = 1\text{ A} \times 4\text{ Ω} = 4\text{ v}$$
$$E_{Req} = I_T R_{eq} = 1\text{ A} \times 8\text{ Ω} = 8\text{ v}$$
$$E_T = E_{R1} + E_{Req} = 4\text{ v} + 8\text{ v} = 12\text{ v}$$

The individual currents for R_2 and R_3 can be found since the IR drop has been found (8 v).

$$I_2 = \frac{E_{R2}}{R_2} = \frac{8 \text{ v}}{10 \text{ }\Omega} = 0.8 \text{ A}$$
$$I_3 = \frac{E_{R3}}{R_3} = \frac{8 \text{ v}}{40 \text{ }\Omega} = 0.2 \text{ A}$$
$$I_T = I_2 + I_3 = 0.8 \text{ A} + 0.2 \text{ A} = 1.0 \text{ A}$$

The results for Figure 6-3 are summarized in Figure 6-3(c). Now that the IR drops and currents are defined, power dissipation can be calculated. Recall that for series or parallel circits, the dissipation of the individual resistors is summed. Therefore, combining series and parallel circuits means that the dissipation of each resistor is summed, regardless of whether they are in series, parallel, or in any combination.

$$P_T = P_1 + P_2 + P_3 + \cdots + P_n \qquad (6\text{-}1)$$

Power dissipation for Figure 6-3(c) is

$$P_1 = E_{R1}I_1 = 4 \text{ v} \times 1 \text{ A} = 4 \text{ w}$$
$$P_2 = E_{R2}I_2 = 8 \text{ v} \times 0.8 \text{ A} = 6.4 \text{ w}$$
$$P_3 = E_{R3}I_3 = 8 \text{ v} \times 0.2 \text{ A} = 1.6 \text{ w}$$
$$P_T = P_1 + P_2 + P_3 = 4 \text{ w} + 6.4 \text{ w} + 1.6 \text{ w} = 12 \text{ w}$$

The equivalent circuit of Figure 6-3(b) can be used as a basis for verifying the total power dissipation.

$$R_T = R_1 + R_{eq} = 4 \text{ }\Omega + 8 \text{ }\Omega = 12 \text{ }\Omega$$
$$P_T = \frac{E_T^2}{R_T} = \frac{(12 \text{ v})^2}{12 \text{ }\Omega} = \frac{144 \text{ v}^2}{12 \text{ }\Omega} = 12 \text{ w}$$

which agrees with the previous calculations.

Figure 6-4 shows a series resistor at each end of two parallel resistors. The value of each resistor is shown, along with the total current. To find E_T (the source voltage), Ohm's law is applied to determine the three individual IR drops, which can then be summed to equal E_T. The IR drops across R_1, across the parallel branch of R_2 and R_3, and across R_4, are calculated in the following manner:

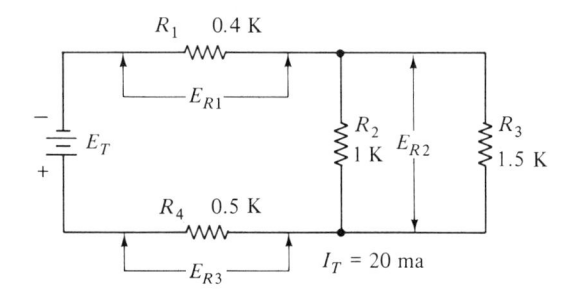

FIGURE 6-4 *Determining the Source Voltage*

[6.2] Series / Parallel Circuits 93

$$E_{R1} = I_T R_1 = 20 \text{ ma} \times 0.4 \text{ k}\Omega = 8 \text{ v}$$
$$E_{R2} = I_T(R_2 \| R_3) = 20 \text{ ma} \times \frac{1 \text{ k}\Omega \times 1.5 \text{ k}\Omega}{1 \text{ k}\Omega + 1.5 \text{ k}\Omega}$$
$$= 20 \text{ ma} \times 0.6 \text{ k}\Omega = 12 \text{ v}$$
$$E_{R4} = I_T R_4 = 20 \text{ ma} \times 0.5 \text{ k}\Omega = 10 \text{ v}$$
$$E_T = E_{R1} + E_{R2} + E_{R4} = 8 \text{ v} + 12 \text{ v} + 10 \text{ v}$$
$$= 30 \text{ v}$$

The currents through R_2 and R_3 are found:

$$I_2 = \frac{E_{R2}}{R_2} = \frac{12 \text{ v}}{1 \text{ k}\Omega} = 12 \text{ ma}$$
$$I_3 = \frac{E_{R2}}{R_3} = \frac{12 \text{ v}}{1.5 \text{ k}\Omega} = 8 \text{ ma}$$
$$I_T = I_2 + I_3 = 12 \text{ ma} + 8 \text{ ma} = 20 \text{ ma} = I_1 = I_4$$

Power dissipation for the entire circuit is found by summing the individual powers. The dissipation for each resistor is

$$P_1 = E_{R1} I_T = 8 \text{ v} \times 20 \text{ ma} = 160 \text{ mw}$$
$$P_2 = E_{R2} I_2 = 12 \text{ v} \times 12 \text{ ma} = 144 \text{ mw}$$
$$P_3 = E_{R2} I_3 = 12 \text{ v} \times 8 \text{ ma} = 96 \text{ mw}$$
$$P_4 = E_{R4} I_T = 10 \text{ v} \times 20 \text{ ma} = 200 \text{ mw}$$
$$P_T = P_1 + P_2 + P_3 + P_4 = 160 \text{ mw} + 144 \text{ mw}$$
$$+ 96 \text{ mw} + 200 \text{ mw} = 600 \text{ mw}$$

The total power can also be found:

$$P_T = E_T I_T = 30 \text{ v} \times 20 \text{ ma} = 600 \text{ mw}$$

Figure 6–5 is an example in which the IR drop across R_1 is 2 v, and R_3 is the unknown. The example is solved:

$$I_T = \frac{E_{R1}}{R_1} = \frac{2 \text{ v}}{20 \text{ }\Omega} = 0.1 \text{ A}$$
$$E_{R5} = I_T R_5 = 0.1 \text{ A} \times 50 \text{ }\Omega = 5 \text{ v}$$

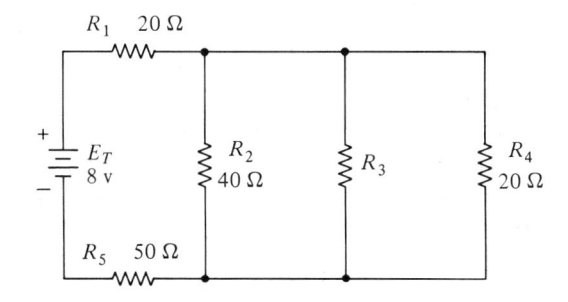

FIGURE 6–5 *Determining an Unknown Resistance*

$$E_{R2} = E_{par} = E_T - (E_1 + E_5) = 8\text{ v} - (2\text{ v} + 5\text{ v})$$
$$= 1\text{ v}$$

Now that the total current and the *IR* drop (E_{par}) of the parallel branch are known, the currents for the branch can be determined.

$$I_2 = \frac{E_{par}}{R_2} = \frac{1\text{ v}}{40\ \Omega} = 0.025\text{ A}$$
$$I_4 = \frac{E_{par}}{R_4} = \frac{1\text{ v}}{20\ \Omega} = 0.05\text{ A}$$
$$I_3 = I_T - (I_2 + I_4) = 0.1\text{ A} - (0.025\text{ A} + 0.05\text{ A})$$
$$= 0.025\text{ A}$$
$$R_3 = \frac{E_{par}}{I_3} = \frac{1\text{ v}}{0.025\text{ A}} = 40\ \Omega$$

The total power is

$$P_T = E_T I_T = 8\text{ v} \times 0.1\text{ A} = 0.8\text{ w}$$

and individual resistors dissipate as follows:

$$P_1 = E_{R1} I_T = 2\text{ v} \times 0.1\text{ A} = 0.2\text{ w}$$
$$P_2 = E_{par} I_2 = 1\text{ v} \times 0.025\text{ A} = 0.025\text{ w}$$
$$P_3 = E_{par} I_3 = 1\text{ v} \times 0.025\text{ A} = 0.025\text{ w}$$
$$P_4 = E_{par} I_4 = 1\text{ v} \times 0.05\text{ A} = 0.05\text{ w}$$
$$P_5 = E_{R5} I_T = 5\text{ v} \times 0.1\text{ A} = 0.5\text{ w}$$
$$P_T = 0.2\text{ w} + 0.025\text{ w} + 0.025\text{ w} + 0.05\text{ w} + 0.5\text{ w}$$
$$= 0.8\text{ w}$$

An example of a voltage divider is shown in Figure 6-6. The entire string of resistors is presented in a vertical line, and each junction has a test point labeled TP-0, TP-1, etc. Therefore, a voltmeter that has its positive lead connected to the common ground (TP-0) would have the voltage readings indicated at each TP in Figure 6-6. The voltages are negative with reference to common. A voltage divider can provide specific voltages to operate devices that require a smaller voltage than the available source voltage (which is -120 v in the figure). Notice the variety of voltages shown: TP-1 $= -27$ v, TP-2 $= -45$ v, TP-3 $= -75$ v, and TP-4 $= -120$ v. The voltages at the junctions are dependent upon the values of resistance that are distributed across the source.

Given the resistances, the voltages may be verified by applying Ohm's law. To do this, the currents must be calculated. Total current is derived through the following sequence:

$$\frac{1}{R_{eq}} = \frac{1}{R_3} + \frac{1}{R_4} + \frac{1}{R_5} = \frac{1}{40\ \Omega} + \frac{1}{60\ \Omega} + \frac{1}{24\ \Omega} = \frac{20}{240\ \Omega}$$

[6.2] Series / Parallel Circuits 95

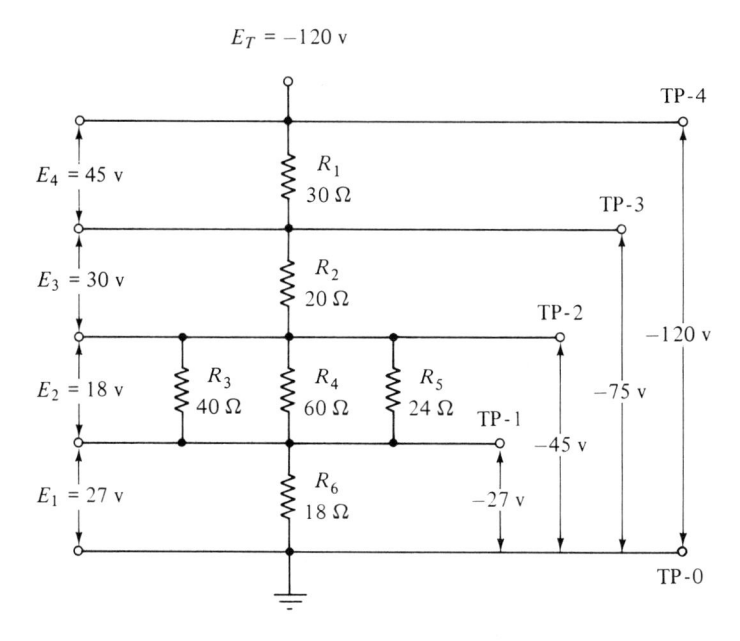

FIGURE 6-6 *Voltage Divider Network*

$$R_{eq} = \frac{240 \, \Omega}{20} = 12 \, \Omega$$

$$I_T = \frac{E_T}{R_1 + R_2 + R_{eq} + R_6}$$

$$= \frac{120 \text{ v}}{30 \, \Omega + 20 \, \Omega + 12 \, \Omega + 18 \, \Omega}$$

$$= \frac{120 \text{ v}}{80 \, \Omega} = 1.5 \text{ A}$$

The *IR* drops between junctions are shown along the left side of the circuit. They are calculated:

$$E_1 = I_T R_6 = 1.5 \text{ A} \times 18 \, \Omega = 27 \text{ v}$$
$$E_2 = I_T R_{eq} = 1.5 \text{ A} \times 12 \, \Omega = 18 \text{ v}$$
$$E_3 = I_T R_2 = 1.5 \text{ A} \times 20 \, \Omega = 30 \text{ v}$$
$$E_4 = I_T R_1 = 1.5 \text{ A} \times 30 \, \Omega = 45 \text{ v}$$
$$E_T = E_1 + E_2 + E_3 + E_4 = 27 \text{ v} + 18 \text{ v} + 30 \text{ v} + 45 \text{ v}$$
$$= 120 \text{ v}$$

The above calculations deal in magnitudes of voltage, and show a total voltage of 120 v. The sign is negative since the source is given as −120 v. Starting at ground (TP-0) the voltages go toward a negative source (-120 v) and become more negative as they approach the source. Each test point is a function of the *IR* drops. For example, TP-1 is 27 v which is the *IR* drop of R_6. It becomes -27 v because it is more negative.

$$\text{TP-2} = -E_1 - E_2 = -27 \text{ v} - 18 \text{ v} = -45 \text{ v}$$
$$\text{TP-3} = -E_1 - E_2 - E_3 = -27 \text{ v} - 18 \text{ v} - 30 \text{ v} = -75 \text{ v}$$

or

$$\text{TP-2} - E_3 = -45 \text{ v} - 30 \text{ v} = -75 \text{ v}$$
$$\text{TP-4} = \text{the source voltage } -120 \text{ v}$$

Now that the conditions shown in Figure 6-6 have been verified, other parameters such as power dissipation may be calculated.

$$P_1 = E_4 I_T = 45 \text{ v} \times 1.5 \text{ A} = 67.5 \text{ w}$$
$$P_2 = E_3 I_T = 30 \text{ v} \times 1.5 \text{ A} = 45 \text{ w}$$
$$P_3 = \frac{E_2^2}{R_3} = \frac{(18 \text{ v})^2}{40 \text{ }\Omega} = \frac{324 \text{ v}^2}{40 \text{ }\Omega} = 8.1 \text{ w}$$
$$P_4 = \frac{E_2^2}{R_4} = \frac{(18 \text{ v})^2}{60 \text{ }\Omega} = \frac{324 \text{ v}^2}{60 \text{ }\Omega} = 5.4 \text{ w}$$
$$P_5 = \frac{E_2^2}{R_5} = \frac{(18 \text{ v})^2}{24 \text{ }\Omega} = \frac{324 \text{ v}^2}{24 \text{ }\Omega} = 13.5 \text{ w}$$
$$P_6 = E_1 I_T = 27 \text{ v} \times 1.5 \text{ A} = 40.5 \text{ w}$$
$$P_T = P_1 + P_2 + P_3 + P_4 + P_5 + P_6$$
$$= 67.5 \text{ w} + 45 \text{ w} + 8.1 \text{ w} + 5.4 \text{ w} + 13.5 \text{ w} + 40.5 \text{ w}$$
$$= 180 \text{ w}$$

Taking the product of the source voltage and total current is another approach to deriving total power:

$$P_T = E_T I_T = 120 \text{ v} \times 1.5 \text{ A} = 180 \text{ w}$$

Figure 6-7(a) illustrates an application involving a voltage divider. The two 10-ohm resistors divide the source voltage in half.

$$I_T = \frac{E}{R_1 + R_2} = \frac{30 \text{ v}}{10 \text{ }\Omega + 10 \text{ }\Omega} = 1.5 \text{ A}$$
$$E_{R1} = I_T R_1 = 1.5 \text{ A} \times 10 \text{ }\Omega = 15 \text{ v}$$
$$E_{R2} = I_T R_1 = 1.5 \text{ A} \times 10 \text{ }\Omega = 15 \text{ v}$$

If the 15 v would remain constant at the junction regardless of the load, the circuit could be easily analyzed. However, the presence of an external load complicates the problem.

For example, suppose the switch is closed, and the external load $R_X = 10 \text{ }\Omega$. The resistance across the bottom junction is now

[6.2] Series / Parallel Circuits

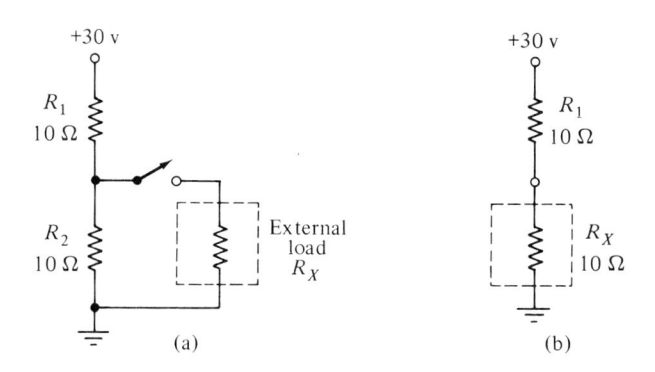

FIGURE 6-7 Variations with Loading

$$R_{par} = \frac{R_2 \text{ or } R_X}{2} = \frac{10\ \Omega}{2} = 5\ \Omega$$

Total current and the junction voltage are

$$I_T = \frac{E}{R_1 + R_{par}} = \frac{30\ v}{10\ \Omega + 5\ \Omega} = 2\ A$$

$$E_{R2} = I_T R_{par} = 2\ A \times 5\ \Omega = 10\ v$$

Without the load, the junction was +15 v with reference to ground. Adding a 10-ohm load drops the junction voltage from +15 v to +10 v. Therefore, when using voltage dividers for external circuits that will load the divider, the current requirement of the load must be calculated into the divider. For instance, in order to drive the external load with +15 v, R_1 must have the same value as the parallel resistance. In this case, $R_{par} = 5$ ohms, thus R_1 must be 5 ohms in order to split the 30 v in half.

Another approach is to include the external load in the voltage divider. Figure 6-7(b) shows the same load of 10 ohms in series with a 10-ohm resistor. The load still has 15 v across it and draws 1.5 A of current.

$$I_T = \frac{E}{R_1 + R_X} = \frac{30\ v}{10\ \Omega + 10\ \Omega} = 1.5\ A$$

$$E_{RX} = I_T R_X = 1.5\ A \times 10\ \Omega = 15\ v$$

Example 6-1 Suppose R_X in Figure 6-7(b) is 5 ohms and requires 0.5 A for proper operation. What value of R_1 will satisfy the requirement, and what would the voltage be across R_X?

Solution

$$R_T = \frac{E}{I_T} = \frac{30\ v}{0.5\ A} = 60\ \Omega$$

$$R_1 = R_T - R_X = 60\ \Omega - 5\ \Omega = 55\ \Omega$$

$$E_{RX} = I_T R_X = 0.5\ A \times 5\ \Omega = 2.5\ v$$

6.3 Parallel/Series

Parallel/series circuits are two or more parallel branches that are connected in series. The branches may consist of any number of parallel resistances. Figure 6-8(a) is a circuit with two parallel branches in series.

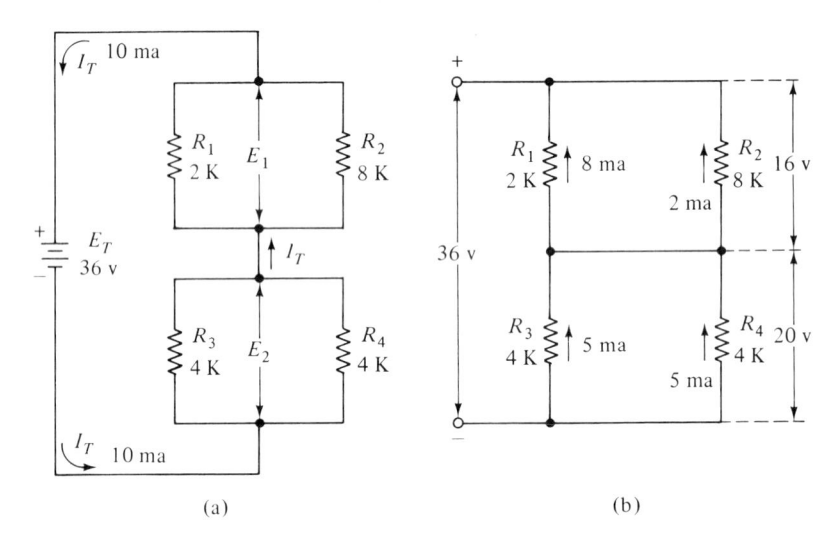

FIGURE 6-8 *Parallel/Series Circuit*

Notice that the total current (I_T) enters the lower branch where it divides. The current rejoins and becomes I_T at the center junction between the branches. I_T again divides, and rejoins at the top to enter the positive terminal of the battery. To determine the individual branch currents, E_1 and E_2 must be derived. To find E_1 and E_2, the parallel resistances must be derived.

For E_1:
$$R_{T1} = \frac{R_1 R_2}{R_1 + R_2} = \frac{2\ k\Omega \times 8\ k\Omega}{2\ k\Omega + 8\ k\Omega} = 1.6\ k\Omega$$
$$E_1 = I_T R_{T1} = 10\ ma \times 1.6\ k\Omega = 16\ v$$

For E_2:
$$R_{T2} = \frac{R_3 \text{ or } R_4}{2} = \frac{4\ k\Omega}{2} = 2\ k\Omega$$
$$E_2 = I_T R_{T2} = 10\ ma \times 2\ k\Omega = 20\ v$$

The *IR* drops and branch currents are shown in Figure 6-8(b).

$$I_1 = \frac{E_1}{R_1} = \frac{16\ v}{2\ k\Omega} = 8\ ma$$
$$I_2 = \frac{E_1}{R_2} = \frac{16\ v}{8\ k\Omega} = 2\ ma$$

[6.3]   Series / Parallel Circuits   99

$$I_3 = \frac{E_2}{R_3} = \frac{20 \text{ v}}{4 \text{ k}\Omega} = 5 \text{ ma}$$

$$I_4 = \frac{E_2}{R_4} = \frac{20 \text{ v}}{4 \text{ k}\Omega} = 5 \text{ ma}$$

The circuit of (b) is exactly the same as the one in (a). However, the connection [in (b)] across the center junction may be confusing. At this point, it does not matter which way current flows through the junction wire (that will be covered later in the text). The connection should simply be visualized as a wire that joins the parallel branches [as shown in (a)].

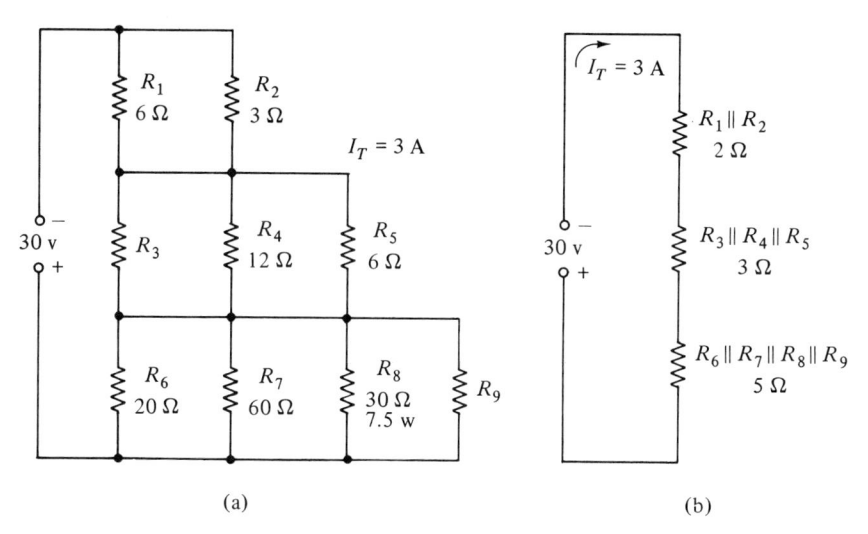

FIGURE 6-9 *Parallel/Series Analysis*

A more involved problem is illustrated in Figure 6-9 (a). The circuit resembles the one in Figure 6-8(b). For ease of analysis, the horizontal lines that connect the junctions should still be visualized as a single terminal as shown in Figure 6-8(a).

Example 6-2   Determine the values of R_3 and R_9 for the given conditions in Figure 6-9(a). Note that $I_T = 3$ A.

Solution   The IR drops are found across R_1, R_3, and R_8. The IR drop (E_{R1}) of R_1 is

$$R_{par} = \frac{R_1 R_2}{R_1 + R_2} = \frac{6 \, \Omega \times 3 \, \Omega}{6 \, \Omega + 3 \, \Omega} = 2 \, \Omega$$

$$E_{R1} = I_T R_{par} = 3 \text{ A} \times 2 \, \Omega = 6 \text{ v}$$

Series / Parallel Circuits [6.3]

The *IR* drop across R_8 is
$$E_{R8} = \sqrt{R_8 P_8} = \sqrt{30\ \Omega \times 7.5\ \text{w}}$$
$$= \sqrt{225\ \text{v}^2} = 15\ \text{v}$$

The *IR* drop across R_3 is
$$E_{R3} = E_T - (E_{R1} + E_{R8}) = 30\ \text{v} - (6\ \text{v} + 15\ \text{v})$$
$$= 9\ \text{v}$$

I_T has been given, and the *IR* drop has been derived across R_3. So the simplest way to find R_3 would be to find the individual currents for R_4 and R_5 and proceed with E/I.

$$I_4 = \frac{E_{R3}}{R_4} = \frac{9\ \text{v}}{12\ \Omega} = 0.75\ \text{A}$$

$$I_5 = \frac{E_{R3}}{R_5} = \frac{9\ \text{v}}{6\ \Omega} = 1.5\ \text{A}$$

$$I_3 = I_T - (I_4 + I_5) = 3\ \text{A} - (0.75\ \text{A} + 1.5\ \text{A})$$
$$= 0.75\ \text{A}$$

$$R_3 = \frac{E_{R3}}{I_3} = \frac{9\ \text{v}}{0.75\ \text{A}} = 12\ \Omega$$

The same method is applied for finding R_9.

$$I_6 = \frac{E_{R8}}{R_6} = \frac{15\ \text{v}}{20\ \Omega} = 0.75\ \text{A}$$

$$I_7 = \frac{E_{R8}}{R_7} = \frac{15\ \text{v}}{60\ \Omega} = 0.25\ \text{A}$$

$$I_8 = \frac{E_{R8}}{R_8} = \frac{15\ \text{v}}{30\ \Omega} = 0.5\ \text{A}$$

$$I_9 = I_T - (I_6 + I_7 + I_8) = 3\ \text{A} - (0.75\ \text{A} + 0.25\ \text{A} + 0.5\ \text{A}) = 1.5\ \text{A}$$

$$R_9 = \frac{E_{R8}}{I_9} = \frac{15\ \text{v}}{1.5\ \text{A}} = 10\ \Omega$$

An equivalent circuit for (a) is shown in (b). The parallel resistances are calculated:

$R_1 \parallel R_2$
$$R_T = \frac{R_1 R_2}{R_1 + R_2} = \frac{6\ \Omega \times 3\ \Omega}{6\ \Omega + 3\ \Omega} = 2\ \Omega$$

$R_3 \parallel R_4 \parallel R_5$
$$\frac{1}{R_T} = \frac{1}{R_3} + \frac{1}{R_4} + \frac{1}{R_5} = \frac{1}{12\ \Omega} + \frac{1}{12\ \Omega} + \frac{1}{6\ \Omega}$$
$$= \frac{4}{12\ \Omega}$$

$$R_T = \frac{12\,\Omega}{4} = 3\,\Omega$$

$R_6 \parallel R_7 \parallel R_8 \parallel R_9$

$$\frac{1}{R_T} = \frac{1}{R_6} + \frac{1}{R_7} + \frac{1}{R_8} + \frac{1}{R_9}$$
$$= \frac{1}{20\,\Omega} + \frac{1}{60\,\Omega} + \frac{1}{30\,\Omega} + \frac{1}{10\,\Omega} = \frac{12}{60\,\Omega}$$
$$R_T = \frac{60\,\Omega}{12} = 5\,\Omega$$

The sum of the equivalent resistances is

$$R = 2\,\Omega + 3\,\Omega + 5\,\Omega = 10\,\Omega$$
$$E_T = I_T R = 3\,\text{A} \times 10\,\Omega = 30\,\text{v}$$

The circuit of Figure 6–9 has been analyzed in several ways. In general, no set procedure is necessary. Any approach that solves the problem is considered appropriate.

6.4 Combination Circuits

Mixing series and parallel resistors adds to the complexity of analysis. The techniques studied so far are applicable to any combination of resistors. In general, a circuit can be redrawn into equivalent circuits to aid in understanding and problem solving. Some of this has already been illustrated.

Example 6–3 Figure 6–10(a) is an example of combining series and parallel circuitry. Determine the source voltage (E_T), total power dissipation, and power dissipated by each resistor.

Solution By reducing the circuit to a single resistance and by applying $I_T' = 3$ ma, E_T can be derived.
Circuit (b) shows $R_4 \parallel R_5$.

$$R = \frac{R_4 R_5}{R_4 + R_5} = \frac{2\,\text{k}\Omega \times 1.2\,\text{k}\Omega}{2\,\text{k}\Omega + 1.2\,\text{k}\Omega} = 0.75\,\text{k}\Omega$$

Circuit (c) combines R_3 with the parallel resistors.

$$R = R_3 + 0.75\,\text{k}\Omega = 2.25\,\text{k}\Omega + 0.75\,\text{k}\Omega = 3\,\text{k}\Omega$$

Circuit (d) parallels R_2 with the R of 3 kΩ.

$$R = \frac{R_2 \times 3\,\text{k}\Omega}{R_2 + 3\,\text{k}\Omega} = \frac{6\,\text{k}\Omega \times 3\,\text{k}\Omega}{6\,\text{k}\Omega + 3\,\text{k}\Omega} = 2\,\text{k}\Omega$$

Circuit (e) sums R_1 with the last R.

$$R_T = R_1 + R = 3\,\text{k}\Omega + 2\,\text{k}\Omega = 5\,\text{k}\Omega$$

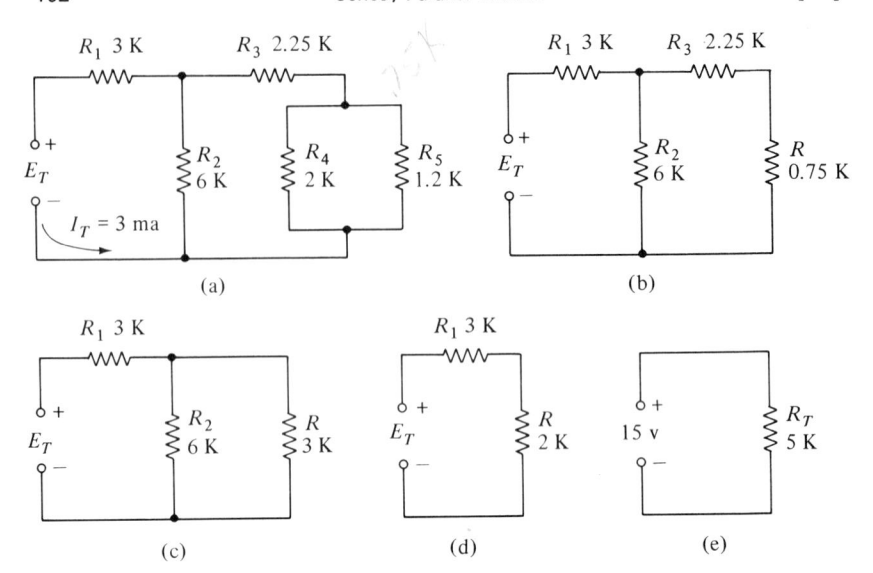

FIGURE 6-10 Combination of Series and Parallel Circuitry

Since the total resistance has been found, E_T can be derived with $I_T = 3$ ma.

$$E_T = I_T R_T = 3 \text{ ma} \times 5 \text{ k} = 15 \text{ v}$$

And P_T is

$$P_T = E_T I_T = 15 \text{ v} \times 3 \text{ ma} = 45 \text{ mw}$$

In circuit (a), the *IR* drops are:

$$E_{R1} = I_T R_1 = 3 \text{ ma} \times 3 \text{ k}\Omega = 9 \text{ v}$$
$$E_{R2} = E_T - E_1 = 15 \text{ v} - 9 \text{ v} = 6 \text{ v}$$
$$I_2 = \frac{E_{R2}}{R_2} = \frac{6 \text{ v}}{6 \text{ k}\Omega} = 1 \text{ ma}$$
$$I_3 = I_T - I_2 = 3 \text{ ma} - 1 \text{ ma} = 2 \text{ ma}$$
$$E_{R3} = I_3 R_3 = 2 \text{ ma} \times 2.25 \text{ k}\Omega = 4.5 \text{ v}$$
$$E_{R4} = E_{R2} - E_{R3} = 6 \text{ v} - 4.5 \text{ v} = 1.5 \text{ v}$$

All of the parameters are now available for calculating the dissipation in each resistor.

$$P_1 = E_{R1} I_T = 9 \text{ v} \times 3 \text{ ma} = 27 \text{ mw}$$
$$P_2 = E_{R2} I_2 = 6 \text{ v} \times 1 \text{ ma} = 6 \text{ mw}$$
$$P_3 = E_{R3} I_3 = 4.5 \text{ v} \times 2 \text{ ma} = 9 \text{ mw}$$
$$P_4 = \frac{E_{R4}^2}{R_5} = \frac{(1.5 \text{ v})^2}{2 \text{ k}\Omega} = 1.125 \text{ mw}$$

[6.4] Series / Parallel Circuits 103

$$P_5 = \frac{E_{R4}^2}{R_5} = \frac{(1.5 \text{ v})^2}{1.2 \text{ k}\Omega} = 1.875 \text{ mw}$$
$$P_T = P_1 + P_2 + P_3 + P_4 + P_5$$
$$= 27 \text{ mw} + 6 \text{ mw} + 9 \text{ mw} + 1.125 \text{ mw} + 1.875 \text{ mw}$$
$$= 45 \text{ mw}$$

Figure 6–7 was a study involving an external load and its effect on a voltage divider. External loads are further studied in Figure 6–11. A source voltage is given along with requirements for three loads. The fact that each has a different current and voltage requirement suggests the use of a voltage divider. To solve the problem, values of resistors have to be found to satisfy the loading requirements.

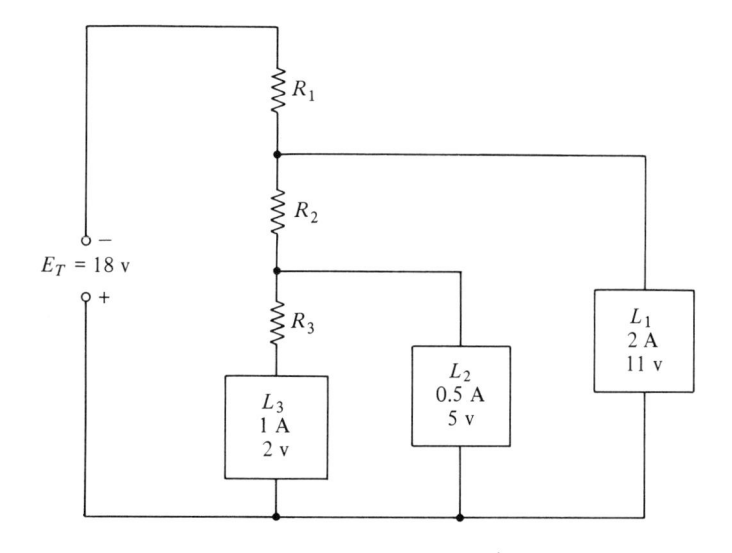

FIGURE 6–11 *Voltage Divider for Three External Loads*

In this situation, loads are usually calculated from the inside out. Load L_3 has a 1-A requirement, which means that R_3 will also carry 1 A. Both R_3 and L_3 will have an *IR* drop of 5 v as required by L_2. Therefore, the total resistance of R_3 and L_3 is

$$R_3 + L_3 = \frac{E_{L2}}{I_3} = \frac{5 \text{ v}}{1 \text{ A}} = 5 \text{ }\Omega$$
$$L_3 = \frac{E_{L3}}{I_3} = \frac{2 \text{ v}}{1 \text{ A}} = 2 \text{ }\Omega$$
$$R_3 = 5 \text{ }\Omega - L_3 = 5 \text{ }\Omega - 2 \text{ }\Omega = 3 \text{ }\Omega$$

The *IR* drop of R_2 is

$$E_{R2} = E_{L1} - E_{L2} = 11 \text{ v} - 5 \text{ v} = 6 \text{ v}$$

And the current through R_2 is
$$I_2 = I_3 + I_{L2} = 1 \text{ A} + 0.5 \text{ A} = 1.5 \text{ A}$$
The value of R_2 is
$$R_2 = \frac{E_{R2}}{I_2} = \frac{6 \text{ v}}{1.5 \text{ A}} = 4 \, \Omega$$
The IR drop across R_1 is the difference between the source voltage and the L_1 voltage requirement:
$$E_{R1} = E_T - E_{L1} = 18 \text{ v} - 11 \text{ v} = 7 \text{ v}$$
The current for R_1 is the sum of the two branches:
$$I_1 = I_2 + I_{L1} = 1.5 \text{ A} + 2 \text{ A} = 3.5 \text{ A}$$
And by Ohm's law,
$$R_1 = \frac{E_{R1}}{I_1} = \frac{7 \text{ v}}{3.5 \text{ A}} = 2 \, \Omega$$

Summarizing,
$$R_1 = 2 \, \Omega$$
$$R_2 = 4 \, \Omega$$
$$R_3 = 3 \, \Omega$$
$$I_T = I_1 = 3.5 \text{ A}$$
$$R_T = \frac{E_T}{I_T} = \frac{18 \text{ v}}{3.5 \text{ A}} = 5.14 \, \Omega$$

The entire circuit can be cross-checked by equating all resistances to a single resistance. Beginning at the inside,

$$L_3 = \frac{E_{L3}}{I_3} = \frac{2 \text{ v}}{1 \text{ A}} = 2 \, \Omega$$
$$L_3 + R_3 = 2 \, \Omega + 3 \, \Omega = 5 \, \Omega$$
$$L_2 = \frac{E_{L2}}{I_{L2}} = \frac{5 \text{ v}}{0.5 \text{ A}} = 10 \, \Omega$$
$$R_{par} = \frac{(L_3 + R_3) \times L_2}{(L_3 + R_3) + L_2}$$
$$= \frac{5 \, \Omega \times 10 \, \Omega}{5 \, \Omega + 10 \, \Omega} = 3.33 \, \Omega$$
$$R_{par} + R_2 = 3.33 \, \Omega + 4 \, \Omega = 7.33 \, \Omega$$
$$L_1 = \frac{E_{L1}}{I_{L1}} = \frac{11 \text{ v}}{2 \text{ A}} = 5.5 \, \Omega$$
$$R = \frac{(R_{par} + R_2) \times L_1}{(R_{par} + R_2) + L_1}$$
$$= \frac{7.33 \, \Omega \times 5.5 \, \Omega}{7.33 \, \Omega + 5.5 \, \Omega} = 3.14 \, \Omega$$
$$R_T = R_1 + R = 2 \, \Omega + 3.14 \, \Omega = 5.14 \, \Omega$$

Series / Parallel Circuits

This agrees with R_T as shown in the summary above.

The final analysis of Figure 6-11 consists of calculating the power dissipated in each of the three resistors. The IR drops and currents have already been calculated, so

$$P_1 = E_{R1}I_1 = 7 \text{ v} \times 3.5 \text{ A} = 24.5 \text{ w}$$
$$P_2 = E_{R2}I_2 = 6 \text{ v} \times 1.5 \text{ A} = 9 \text{ w}$$
$$P_3 = E_{R3}I_3 = 3 \text{ v} \times 1 \text{ A} = 3 \text{ w}$$

Note: $E_{R3} = E_{L2} - E_{L1}$ and $I_3 = I_{L3}$

$$P_T = P_1 + P_2 + P_3 = 24.5 \text{ w} + 9 \text{ w} + 3 \text{ w}$$
$$= 36.5 \text{ w}$$

And the total power for the circuit is

$$P_{\text{tot}} = E_T I_T = 18 \text{ v} \times 3.5 \text{ A} = 63 \text{ w}$$

The power dissipated in the loads is

$$P_L = P_{\text{tot}} - P_T = 63 \text{ w} - 36.5 \text{ w} = 26.5 \text{ w}$$

The load dissipation is

$$P_{L1} = E_{L1}I_{L1} = 11 \text{ v} \times 2 \text{ A} = 22 \text{ w}$$
$$P_{L2} = E_{L2}I_{L2} = 5 \text{ v} \times 0.5 \text{ A} = 2.5 \text{ w}$$
$$P_{L3} = E_{L3}I_{L3} = 2 \text{ v} \times 1 \text{ A} = 2 \text{ w}$$
$$P_L = 22 \text{ w} + 2.5 \text{ w} + 2 \text{ w} = 26.5 \text{ w}$$

QUESTIONS

6-1 Refer to Figure 6-3(a). If $R_1 = 8 \text{ }\Omega$, and $E_T = 8 \text{ v}$, what is the current through R_2?

6-2 For Question 6-1, what is the power dissipated by each resistor?

6-3 In Figure 6-4, if $R_1 = 0.1 \text{ k}\Omega$, what is the value of E_T?

6-4 In Figure 6-5, if $R_5 = 2 \text{ }\Omega$, $E_T = 15 \text{ v}$, and $I_T = 0.5 \text{ A}$, what is the resistance of R_3?

6-5 Refer to Figure 6-6. If $E_T = -30 \text{ v}$, and $R_1 = 10 \text{ }\Omega$, what are the values of E_1, E_2, E_3, and E_4?

6-6 For Question 6-5, what is the total power dissipated by the circuit?

6-7 For Question 6-5, what is the voltage between TP-0 and TP-3?

6-8 Refer to Figure 6-7. If $I_T = 0.75 \text{ A}$, and $R_1 = 32 \text{ }\Omega$, what is the current drain of the external load?

6-9 In Figure 6-8, assume $R_1 = 12$ kΩ. What is the total current in the circuit, and what is the current through R_2?

6-10 In Figure 6-10, if $R_3 = 5.25$ kΩ, what is the total power dissipated by the circuit?

6-11 Refer to Figure 6-11. The requirement for L_3 is 2 A. What are the values of R_1, R_2, and R_3?

6-12 Refer to Figure 6-12. If the current through R_1 is 1.6 ma, what is the value of R_3?

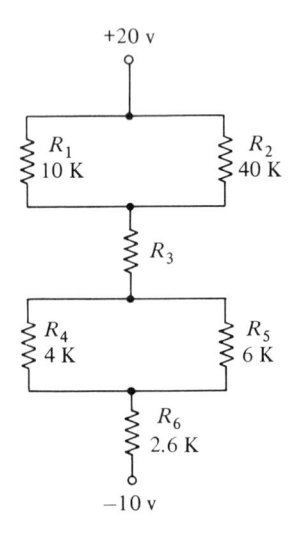

FIGURE 6-12

6-13 For Question 6-12, what is the power dissipated by R_4 and by R_5?

6-14 A particular load requires 1.2 A and 3 v for proper operation. If the source voltage is 18 v, what value of resistor is required in series with the load?

7 Alternating Voltage, Alternating Current, and Pulsating DC

The last few chapters have been devoted to direct current (dc) circuits. The source voltages were specified in terms of dc with positive and negative polarities. In each case the source was either a battery or a dc power supply with the same characteristics as a battery. The source was maintained at a steady level and represented a potential difference between the positive and negative terminals. When the voltage was connected across one or more resistors, a dc or steady current was developed.

This chapter introduces alternating voltage and current and pulsating dc levels. Alternating voltage swings between two levels. The voltage is usually visualized as swinging above (plus) and below (minus) a reference level. The reference may be zero or some dc level above or below zero. Pulsating dc occurs when the dc level shifts at prescribed intervals. The pulses may be narrow or wide, depending upon the application. They can also be positive or negative with respect to a reference level.

7.1 Alternating Voltage

The theory of induced current was studied in Chapter 2. When a wire is passed through a magnetic field, a current is induced in the wire. The magnitude of the current is a function of (1) the strength of the magnetic field and (2) the speed at which it passes through the field. In practical situations, a coil is used in placed of a single wire. In these cases, the magnitude of current also depends upon the number of turns in the coil. Since a wire has some amount of resistance, a voltage potential is developed across the ends of the wire or coil.

The drawing in Figure 7-1 demonstrates the principle of induced voltage. A simple loop represents the coils that would normally be used. The magnetic field is created by the north and south poles of the permanent

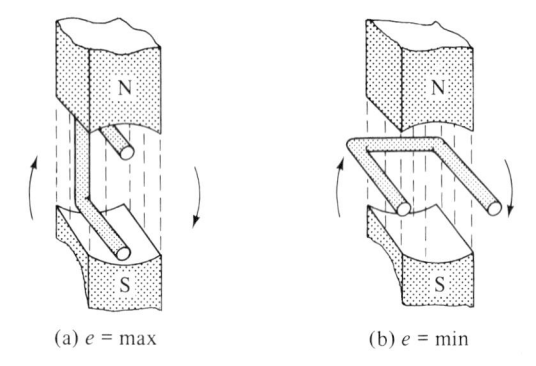

(a) e = max   (b) e = min

FIGURE 7-1 *Induced Voltage Principle*

magnet. In drawing (a) the loop cuts through the magnetic lines of flux. When the loop is moving perpendicular to the flux lines, it cuts the maximum number of lines per unit of time. At this point, the voltage (e) induced in the loop is at a maximum; e = maximum.

In drawing (b) the coil has rotated a quarter turn (90°). The loop cuts the minimum number of flux lines per unit of time; e = minimum. As the loop rotates and approaches the vertical position, the potential increases. When the loop passes the vertical position, the potential begins to decrease. Notice that for each complete turn (cycle) of the loop, the loop is in a vertical position twice and a horizontal position twice. Since the permanent magnet is polarized, a complete revolution causes the polarity of the voltage to change. This is the result of one side of the loop crossing left to right at the top while the other side of the loop crosses right to left at the bottom.

A complete turn or cycle of the loop is illustrated in Figure 7-2. The voltage potential across the ends of the loop is plotted as a function of time, but is labeled in degrees of rotation beginning at 0°. A quarter turn of the loop is represented by 90°, which is followed successively by 180°, 270°, and 360°, which is one complete cycle.

When the loop turns clockwise one-quarter turn (90°), the potential increases to a maximum. After the second quarter cycle, from 90° to 180°, the potential decreases to zero. The next quarter cycle, 180° to 270°,

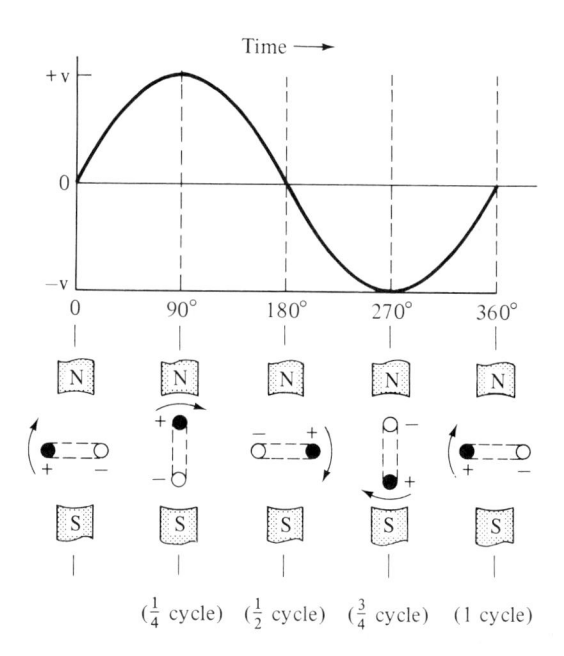

[7.2] Alternating Voltage, Alternating Current, and Pulsating DC

FIGURE 7–2 *Voltage versus Angle of Rotation*

shows the potential increasing in the opposite direction. The change in polarity is indicated by the direction of the positive side of the loop; it has changed direction through the magnetic flux. The last part of the cycle, 270° to 360°, shows the potential decreasing toward the zero reference line. The cycle would repeat itself as long as the loop turned.

In summary, notice that minimum voltage occurs at 0°, 180°, and 360° (minimum voltage occurs every 180°). Maximum voltage occurs at 90° and 270° (every 180°). Therefore, minimum and maximum voltage are displaced by 90°. In the drawing, the cycle began at 0°. In actual practice a cycle could begin at any point, 0, 1, 2 degrees, etc. The important thing is that the minimum and maximum voltage points are separated by 180°.

In Figure 7–2 time moves from left to right in the same direction as the loop. For An arbitrary amount of time, one cycle is shown. If the loop made a complete circle (360°) in one second, the frequency would be one cycle per second. If the loop turned 60 times in one second, the frequency would be 60 cycles per second (abbreviated 60 cps).

7.2 Angular Rotation

The drawings in Figures 7–1 and 7–2 show that the loops rotate a full 360° in a clockwise direction. The loop in Figure 7–3 begins at the left side and revolves in a clockwise direction to agree with the rotating loops in the

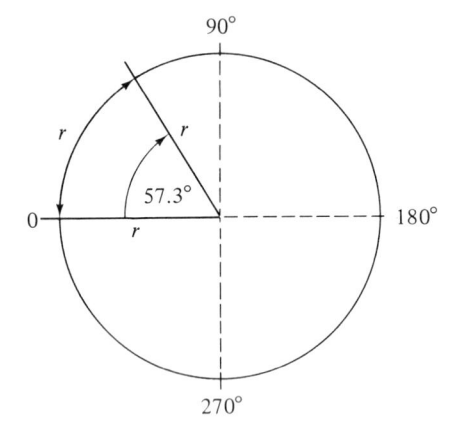

FIGURE 7-3 Angular Rotation

other figures. Rotation in a counter-clockwise direction produces angles that are the reverse.

The circle in Figure 7-3 illustrates a unit of measure called the **radian**. A radian is the angular section of a circle in which the length of the arc formed by the angle is equal to the radius of the circle. The section begins at 0° ends at 57.3°, the point where the length of the arc equals the radius. The circumference of a circle is πd (d for diameter). Since the diameter is twice the radius (r), the circumference is $2\pi r$. Since the radius is equal to one radian (rad), the circumference can also be written as 2π rad. Table 7-1 shows the relationship of degrees to radians in a circle.

TABLE 7-1

DEGREES	0	90	180	270	360
RADIANS	0	$\pi/2$	π	$3\pi/2$	2π

The amount of induced voltage from a rotating loop is proportional to the sine of the angle of rotation. The sine of the angle 0° is 0, and the sine of the angle 90° is 1. The minimum voltage is at 0°, and the maximum voltage is at 90°. As rotation continues from 90° to 180°, the voltage becomes proportionally smaller. Note that the sine of the angle for 180° is 0; for 270° it is −1, and for 360° it is 0. Table 7-2 compares rotational angles and their sines. Notice that half way through a quarter cycle (45°) the voltage reaches 70.7 percent of its maximum value. As a result, the waveform is steeper within the 45° on each side of its zero crossing. This steep slope is easily seen in Figure 7-2.

Alternating Voltage, Alternating Current, and Pulsating DC

TABLE 7-2

ANGLE (∠)	SINE ∠	VOLTAGE
0°	0	Zero
15°	0.26	
30°	0.50	50% of maximum positive value
45°	0.707	70.7% of maximum
60°	0.866	
75°	0.966	
90°	1.00	Maximum positive
135°	0.707	
180°	0	Zero
225°	−0.707	
270°	−1.00	Maximum negative
315°	−0.707	
360°	0	Zero

Because the magnitude of voltage is proportional to the sine of the angle, the waveform is referred to as a sinusoidal waveform or sine wave. A true sine wave can be reproduced graphically from the trigonometric laws of sines. The instantaneous value for any part of a sine wave is calculated using the following formula:

$$e_i = e_m \sin \theta \qquad (7\text{-}1)$$

in which e_i is the instantaneous value in volts; e_m is the maximum voltage of the sine wave, and θ (theta) is the angle of rotation. Several examples of Equation (7-1) follow for a maximum voltage of 10 v.

0°	$e_i = e_m \sin 0° = 10 \text{ v} \times 0 = 0 \text{ v}$
30°	$e_i = e_m \sin 30° = 10 \text{ v} \times 0.5 = 5 \text{ v}$
45°	$e_i = e_m \sin 45° = 10 \text{ v} \times 0.707 = 7.07 \text{ v}$
60°	$e_i = e_m \sin 60° = 10 \text{ v} \times 0.866 = 8.66 \text{ v}$
90°	$e_i = e_m \sin 90° = 10 \text{ v} \times 1.0 = 10 \text{ v}$
135°	$e_i = e_m \sin 135° = 10 \text{ v} \times 0.707 = 7.07 \text{ v}$
180°	$e_i = e_m \sin 180° = 10 \text{ v} \times 0 = 0 \text{ v}$
270°	$e_i = e_m \sin 270° = 10 \text{ v} \times -1.0 = -10 \text{ v}$
300°	$e_i = e_m \sin 300° = 10 \text{ v} \times -0.866 = -8.66 \text{ v}$

7.3 AC Current

When a sine wave of ac voltage is connected to a load resistance, the ac current through the resistance is directly proportional to the ac voltage. That is, when the voltage is zero, the current is zero, and when the voltage is maximum, the current is maximum. The ac voltage causes an ac current waveform that is also a sine wave. It is important to note that Ohm's law applies to ac as well as dc.

Figure 7-4(a) is a circuit that shows an ac voltage source that is connected to a 10-ohm resistor. Drawings (b) and (c) are voltage and current waveforms respectively. The sine waves are plotted on similar coordinates for ease of comparison. Notice that the minimum and maximum values match each other from 0° to 360°. When the maximum voltage (*e*) is 10 v, and the maximum current (*i*) is 1 A,

$$R = \frac{e_m}{i_m} = \frac{10 \text{ v}}{1 \text{ A}} = 10 \text{ }\Omega$$

which agrees with the value of *R* in diagram (a).

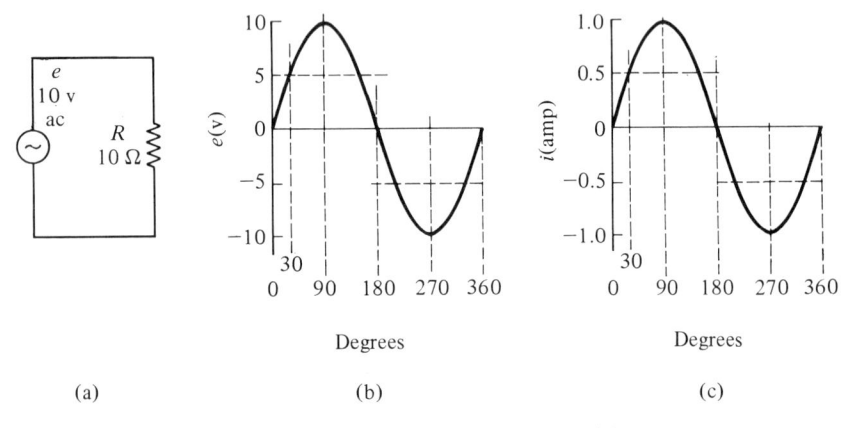

(a)   (b)   (c)

FIGURE 7-4 *Ohm's Law for ac Voltage and Current*

Notice that for 30° the voltage is 5 v, and the current is 0.5 A, causing *R* to be

$$R = \frac{e_i}{i_i} = \frac{5 \text{ v}}{0.5 \text{ A}} = 10 \text{ }\Omega$$

The instantaneous value of current for any part of a sine wave is calculated using the following formula:

$$i_i = i_m \sin \theta \qquad (7\text{-}2)$$

In the above formula i_i is the instantaneous value of current; i_m is the maximum value of current, and θ is the angle of rotation. Several examples follow which show the correlation between formulas (7-1) and (7-2) and Ohm's law.

60° $i_i = i_m \sin 60° = 1.0 \text{ A} \times 0.866 = 0.866 \text{ A}$

$$R = \frac{e_i}{i_i} = \frac{8.66 \text{ v}}{0.866 \text{ A}} = 10 \text{ }\Omega$$

90° $i_i = i_m \sin 90° = 1.0 \text{ A} \times 1.0 = 1.0 \text{ A}$

$R = \dfrac{e_i}{i_i} = \dfrac{10 \text{ v}}{1.0 \text{ A}} = 10 \text{ }\Omega$

135° $i_i = i_m \sin 135° = 1.0 \text{ A} \times 0.707 = 0.707 \text{ A}$

$R = \dfrac{e_i}{i_i} = \dfrac{7.07 \text{ v}}{0.707 \text{ A}} = 10 \text{ }\Omega$

In each case, the value of R is a constant 10 ohms, which tends to prove the validity of the expressions for instantaneous voltage and current.

Another relation called **phase** is evident from Figure 7-4(b) and (c). Notice that when the voltage begins to rise in a positive direction, the current follows in the same direction. When the current waveform duplicates the voltage waveform at each point on the horizontal axis, and both are sine waves, the current is in phase with the voltage across a resistor.

7.4 Amplitude Measurements

The amplitude of sinusoidal voltage and current waveforms can be specified in several ways. For example, Figure 7-4 voltage was discussed in terms of 0 to 10 v, or from the zero reference to the maximum point of the positive excursion. The maximum value is called the **peak voltage**, which is the peak amplitude above the reference line. The maximum positive and negative values (± 10 v) as a total would be peak-to-peak (*p-p*). In the figure voltage is 20 v *p-p*. In specifying an electrical quantity, the quantity also must be defined in terms of peak or *p-p*. A sine wave displayed on an oscilloscope will show the complete waveform in *p-p* terms.

The term average (ave) is sometimes used to indicate the average value of the peak waveform (half-cycle).

$$\text{ave} = 0.637 \times \text{peak} \qquad (7\text{-}3)$$

Ave is the arithmetical average of the sine values of a half-cycle (0–180°).

$$\dfrac{\sin 0° + \sin 1° + \cdots + \sin 180°}{\text{number of sine values}} = \text{ave} = 0.636 \times \text{peak}$$

The average value of a full cycle (0–360°) is zero because the positive and negative portions of the sine wave would cancel since they have identical forms.

Figure 7-5 is a drawing of one cycle, and it shows the relationship of peak-to-peak, peak, and ave. The vertical axis is plotted for the positive half-cycle in terms of 0 to 100 percent. Ave then is 63.7 percent of the total amplitude.

In the same figure, another value shown in called rms, which is 70.7 percent of the total amplitude. RMS is the abbreviation for **root mean square**. RMS is the most common way to specify alternating voltage and

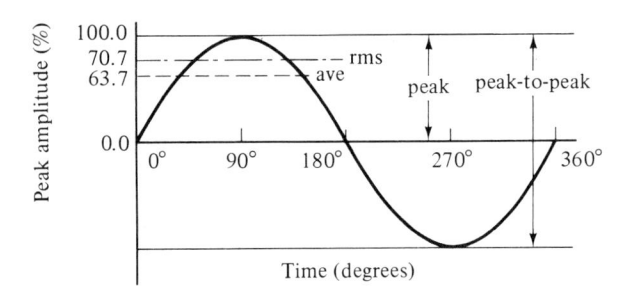

114   Alternating Voltage, Alternating Current, and Pulsating DC   [7.4]

FIGURE 7-5 *Sine Wave Conversion Factors*

current. Most ac meters are calibrated in rms values. Ordinary 117-v house voltage is an rms value. RMS values are equivalent to dc values. A voltage of 10 v rms is just as effective as 10 v dc, and for that reason rms is sometimes referred to as the **effective value**. In terms of peak values,

$$\text{rms} = 0.707 \times \text{peak} \qquad (7\text{-}4),$$

To find an rms value, the sines of the angles between 0° and 180° are squared, summed, and divided by the number of angles. The square root of the quotient is the rms value.

$$\sqrt{\frac{(\sin 0°)^2 + (\sin 1°)^2 + \cdots + (\sin 180°)^0}{\text{number of sine values}}} = \text{rms} = 0.707 \times \text{peak}$$

RMS values can be used to calculate power which is equivalent to dc power. For example,

$$P = E_{\text{rms}} I_{\text{rms}} = E_{\text{dc}} I_{\text{dc}}$$

Notice that both voltage and current are shown in rms terms. Therefore, Ohm's law for dc power is applicable to ac power when the ac is specified in rms values.

TABLE 7-3

Given	To Obtain:			
	ave	rms	peak	p-p
ave		1.11 × ave	1.57 × ave	3.14 × ave
rms	0.9 × rms		1.414 × rms	2.828 × rms
peak	0.637 × peak	0.707 × peak		2 × peak
p-p	0.318 × p-p	0.354 × p-p	0.5 × p-p	

[7.4] Alternating Voltage, Alternating Current, and Pulsating DC

Table 7-3 is a convenient reference for converting one form to another. The four forms, ave, rms, peak, and p-p, are listed both vertically and horizontally. To obtain rms from peak, locate rms at the top and peak on the side. The formula for rms (0.707 × peak) is found in the second column, straight across to the right.

Example 7-1 An ac voltmeter reads 120 v ac at a receptacle for an appliance. If an oscilloscope were connected to the outlet, what would be the peak-to-peak value?

Solution The meter reads in terms of rms. The oscilloscope displays the total waveform. From Table 7-3,

$$p\text{-}p = 2.828 \times \text{rms} = 2.828 \times 120 \text{ v} = 340 \text{ v } p\text{-}p$$

Example 7-2 What would an equivalent dc circuit look like for Figure 7-4(a)?

Solution In the figure, $e = 10$ v peak, and $i = 1.0$ A peak [see drawings (b) and (c)]. The rms voltage and current are

$$e_{\text{rms}} = 0.707 \times \text{peak} = 0.707 \times 10 \text{ v} = 7.07 \text{ v rms}$$
$$i_{\text{rms}} = 0.707 \times \text{peak} = 0.707 \times 1.0 \text{ A} = 0.707 \text{ A rms}$$

A dc voltage source of 7.07 v would produce the equivalent dc current of 0.707 A through the 10-ohm resistor.

$$I = \frac{E}{R} = \frac{7.07 \text{ v}}{10 \text{ }\Omega} = 0.707 \text{ A}$$

Example 7-3 The peak value of an alternating voltage across a resistor is 20 v. The average alternating current is 5 A. What is the power dissipated by the resistor, and what is the resistance of the resistor?

Solution To find the dc equivalents, the voltage and current must be converted to rms values. From Table 7-3,

$$e_{\text{rms}} = 0.707 \times \text{peak} = 0.707 \times 20 \text{ v} = 14.14 \text{ v}$$
$$i_{\text{rms}} = 1.11 \times \text{ave} = 1.11 \times 5 \text{ A} = 5.55 \text{ A}$$
$$P = e_{\text{rms}} \times i_{\text{rms}} = 14.14 \text{ v} \times 5.55 \text{ A} = 78.5 \text{ w}$$
$$R = \frac{e_{\text{rms}}}{i_{\text{rms}}} = \frac{14.14 \text{ v}}{5.55 \text{ A}} = 2.55 \text{ }\Omega$$

In Section 7-3, resistance was calculated from several angles of voltage and current sine waves. As long as the angles were the same for both waveforms, the resistance was constant (10 ohms for the example in Figure 7-4). It is important to stress that the two waveforms are directly proportional and of the same phase. RMS is also a proportionality (70.7 percent of peak value). Thus, resistance can be calculated from any angle of the waveforms as long as the terms are proportional. The rms values produce power values that are equivalent to dc power values.

7.5 Phase Angle

In previous discussions the waveforms of voltage and current were in phase. This occurs when the phase angle is zero, and the two waveforms begin at the same point, have the same time base, and complete a cycle in the same amount of time. Figure 7–6(a) is an example in which two

A leads B 180°

(a)

A lags B 90°

(b)

A leads B 90°

(c)

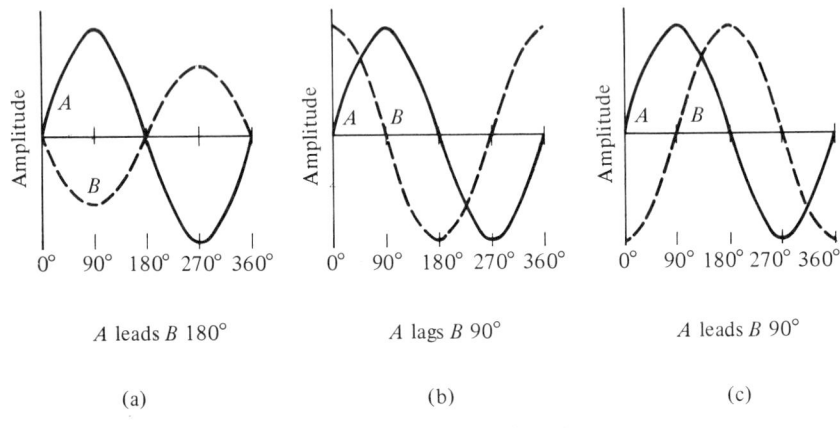

FIGURE 7-6 *Phase Relationships*

sinusoidal waveforms complete one cycle in the same time but are the opposite of each other. Wave A begins in the positive direction and reaches maximum positive amplitude at 90°. B begins in the negative direction and reaches maximum negative amplitude at 90°. The opposite is true at 270°, and both waveforms cross zero at 180°. Notice that wave A begins its positive excursion 180° before wave B begins its positive excursion. A leads B by 180°, or B lags A by 180°. Since the phase angle is 180°, the two waveforms are said to be 180° **out of phase**.

In drawing (a), wave A is shown with a larger amplitude than B. As long as both are sinusoidal and the time per cycle is the same, amplitude differences do not change the phase angle.

In drawing (b), wave B is at maximum amplitude when wave A is at zero, or B starts 90° before A, and every 90° one wave is at zero while other is at maximum (positively or negatively). Therefore, B leads A by the 90° or A lags B by 90°. The actual generation of B can be compared to the loop in Figure 7–2 beginning the cycle (0°) in its vertical position.

Waveform B is a cosine wave since it follows the rules of cosines. For example, $\sin 0° = 0$, and $\cos 0° = 1$, which agrees with Figure 7–6(b). Further, $\sin 90° = 1$, and $\cos 90° = 0$. Several comparisons are made in Table 7–4.

TABLE 7–4

Angle ($\angle$)	Sin $\angle$	Cos $\angle$
0°	0	1.00
15°	0.26	0.966
30°	0.50	0.866
45°	0.707	0.707
60°	0.866	0.50
75°	0.966	0.26
90°	1.00	0

The waveforms in Figure 7–6(c) illustrate another cosine relationship. Wave B is 90° later than wave A, or A leads B by 90°. The same proportions hold for (c) as hold for the previous cosine in (b); when one is at zero, the other is at its maximum value. The relationships in Table 7–4 are also true for (c), except that the cosine values begin as minus values during the first 90°.

7.6 Frequency

Frequency is defined as the number of cycles completed during a period of one second. Up to this point, waveforms have been defined in terms of angular degrees and radians for one complete cycle. These terms do not change. Each cycle is still measured in terms of its angle. An example of frequency is the 60-cycle power lines that supply ordinary house current. The 60 cycles means 60 cycles per second (60 cps).

Recently the term cps has been replaced by the word **Hertz** (Hz); 60 cps = 60 Hz. Frequency will now be referred to in units of Hz.

Rotating power generators for house current operate at 60 Hz. Other types of rotating generators operate up to several thousand Hz. Electrical circuits are capable of generating waveforms at frequencies in the millions per second. Recall that resistances were referred to in ohms (units), kilohms ($\times 10^3$), and megohms ($\times 10^6$). Frequency follows the same pattern, but with an additional unit known as the **giga** ($\times 10^9$). The units are compared below.

One cycle per second = 1 Hz
One kilocycle per second = 1 kHz
One megacycle per second = 1 MHz

118   Alternating Voltage, Alternating Corrent, and Pulsating DC   [7.6]

One gigacycle per second = 1 GHz

Figure 7-7 is an example of frequency. The upper waveform (A) is a single cycle during the time frame of 1 sec, and is 1 Hz. The lower waveform (B) shows six revolutions or six cycles during the same one-second

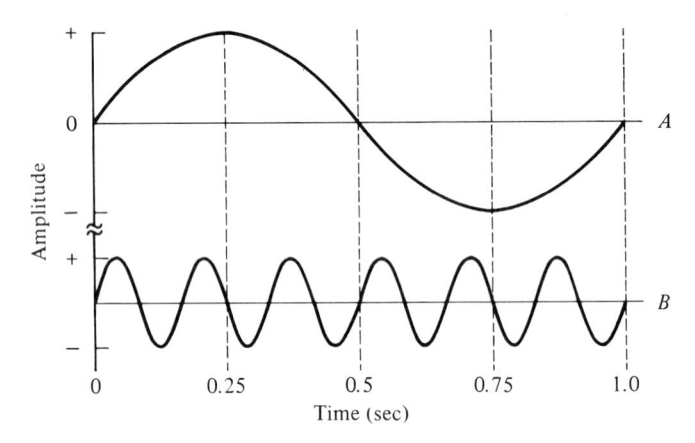

FIGURE 7-7 *Frequency of a Sine Wave Signal*

interval and is 6 Hz. Note that amplitude does not determine the frequency of a waveform. Several examples of frequency conversions follow.

Example 7-4   If 20 complete cycles occur during a period of 1 msec (1 sec × 10^{-3} = 1 msec), what is the frequency?

Solution   Since frequency (f) is cycles per unit time,

$$f = \frac{20 \text{ cycles}}{1 \text{ msec}} = \frac{20 \text{ cycles}}{1 \times 10^{-3} \text{ sec}} = 20 \times 10^3 \text{ Hz}$$
$$= 20 \text{ kHz}$$

Example 7-5   If 75 complete cycles occur during a period of one μsec (1 sec × 10^{-6} = 1 μsec), what is the frequency?

Solution
$$f = \frac{75 \text{ cycles}}{1 \text{ }\mu\text{sec}} = \frac{75 \text{ cycles}}{1 \times 10^{-6} \text{ sec}}$$
$$= 75 \times 10^6 \text{ Hz} = 75 \text{ MHz}$$

Example 7-6   What is the frequency if 300 cycles occur during a period of 60 μsec?

Solution
$$f = \frac{300 \text{ cycles}}{60 \text{ }\mu\text{sec}} = \frac{300 \text{ cycles}}{60 \times 10^{-6} \text{ sec}}$$
$$= 5 \times 10^6 \text{ Hz} = 5 \text{ MHz}$$

[7.6] Alternating Voltage, Alternating Current, and Pulsating DC 119

Time and frequency are also related in terms of a period. A period is the time it takes to complete one cycle. If the period for a cycle is 1 msec, the frequency is

$$f = \frac{1 \text{ cycle}}{1 \text{ msec}} = \frac{1 \text{ cycle}}{1 \times 10^{-3} \text{ sec}}$$
$$= 1 \times 10^3 \text{ Hz} = 1 \text{ kHz}$$

Since the period is the time for a frequency of one cycle, the frequency is actually the reciprocal of time:

$$f = \frac{1}{t} \tag{7-5}$$

In the formula f is in Hz, and t is in sec. By rearranging equation (7-5), the time for a single cycle is found:

$$t = \frac{1}{f} \tag{7-6}$$

Examples of frequency and period conversion follow.

Example 7-7 The time for one complete cycle is 0.1 μsec. What is the frequency?

Solution $\quad f = \dfrac{1}{t} = \dfrac{1}{0.1 \times 10^{-6} \text{ sec}} = 10 \times 10^6 \text{ Hz} = 10 \text{ MHz}$

Example 7-8 What is the period of one cycle for a frequency of 40 kHz?

Solution $\quad t = \dfrac{1}{f} = \dfrac{1}{40 \times 10^3 \text{ Hz}} = 25 \times 10^{-6} \text{ sec} = 25 \text{ μsec}$

Wavelength is defined as the length of one complete cycle at a specific frequency. For example, radio waves travel through air at a velocity equal to the speed of light (186,000 miles per second). The wavelength (abbreviated lambda λ) depends upon both frequency and velocity.

$$\lambda = \frac{\text{velocity}}{\text{frequency}} = \frac{v}{f} \tag{7-7}$$

When lambda is in metric values (normally centimeters or meters), velocity is in metric values (cm per sec or m per sec), and frequency is in Hz. The speed of light converts metrically as follows:

$$186{,}000 \text{ mi/sec} = 3 \times 10^8 \text{ m/sec} = 3 \times 10^{10} \text{ cm/sec}$$

In radio and television work, wavelength is usually measured in meters. Formula (7-7) is then

$$\lambda = \frac{3 \times 10^8 \text{ m/sec}}{f \text{ (Hz)}} \tag{7-8}$$

One important rule is that the wavelength and velocity must be in equivalent units (cm for cm, or m for m). Examples of wavelength follow.

Example 7-9 The FM radio band is from 88 to 108 MHz. What is the wavelength of the midband point?

Solution Since 100 MHz is close to midband (exact center of the band is 98 MHz),

$$\lambda(m) = \frac{v\,(m)}{f\,(Hz)} = \frac{3 \times 10^8 \text{ m/sec}}{100 \text{ MHz}}$$
$$= \frac{3 \times 10^8 \text{ m}}{100 \times 10^6} = 3 \text{ m}$$

Figure 7-8 shows a sine wave and its relationship to wavelength. Time is plotted from the period formula:

$$t = \frac{1}{f} = \frac{1}{100 \text{ MHz}} = \frac{1}{100 \times 10^6}$$
$$= 0.01 \text{ }\mu\text{sec} = 10 \text{ nsec}$$

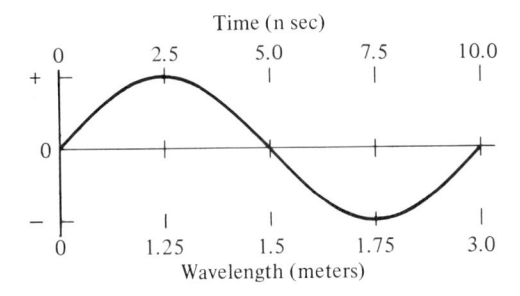

FIGURE 7-8 *Wavelength versus Frequency*

Example 7-10 What is the wavelength in meters and feet for a frequency of 15 MHz?

Solution
$$\lambda\,(m) = \frac{3 \times 10^8 \text{ m/sec}}{15 \text{ MHz}} = \frac{3 \times 10^8 \text{ m}}{15 \times 10^6}$$
$$= 0.2 \times 10^2 \text{ m} = 20 \text{ m}$$

To convert to inches,

1 in. = 2.54 cm   1 cm = 0.3937 in.
1 m = 39.37 in.

∴ λ (in.) = 20 m = 20 × 39.37 in. = 787.4 in.

and λ (ft) = $\frac{\text{in.}}{12} = \frac{787.4}{12} = 6.55$ ft

[7.7]   Alternating Voltage, Alternating Current, and Pulsating DC   121

Example 7-11   What is the frequency for a wavelength of 50 cm?

Solution
$$\lambda = \frac{v}{f} \quad \therefore \quad f = \frac{v}{\lambda} = \frac{3 \times 10^{10} \text{ cm/sec}}{50 \text{ cm}}$$
$$= 0.6 \times 10^9 \text{ Hz} = 600 \text{ MHz}$$

Sound waves are much slower than radio waves because they are transmitted by mechanical vibration rather than by electrical signals. The speed of sound waves in air is 1130 feet per second. Therefore, the wavelength for sound waves is determined:

$$\lambda = \frac{1130 \text{ ft/sec}}{f(\text{Hz})} \tag{7-9}$$

Example 7-12   If the frequency of a soundwave is 10 kHz, what is the wavelength in feet?

Solution
$$\lambda = \frac{1130 \text{ ft/sec}}{1 \text{ kHz}} = \frac{1130 \text{ ft}}{1 \times 10^3} = 1.13 \text{ ft}$$

Example 7-13   If the wavelength is 22.6 feet, what is the frequency of the sound waves?

Solution
$$f = \frac{1130 \text{ ft/sec}}{22.6 \text{ ft}} = \frac{1130}{22.6} = 50 \text{ Hz}$$

7.7 Pulsating DC

Pulsating dc is closely related to sine wave analysis. For example, when alternating voltage is at a dc level that varies in a sinusoidal manner, pulses are at dc levels that vary in a sharper manner, often a step function. One way that pulses can be generated is by opening and closing a switch as in Figure 7-9. The circuit in (a) has a dc source voltage of 10 v. When the switch is closed, the 10 v is applied across the 10-ohm resistor. The output is 10 v when the switch is closed and is zero when the switch is open.

Figure 7-9(b) is a sample of pulsating dc at the output of circuit (a).

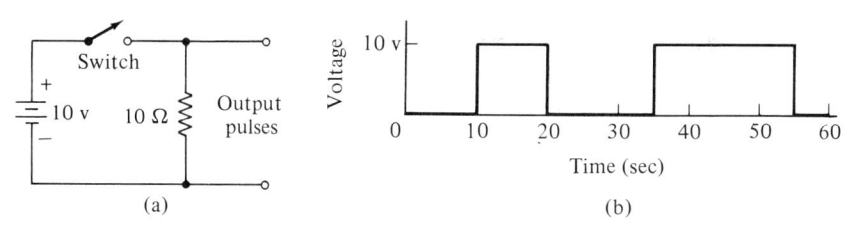

FIGURE 7-9 *Pulsating dc Levels*

122   Alternating Voltage, Alternating Current, and Pulsating DC   [7.7]

Assume the switch is open for 10 sec; the dc output is zero. After the initial 10 sec, the switch is closed for 10 sec (10 to 20 sec), and the output is 10 v. At the 20-second point, the switch is opened for 15 sec (20 to 35 sec). At 35 sec, the switch is closed for 20 sec (35 to 55 sec). Finally, at 55 sec, the switch is opened. This example shows pulses that are really quite slow. In modern electronics, a typical time scale is often shown in milliseconds, microseconds, or even nanoseconds.

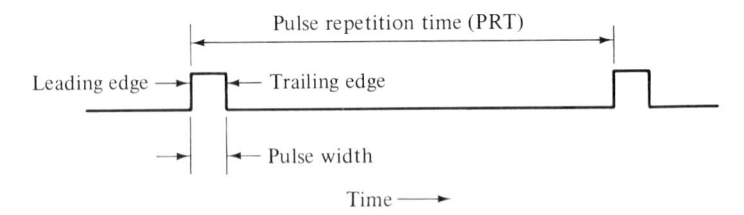

FIGURE 7-10 *Definitions for Pulsating dc*

Several terms associated with pulses are illustrated in Figure 7-10. Each pulse has a **leading** and a **trailing edge**. **Pulse width** (T_w) is the width of the pulse on a time scale (10 sec, 10 ms, 10 μs, etc.). The two pulses in Figure 7-9(b) have pulse widths of

$$T_w = 20 \text{ sec} - 10 \text{ sec} = 10 \text{ sec}$$
$$T_w = 55 \text{ sec} - 35 \text{ sec} = 20 \text{ sec}$$

Another term is the **pulse repetition time** (PRT). PRT is the time from the leading edge of one pulse to the leading edge of the next pulse. If a series of pulses all have the same PRT, the frequency is found:

$$f = \frac{1}{\text{PRT}} \qquad (7\text{--}10)$$

If the pulses are specified in frequency, PRT is found:

$$\text{PRT} = \frac{1}{f} \qquad (7\text{--}11)$$

Example 7-14 What is the frequency for a pulse repetition time of 1 μsec? of 20 msec?

Solution For PRT = 1 μsec,

$$f = \frac{1}{\text{PRT}} = \frac{1}{1 \text{ }\mu\text{sec}} = \frac{1}{1 \times 10^{-6}} = 1 \text{ MHz}$$

and for PRT = 20 msec,

$$f = \frac{1}{20 \text{ msec}} = \frac{1}{20 \times 10^{-3}} = 50 \text{ Hz}$$

Example 7-15 What is the pulse repetition time for pulses being generated at a rate of 25 kHz? of 400 MHz?

Solution
$$\text{PRT} = \frac{1}{f} = \frac{1}{25 \times 10^3} = 0.04 \times 10^{-3} = 40 \ \mu\text{sec}$$

$$\text{PRT} = \frac{1}{400 \times 10^6} = 2.5 \times 10^{-9} = 2.5 \ \text{nsec}$$

Example 7-16 What is the frequency for a train of pulses that occur at equal intervals if the pulses are each 2 μsec wide, and if the time from the trailing edge to the leading edge of the next is 18 μsec?

Solution PRT equals the time from one leading edge to the next. In this example, the pulse width must be added to the specified 18 μsec:

$$\text{PRT} = 18 \ \mu\text{sec} + 2 \ \mu\text{sec} = 20 \ \mu\text{sec}$$

$$f = \frac{1}{\text{PRT}} = \frac{1}{20 \ \mu\text{sec}} = \frac{1}{20 \times 10^{-6}}$$
$$= 0.05 \times 10^6 = 50 \ \text{kHz}$$

When working with pulsating dc, it is often convenient to consider a term called the **duty cycle**. Duty cycle defines the ratio of on and off times for a repetitious series of pulses. The ratio is expressed in percent; the percentage is for the portion of the PRT that a pulse is in the on state, or simply the pulse width (T_w) divided by the PRT:

$$\text{Duty Cycle} = \frac{T_w}{\text{PRT}} \qquad (7\text{--}12)$$

Figure 7-11(a) is an example of a 50-percent duty cycle. Notice that the pulses are on (positive) for 10 msec and off for 10 msec. The duty cycle is

$$\text{Duty Cycle} = \frac{T_w}{\text{PRT}} = \frac{10 \ \text{msec}}{20 \ \text{msec}} = 0.5 = 50 \ \%$$

Figure 7-11(b) is an example of a lower duty cycle. The pulses are on

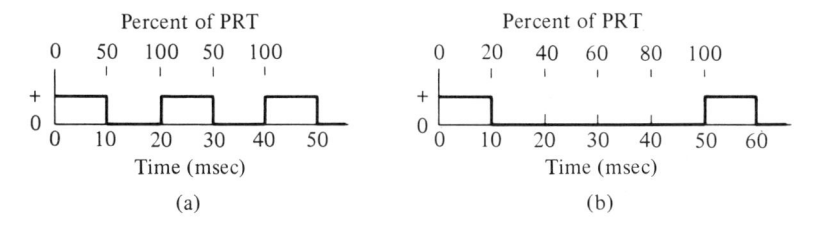

FIGURE 7-11 *Duty Cycle*

for 10 msec (0 to 10 msec) and off for 40 msec (10 to 50 msec). The pulse width (T_w) is 10 msec, and the PRT is 50 msec.

$$\text{Duty Cycle} = \frac{T_w}{\text{PRT}} = \frac{10 \text{ msec}}{50 \text{ msec}} = 0.2 = 20\%$$

Example 7-17 What is the duty cycle for a frequency of 100 kHz and a pulse width of 1 μsec?

Solution The pulse repetition time is

$$\text{PRT} \frac{1}{f} = \frac{1}{100 \text{ kHz}} = \frac{1}{0.1 \times 10^6}$$
$$= 10 \times 10^{-6} \text{ sec} = 10 \text{ } \mu\text{sec}$$

Since pulse width is 1 μsec, duty cycle is

$$\text{Duty Cycle} = \frac{T_w}{\text{PRT}} = \frac{1 \text{ } \mu\text{sec}}{10 \text{ } \mu\text{sec}} = 0.1 = 10\%$$

Example 7-18 If the wavelength is 20 cm, and the pulse width is 0.2 nsec, what is the duty cycle?

Solution Wavelength has to be converted to frequency, frequency to PRT, then PRT to duty cycle:

$$f = \frac{3 \times 10^{10} \text{ cm/sec}}{\lambda} = \frac{3 \times 10^{10}}{20}$$
$$= 1.5 \times 10^9 = 1.5 \text{ GHz}$$
$$\text{PRT} = \frac{1}{f} = \frac{1}{1.5 \times 10^9} = 0.67 \times 10^{-9} = 0.67 \text{ nsec}$$
$$\text{Duty Cycle} = \frac{T_w}{\text{PRT}} = \frac{0.2 \text{ nsec}}{0.67 \text{ nsec}} = 0.3 = 30\%$$

In practical circuit applications, pulses are rarely as precise as the step functions discussed here. DC levels do not ordinarily change so abruptly. When a voltage or current is switched, it takes awhile to reach its full on or off level. In fact, pulses tend to rise and fall in a nonlinear manner. Specific examples of pulse shaping are explained in later chapters after preliminary work is accomplished with the more complex circuits.

In Figure 7-12 the pulse rises slowly, then increases its slope as it passes the 10-percent point, then slowly levels off as it passes the 90-percent point. Because the slope tends to be erratic prior to the first 10 percent and past the final 10 percent points, **rise time** (T_R) is accepted as the time it takes a pulse to rise from 10 percent to 90 percent of its total amplitude. For the same reason, **fall time** (T_F) is defined from 90 percent to 10 percent of the total amplitude.

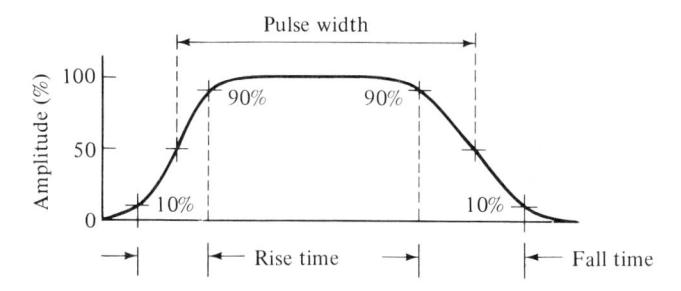

Alternating Voltage, Alternating Current, and Pulsating DC   125

FIGURE 7–12 *Definitions for a Single Pulse*

The **pulse width** (T_w) is normally defined at the 50-percent points on the curve. There are, however, cases where pulse widths are defined at other points. For instance, the pulse may swing between dc levels, and one certain level is considered important to circuit operation. Thus, pulse width can be defined at other points providing the points are specified.

QUESTIONS

7-1   How many degrees correspond to $\pi/5$ radians?

7-2   How many radians correspond to 210°?

7-3   If a sine wave of 12 v peak is applied across a resistor of 30 ohms, what is the peak current?

7-4   In Question 7-3 above, what is the peak power dissipated in the resistor?

7-5   What is the rms value of a peak-to-peak sine wave of 40 mv?

7-6   If the effective value of a sine wave current is 100 ma through a 1-kΩ resistor, what is the peak-to-peak voltage?

7-7   A sine wave has a maximum excursion of ± 15 v. What is its instantaneous amplitude (e_i) at 28.7°? At 280°?

7-8   If a sine wave has an instantaneous i of 13 ma at 135°, what is the maximum current?

7-9   If a sinusoidal waveform reaches a maximum negative value at 130°, what is its phase angle relative to the sine wave in Figure 7-4(b)?

7-10 If the phase angle of a sinusoidal waveform is 45° in relation to the sine wave in Figure 7-4(b), when does it reach its maximum negative value?

7-11 What is the frequency for a period of 200 nsec per cycle?

7-12 What is the period of one cycle at 40 kHz?

7-13 What is the wave length in centimeters for 15 MHz?

7-14 What is the frequency for a wave length of 25 cm?

7-15 What is the wavelength of a sound wave at 2.32 kHz?

7-16 What is the frequency for a sound wave with a wave length of 2 ft?

7-17 What is the PRT for 125 MHz?

7-18 What is the frequency for a PRT of 12.5 μsec?

7-19 If the frequency is 2 MHz, and duty cycle is 20 percent, what is the pulse width of a pulsating dc level?

7-20 What is the duty cycle for 5-μsec dc pulses at a frequency of 20 kHz?

8 Capacitance

This chapter explains capacitance and its relationship with electrical circuits. In general, a capacitor stores an electrical charge. By the very nature of its storage capacity, the capacitor transfers alternating or pulsating voltages and currents from one side to the other. Because the basic capacitor is formed by two plates separated by an insulating dielectric, it blocks direct current. A capacitor can therefore discriminate between ac and dc, making it possible to "guide" electrical signals through a circuit.

8.1 Charging a Capacitor

A **capacitor** can be defined as two metal plates (or bodies) that are separated and capable of storing a charge. When the plates have an accumulated charge of opposite polarity, an electrostatic field exists between them. Figure 8-1(a) illustrates an electrostatic field between opposite charges. The lines of force are mapped from minus to plus. Those in the center are strongest because the distance is closest. The field weakens as the flux lines move out from the center. The extreme lines are not shown connected because they are ordinarily so weak they cannot perform work of any significance.

Drawings (b) and (c) illustrate two metal plates and their flux lines. (b) shows the plates separated by air, and the flux lines are similar to those

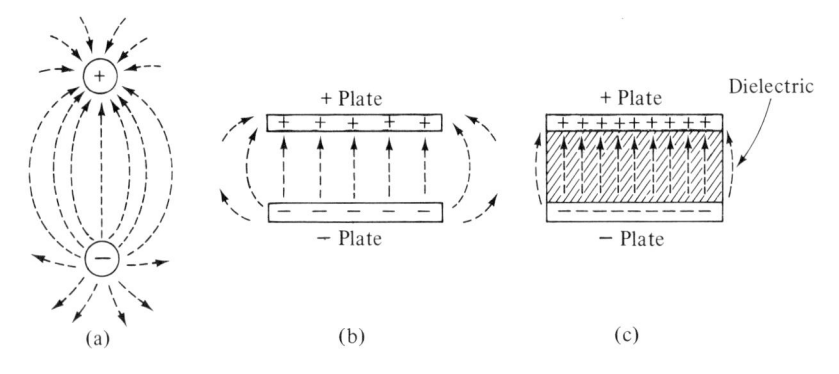

FIGURE 8-1 *Electrostatic Fields*

in (a). They spread at the ends and become weaker as they move outward. In (c) an insulating dielectric other than air is placed between the plates. The dielectric tends to concentrate the field and also provides a way to accurately separate the plates. Examples of dielectric insulating materials are mica, paper, and ceramic. Metal plates are capable of holding charges as long as they are insulated from each other.

The field created by charged plates of opposite polarity is defined in terms of force.

$$F = \frac{q_1 q_2}{d^2} \qquad (8\text{-}1)$$

In the above formula F is force in dynes, q_1 and q_2 are charges in electrostatic units (esu), and d is distance between the plates in centimeters. The equation assumes that the plates are separated by air, which has a **dielectric constant** (k) of one.

Various insulating dielectrics are defined in terms of a constant. Air (or a vacuum) is defined as $k = 1$ since it is used as a reference point. Table 8-1 lists typical dielectric constants for some of the common dielectric materials.

The values listed in Table 8-1 are only approximate since the constants are dependent upon the precise composition of the dielectrics. For

TABLE 8-1

Dielectric Constants (k)			
Air or Vacuum	1.0	Oil	2.8
Ceramic	100	Paper	2.5
Glass	7.0	Paraffin	2.0
Mica	5.8	Quartz	5.0

[8.1] Capacitance

inclusion of any dielectric, formula (8–1) becomes

$$F = \frac{q_1 q_2}{kd^2} \quad (8\text{–}2)$$

The q values are still in esu, but a more practical unit is the *coulomb*:

$$1 \text{ coulomb} = 3 \times 10^9 \text{ esu}$$

By substituting coulombs for esu, formula (8–2) becomes

$$F = (9 \times 10^{18}) \frac{q_1 q_2}{kd^2} \quad (8\text{–}3)$$

where F is in dynes, q_1 and q_2 are in coulombs, d is in centimeters and k is the dielectric constant from Table 8–1.

Example 8–1 Calculate the force between two plates that are separated by 2 cm of glass. Each plate is charged with 30 μcoulombs.

Solution The dielectric constant for glass is 7.0. Substituting the given values into equation (8–3) yields

$$F = (9 \times 10^{18}) \frac{30 \times 10^{-6} \times 30 \times 10^{-6}}{7 \times (2)^2}$$

$$= \frac{9 \times 10^{18} \times 900 \times 10^{-12}}{28} = \frac{8100 \times 10^6}{28}$$

$$= 289 \times 10^6 \text{ dynes}$$

Example 8–2 One plate of a capacitor has a charge of 15 μcoulombs, and the other plate has a charge of 40 μcoulombs. The force between the plates is 135×10^6 dynes with a dielectric of paper. How far apart are the plates?

Solution Equation (8–3) is manipulated in terms of distance squared.

$$d^2 = (9 \times 10^{18}) \frac{q_1 q_2}{Fk}$$

$$= \frac{9 \times 10^{18} \times 15 \times 10^{-6} \times 40 \times 10^{-6}}{135 \times 10^6 \times 2.5}$$

$$= \frac{5400 \times 10^6}{337.5 \times 10^6} = 16 \text{ cm}^2$$

$$d = \sqrt{d^2} = \sqrt{16 \text{ cm}^2} = 4 \text{ cm}$$

A basic capacitor has been defined as two plates capable of storing charges or electrical energy. When a capacitor is formed, the surfaces are spaced so that the distance between them is constant over both the surfaces. A crude capacitor can be formed with two sheets of metal held apart by several sheets of paper and then taped together to keep the metal plates from moving around.

To charge the plates of a capacitor, some means of external energy is required. Figure 8-2 is a circuit that illustrates the charging of a capacitor.

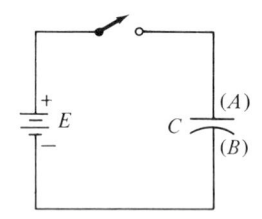

FIGURE 8-2 *Basic Charge Circuit*

Initially, the capacitor is neutral; each plate has an equal number of electrons, and the potential difference between the plates is zero. When the switch is closed, the positive terminal of the battery attracts electrons from the upper plate (A). Electrons flow toward the positive terminal, through the battery to the lower plate (B). Electrons continue to flow and accumulate on the lower plate until the potential difference across the plates is equal to the applied voltage. If the series resistance in the circuit were zero, the capacitor would charge to the applied voltage instantly. The effect of series resistance is explained in Chapter 10.

After the capacitor is charged, the switch can be opened, and the capacitor will store the applied voltage. That is, plate (B) has the excess accumulation of electrons; plate (A) has a deficiency of electrons, and when the switch is open, the charges do not have a path for equalizing. Thus, the potential difference across the plates will remain. In a practical sense, the dielectric has some amount of resistance or leakage, so eventually the charges will neutralize.

A charged capacitor can be discharged immediately by shorting the plates or by connecting a wire between the plates. Without series resistance, the discharge is instantaneous.

Figure 8-3 shows a sequence of charges and discharges. In (a), the capacitor is neutralized—an equal number of positive and negative charges are on each plate. In (b), a 12-v battery is connected across the capacitor. The positive terminal attracts electrons from the upper plate, and electron current flows counterclockwise in the circuit. When the difference in charges equals the applied potential of 12 v, the charge cycle stops, and no additional current flows.

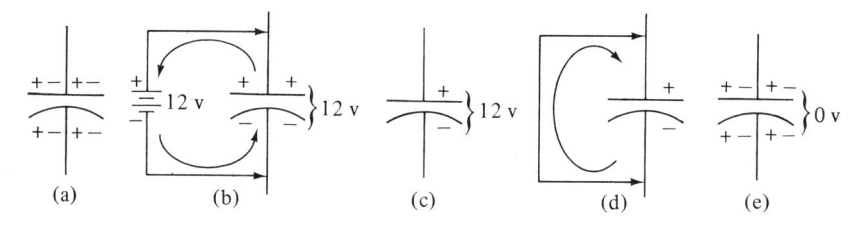

FIGURE 8-3 *Charge and Discharge Cycle*

In (c), the capacitor is shown alone with a stored charge of 12 v; the polarity is positive on top and negative on the bottom to match the current flow from diagram (b). The capacitor is being discharged in (d) when current flows clockwise; the positive plate attracts free electrons from the lower plate. The discharge continues until the charges balance. In (e), the capacitor is neutralized and has a zero potential across the plates.

8.2 Measurement of Capacitance

Capacitance is measured in terms of the **farad**, named after Michael Faraday (1791–1867). One farad (f) is defined as a stored charge of 1 coulomb with a potential voltage difference of 1 v:

$$C = \frac{Q}{E} \text{ or } 1\text{ f} = \frac{1 \text{ coulomb}}{1 \text{ v}} \tag{8-4}$$

Capacity C is in farads, Q is in coulombs, and E is in volts. One further point from Chapter 1,

$$Q = It \text{ or } 1 \text{ coulomb} = 1 \text{ amp} \times 1 \text{ sec} \tag{8-5}$$

Typical units of capacitance in practical applications are in the range of 10^{-6} or 10^{-12} f. Examples of abbreviated forms are

$$1 \times 10^{-6} \text{ f} = 1 \text{ microfarad} = 1 \text{ } \mu\text{f}$$
$$1 \times 10^{-12} \text{ f} = 1 \text{ picofarad} = 1 \text{ pf}$$

The pf was formerly referred to as $\mu\mu$f. Several examples follow using formulas (8-4) and (8-5).

Example 8-3 What is the capacitance for a charge of 300 μcoulombs and a potential of 75 v?

Solution Using formula (8-4) gives

$$C = \frac{Q}{E} = \frac{300 \times 10^{-6}}{75} = 4 \times 10^{-6} \text{ f} = 4 \text{ } \mu\text{f}$$

Example 8-4 What is the potential voltage for a charge of 20 μcoulombs and a capacitance of 0.5 μf?

Solution   Rearranging formula (8-4) yields

$$E = \frac{Q}{C} = \frac{20 \times 10^{-6}}{0.5 \times 10^{-6}} = 40 \text{ v}$$

Example 8-5   How much charge is stored if a constant current of 10 μa charges a capacitor for 12 sec?

Solution   Using formula (8-5) gives

$$Q = It = 10 \times 10^{-6} \times 12 = 120 \times 10^{-6}$$
$$= 120 \text{ } \mu\text{coulombs}$$

Example 8-6   How long does it take to charge a capacitor to 75 μ-coulombs with a constant current of 15 ma?

Solution   Rearranging formula (8-5) produces

$$t = \frac{Q}{I} = \frac{75 \times 10^{-6}}{15 \times 10^{-3}} = 5 \times 10^{-3} \text{ sec} = 5 \text{ msec}$$

Example 8-7   A 0.05-μf capacitor charges to 10 v in 100 msec. How much is the charging current? Assume a constant current source.

Solution   Rearranging formula (8-4) and substituting the given values gives

$$Q = CE = 0.05 \times 10^{-6} \times 10 = 0.5 \times 10^{-6} \text{ coulombs}$$
$$= 0.5 \text{ } \mu\text{coulombs}$$

Formula (8-5) becomes

$$I = \frac{Q}{t} = \frac{0.5 \times 10^{-6}}{100 \times 10^{-3}} = 5 \times 10^{-6} \text{ amp} = 5 \text{ } \mu\text{a}$$

If the area of the capacitor plates is increased, a larger charge can be stored. As the size increases, the intensity of the flux field increases. Therefore, capacitance is directly proportional to the area of the plates. Conversely, capacitance is inversely proportional to the distance between the plates. Recall formulas (8-1) and (8-2); the force increased by the inverse square of the distance. Capacitance is a simple inverse function. A formula for calculating capacitance is

$$C = \frac{kA \times 1.11 \times 10^{-12}}{4\pi d} \qquad (8\text{-}6)$$

where C is in farads, k is the dielectric constant (refer to Table 8-1), A is the area of one of the plates in cm^2, and d is the distance between the plates in cm. The constant 1.11×10^{-12} converts from esu to farads.

In formula (8-6), notice the relationship between area (*A*) and capacitance (*C*), and notice that *A* is directly proportional to *C* and that distance (*d*) is inversely proportional to *C*. Two examples of formula (8-6) follow.

Example 8-8 What is the capacitance for plates that are 3 cm² and separated by 0.4 cm of mica?

Solution Table 8-1 shows a dielectric constant (*k*) of 5.8 for mica. Substituting this into the equation yields

$$C = \frac{5.8 \times 3 \times 1.11 \times 10^{-12}}{4\pi \times 0.4} = \frac{19.3 \times 10^{-11}}{5.02}$$
$$= 3.84 \times 10^{-12} \text{ f} = 3.84 \text{ pf}$$

Example 8-9 What size plates are required to achieve 10 pf with a sheet of ceramic 2 cm thick?

Solution Ceramic has a dielectric constant of 100 (from Table 8-1). Rearranging formula (8-6) and substituting given values, gives

$$A = \frac{4\pi dC}{k \times 1.11 \times 10^{-12}} = \frac{4\pi \times 2 \times 10 \times 10^{-12}}{100 \times 1.11 \times 10^{-12}}$$
$$= \frac{251 \times 10^{-12}}{111 \times 10^{-12}} = 2.36 \text{ cm}^2$$

The dielectric constant (*k*) is also directly proportional to capacitance [see formula (8-6)]. As the dielectric becomes more dense, the flux between plates is concentrated to higher degrees. Figure 8-1 illustrated the concentration of a field between charged plates. An important point, though, is that the dielectric constant (*k*) increases the capacitance, but the force (*F*) is inversely proportional to the constant [refer to formula (8-2)]. So capacitance is simply the capacity of a capacitor in terms of accumulated charges or its ability to accept charges.

Various dielectrics and their constants have been discussed and used in calculating force and capacitance. Both force and capacitance are related to the charge (*Q*) on the plates, and charged plates result in a difference in potential between the plates. If the potential were high enough, the charge could jump the gap (arc) between plates. Thus, another reason for using a dielectric between plates is that it allows higher potentials to accumulate without arcing. For example, the dielectric strength of air is approximately 20 v per mil (1 × 10⁻³ in). The strengths of dielectrics listed in Table 8-1 range from about 200 v per mil for glass to more than 1200 v per mil for mica and paper. Characteristics of dielectrics other than air or vacuum are:

1. lower force (F)
2. concentrated flux field
3. higher capacity for storing charges
4. higher potentials per unit distance between plates

Stray capacitance in a circuit is the effect of wiring and various components. Capacitance is the effect of two metal (or current carrying) objects with an area A and separated by a distance d [see formula (8–6)]. The capacitance can be wire-to-wire, wire-to-chassis, wire-to-component, component-to-component, or parts within a component. Stray capacitance values are normally quite small, ranging from less than 1 pf to 15 or 20 pf. Subsequent sections in this chapter will show that stray capacitance is insignificant for low frequency applications and is quite significant at high frequencies.

Leakage resistance for a capacitor is the leakage that occurs through the insulating dielectric and the insulation between the leads. There is always some leakage regardless of the dielectric selected. When a voltage source is removed from a capacitor, the plates will eventually become neutralized. Electrolytics have leakage resistances as low as 250 kΩ for large capacitors. Leakage is typically 10 to 100 M ohms for dielectrics of paper, mica, or ceramic.

8.3 DC and AC Effects

When a capacitor is placed in a circuit, it acts as an open circuit to dc voltage. The reasoning is somewhat obvious since the plates are insulated by air or some other dielectric. The dc will charge the capacitor to a potential that is equal to the applied voltage, and current will stop flowing. Because a discharge path does not exist, the capacitor will hold the charge —positive on one plate and negative on the other plate.

Conversely, a capacitor acts as a near short-circuit to ac voltage and pulsating dc. For example, ac changes polarity at a given rate per second. On the positive cycle, the capacitor charges in one direction, and on the negative cycle, the capacitor charges in the opposite direction. Current changes direction so that the charge on each plate reverses once for each cycle.

Figure 8–4 illustrates the charge and discharge process for an ac signal. One complete cycle is shown from 0 to 360°. The diagrams across the top of the waveform show the instantaneous charge condition for every 90°. The diagrams across the bottom show the current flow during the 90° increments. At 0° the capacitor is neutralized with equal charges on each plate. During the first 90° the voltage is rising in a positive direction; the potential is increasing. The electrons are attracted from the upper plate,

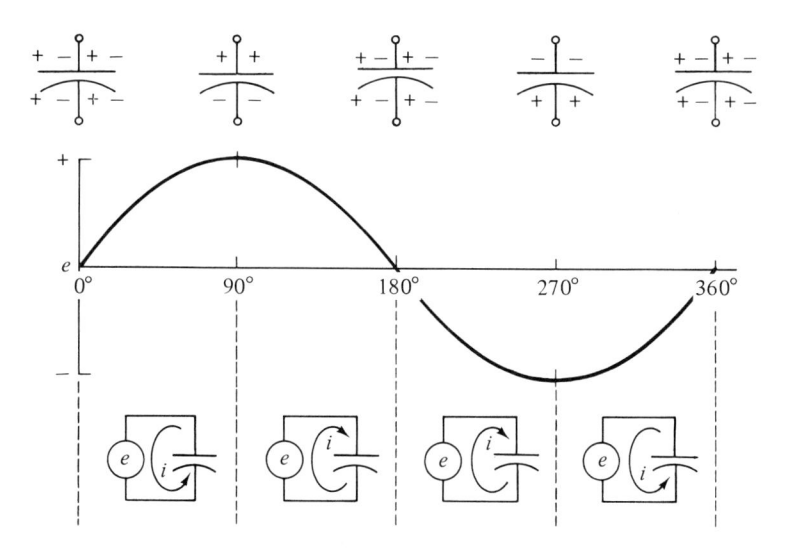

FIGURE 8-4 *Charge Reaction to Sine Wave of Voltage*

causing current to flow in a counterclockwise direction. The process results in negative charges on the lower plate. At the 90° point the cycle is momentarily at its peak, and the capacitor has a maximum charge equivalent to the applied peak voltage.

The potential decreases to zero from 90° to 180°. Since the potential is decreasing, the source provides a discharge path and allows the charge on the capacitor to decrease. Thus, current flows in the opposite direction (clockwise). As soon as the sine wave reaches 180° (zero potential), the charges on the capacitor are equalized, following the rule that the charge potential is equal to the applied voltage. From 180° to 270° the current continues in the same direction. The previous quarter-cycle simply equalized the charge (replaced charged electrons). Now that the cycle is increasing in the negative direction, it continues moving electrons in the same direction (clockwise) until the capacitor is fully charged in the opposite polarity at 270°.

From 270° to 360°, the potential decreases to zero. The current reverses direction and flows counterclockwise as the charges move toward equilibrium. The charges are balanced, and the potential across the capacitor is zero at 360°.

Although current did not actually flow through the capacitor, current did flow through the circuit—first in one direction and then in the other. Current flow in the circuit is what actually accomplishes work. The capacitor provides a means for the electrons to accumulate on alternate plates.

136   Capacitance   [8.3]

If the circuit were open and without a capacitor, current could not flow because there would not be a place for charges to accumulate.

Pulsating dc works in much the same way as ac, except the charge action is more abrupt. The pulse in Figure 8-5 is spread to illustrate the charge and dischage paths. From 0 to 30 msec, the charges remain neutralized.

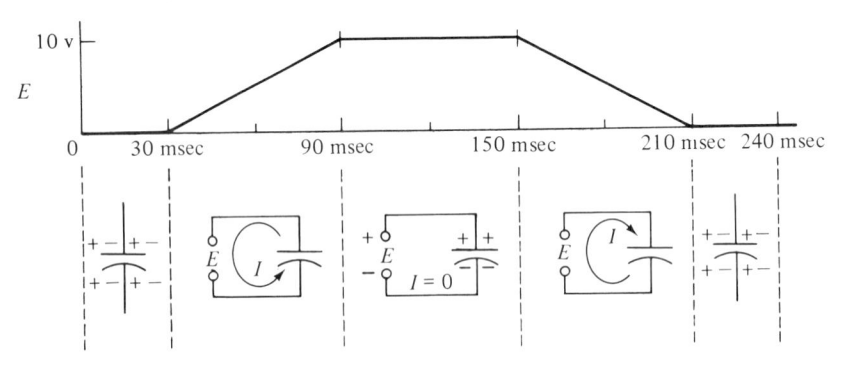

FIGURE 8-5 *Charge Reaction to dc Pulse*

From 30 msec to 90 msec, the pulse rises from 0 v to 10 v, and the capacitor charges from a zero potential to a 10-v potential. Notice that current flow is counterclockwise. The capacitor remains charged between 90 and 150 msec. The voltage decreases to zero from 150 to 210 msec. During the decreasing potential, the current flows in the oposite direction as the electrons flow to the opposite plate. Beyond 210 msec the capacitor remains neutralized.

Notice that current flows in opposite directions at 30-90 msec and at 150-210 msec. Current does not flow while $E = 10$ v (between 90 and 150 msec) because the capacitor is fully charged to equal the applied voltage. The capacitor behaves as though a dc battery with a potential of 10 v were being applied across the plates.

The circuits would operate in exactly the opposite way for a negative pulse. The capacitor would still be neutralized when $E = 0$. The current would flow in the other direction (clockwise) while the level shifted toward -10 v, and at $E = -10$ v the capacitor would remain charged negatively on the upper plate and positively on the lower plate. The current would flow counterclockwise while the level shifted from -10 v to zero.

Figure 8-6(a) is a circuit that combines two source voltages. The dc battery maintains a reference charge potential of 15 v across the plates of the capacitor. The ac source (e) either adds or subtracts from the reference voltage of 15 v. In (b) the circuit voltages are shown in graphic form. The reference level is a dotted line at the 15-v point. A sine wave is shown

[8.3]   Capacitance

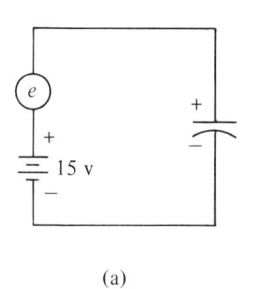

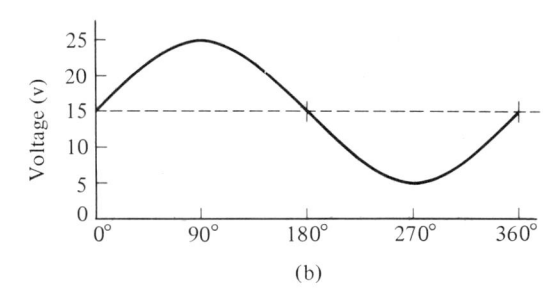

FIGURE 8-6   *ac Voltage on dc Reference Level*

swinging plus and minus around the reference. During the first 90° the sine wave potential increases in a positive direction and causes additional current to flow which increases the total potential across the capacitor. From 90° to 180° the sine wave decreases to the reference level and redistributes the added electron charge that was accumulated during the first 90°.

From 180° to 270° the sine wave potential opposes the dc battery and redistributes more electrons and reduces the net charge potential across the capacitor. From 270° to 360° the charge increases to the reference level of 15 v (at 360°).

Figure 8-6 illustrates the effect of changing voltage around a reference level. The theory holds true for any dc level, whether positive or negative. Pulsating dc has a similar reaction. Pulses follow the same rules used in Figure 8-5. Either positive or negative pulses operate by adding or subtracting from the reference level, whether the level is positive or negative.

The efficiency with which the charging and discharging of a capacitor follows the applied voltage is something less than 100 percent. For example, the dielectric is unable to immediately follow instantaneous ac changes. At higher frequencies in and above the radio bands (0.5 MHz and up) the efficiency falls off more and more. The inefficiency is known as the **absorption loss**. The absorption loss is actually the difference between the applied ac and the ac stored in the capacitor at any instant in time.

In Chapter 2, losses in magnetics were described as the **hysteresis loss**. Absorption loss follows the same principle as hysteresis loss.

Heat loss in the dielectric is another type of loss in capacitors. Heat loss is measured in terms of **power factor**, which is the ratio of the input power that is dissipated by heat within the dielectic of a capacitor. For example, a power factor of 0.001 means 1/1000 of the input power is dissipated as heat loss. Power factor may be expressed as a decimal or percentage. Examples for paper, mica, and ceramic dielectrics range from 0.0001 to 0.01 for frequencies in the RF band.

8.4 Capacitors in Series

When capacitors are connected in series, the effect is an increase in the thickness of the dielectric insulation between them. Increasing the distance decreases the capacitance [see formula (8-6)].

$$C = \frac{kA 1.11 \times 10^{-12}}{4\pi d}$$

A sequence of charging and discharging two capacitors in series is shown in Figure 8-7. Diagram (a) has the capacitors neutralized. In (b) the capacitors charge with an electron current flow from the upper plate of C_1 to

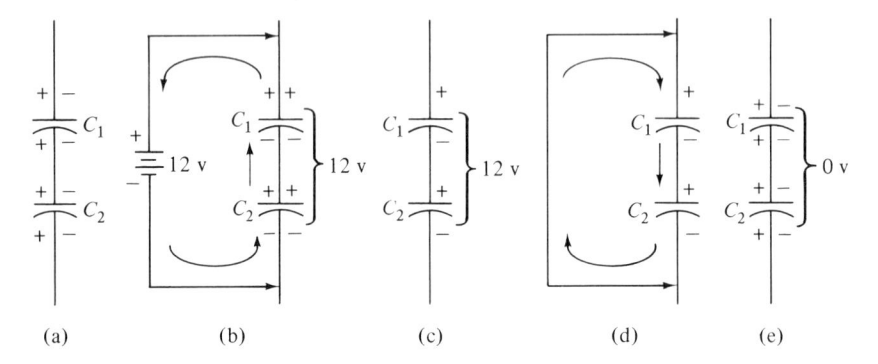

FIGURE 8-7 *Charge and Discharge of Series Capacitors*

the positive terminal of the battery. The excess electrons accumulate on the lower plate of C_2. As the electrons charge the lower plate, the field they create repels electrons from the upper plate of C_2. Electron current flows between the capacitors while the electrons accumulate on the lower plate of C_1. Even though current does not flow through the capacitors, the reaction is the same during a charge (or discharge) cycle. Charging continues until the accumulated charges provide a potential voltage difference of 12 v (the applied voltage) from the top plate of C_1 to the bottom plate of C_2.

The capacitors in (c) show a stored charge equivalent of 12 v across the two capacitors. In (d) a discharge path is provided by shorting the capacitors with a wire. As the electrons redistribute on the top of C_1, the field they create repels excess electrons from the lower plate (of C_1) so they return to the upper plate of C_2. The discharge continues until the plates are neutralized with an equal charge on all plates as in (e).

The charging and discharging of more than two capacitors follows exactly the same analogy. *The charges move in a chain reaction capacitor to capacitor.* Current flows through each of the connecting wires during the charge or discharge cycle.

[8.4] Capacitance

An ac or pulsating dc source operates as described earlier for Figures 8-4, 8-5, and 8-6. The ac or dc pulses cycle in the same manner, and when two or more capacitors are in series, the theory of repelling electrons on the opposite plate still holds true.

The total capacitance (in farads) for series capacitors is calculated in the same way total resistance is calculated for parallel resistors. Therefore, the parallel resistance formulas apply to series capacitance.

$$C_T = \frac{1}{\frac{1}{C_1} + \frac{1}{C_2} + \cdots + \frac{1}{C_n}} \tag{8-7}$$

$$\frac{1}{C_T} = \frac{1}{C_1} + \frac{1}{C_2} + \cdots + \frac{1}{C_n} \tag{8-8}$$

The short-form for two capacitors in series is

$$C_T = \frac{C_1 C_2}{C_1 + C_2} \tag{8-9}$$

When the values are the same for each series capacitor,

$$C_T = \frac{C_1}{2} = \frac{C_2}{2} \tag{8-10}$$

For $C_1 = C_2 = C_3$,

$$C_T = \frac{C_1}{3} = \frac{C_2}{3} = \frac{C_3}{3} \tag{8-11}$$

For four equal values, any one of them is divided by 4, five equal values are divided by 5, and so on.

The total capacitance for Figure 8-8(a) is calculated from each of the formulas (8-7) through (8-10).

$$C_T = \frac{1}{\frac{1}{C_1} + \frac{1}{C_2}} = \frac{1}{\frac{1}{2 \times 10^{-6}} + \frac{1}{2 \times 10^{-6}}} \tag{8-7}$$

$$= \frac{1}{1 \times 10^6} = 1 \times 10^{-6} \text{ f} = 1 \text{ }\mu\text{f}$$

$$\frac{1}{C_T} = \frac{1}{C_1} + \frac{1}{C_2} = \frac{1}{2 \times 10^{-6}} + \frac{1}{2 \times 10^{-6}} \tag{8-8}$$

$$= \frac{2}{2 \times 10^{-6}}$$

$$C_T = \frac{2 \times 10^{-6}}{2} = 1 \times 10^{-6} \text{ f} = 1 \text{ }\mu\text{f}$$

$$C_T = \frac{C_1 C_2}{C_1 + C_2} = \frac{2 \times 10^{-6} \times 2 \times 10^{-6}}{2 \times 10^{-6} + 2 \times 10^{-6}} \tag{8-9}$$

$$= \frac{4 \times 10^{-12}}{4 \times 10^{-6}} = 1 \times 10^{-6} \text{ f} = 1 \text{ }\mu\text{f}$$

$$C_T = \frac{C_1}{2} = \frac{C_2}{2} = \frac{2 \times 10^{-6}}{2} = 1 \times 10^{-6} \text{ f} \qquad (8\text{--}10)$$
$$= 1 \text{ }\mu\text{f}$$

Each of the four formulas resulted in a total capacitance of 1 μf. In similar problems, the choice of a formula is optional.

The total capacitance for the capacitors in Figure 8-8(b) is calculated first by using formula (8-9) for two capacitors:

$$C_T = \frac{C_1 C_2}{C_1 + C_2} = \frac{6 \times 10^{-6} \times 3 \times 10^{-6}}{6 \times 10^{-6} + 3 \times 10^{-6}}$$
$$= \frac{18 \times 10^{-12}}{9 \times 10^{-6}} = 2 \times 10^{-6} \text{ f} = 2 \text{ }\mu\text{f}$$

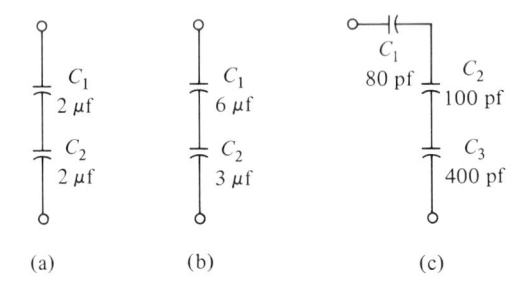

(a)   (b)   (c)

FIGURE 8-8 *Capacitance of Series Capacitors*

For verification, formula (8-7) gives

$$C_T = \frac{1}{\frac{1}{C_1} + \frac{1}{C_2}} = \frac{1}{\frac{1}{6 \times 10^{-6}} + \frac{1}{3 \times 10^{-6}}}$$
$$= \frac{1}{\frac{1}{2 \times 10^{-6}}} = 2 \times 10^{-6} \text{ f} = 2 \text{ }\mu\text{f}$$

An example for three capacitors in series is drawn in Figure 8-8(c). Formula (8-7) or (8-8) applies to this problem. Formula (8-8) will be used here since (8-7) was used in the last example.

$$\frac{1}{C_T} = \frac{1}{C_1} + \frac{1}{C_2} + \frac{1}{C_3} = \frac{1}{80 \times 10^{-12}} + \frac{1}{100 \times 10^{-12}}$$
$$+ \frac{1}{400 \times 10^{-12}} = \frac{5}{400 \times 10^{-12}} + \frac{4}{400 \times 10^{-12}}$$
$$+ \frac{1}{400 \times 10^{-12}} = \frac{10}{400 \times 10^{-12}} = \frac{1}{40 \times 10^{-12}}$$
$$C_T = 40 \times 10^{-12} \text{ f} = 40 \text{ pf}$$

Capacitance

In summary, any number of capacitors in series can be reduced to a single equivalent capacitor by selecting an appropriate formula and calculating the net capacitance.

An additional consideration is shown in Figure 8–9(a). When two capacitors are in series and are charged to the applied voltage, the voltage is divided between the capacitors. In the example, $C_1 = C_2 = 2\ \mu f$. The 12 v are divided equally between the capacitors: 6 v across C_1, and 6 v across C_2.

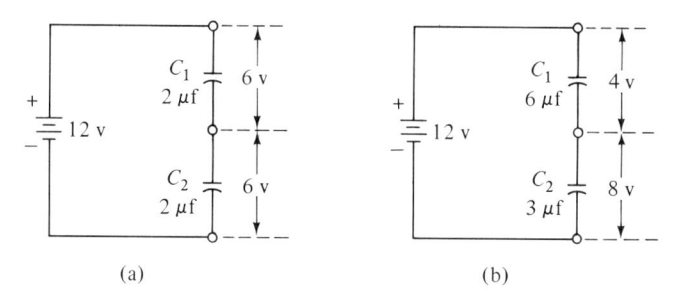

FIGURE 8–9 *Voltage Potential across Series Capacitors*

When capacitors are in series, the accumulated charge on each plate is the same for each capacitor regardless of its capacitance. From formula (8–5),

$$Q = It$$

where Q is the charge in coulombs and is dependent strictly upon current and time, without mention of capacitance. From formula (8–4),

$$C = \frac{Q}{E} \qquad (8\text{–}12)$$

or

$$E = \frac{Q}{C}$$

Since $C_1 = C_2$ in Figure 8–9(a), $E_1 = E_2$ because the Q is the same for each capacitor in a series circuit.

Figure 8–9(b) illustrates two series capacitors of different values. Because the charge Q is the same on the plates of each capacitor, the voltage is inversely proportional to the capacitance as shown in formula (8–12). In the figure C_1 is 9 μf, and C_2 is 3 μf. The 12-v source is split between the two capacitors. The ratio of capacitance is 2:1, and the inverse function means the voltage is split 2:1 with twice as much (8 v) across the smaller capacitor (3 μf). Verification of the proportion can be calculated using formula (8–12). The total capacitance of Figure 8–9(b) is 2 μf as calculated from Figure 8–8(b). The charge is calculated:

142 Capacitance [8.4]

$$Q = CE = 2 \times 10^{-6} \times 12 = 24 \times 10^{-6} \text{ coulombs}$$

For C_1, $\quad E_1 = \dfrac{Q}{C_1} = \dfrac{24 \times 10^{-6} \text{ coulombs}}{6 \times 10^{-6} \text{ f}} = 4 \text{ v}$

For C_2, $\quad E_2 = \dfrac{Q}{C_2} = \dfrac{24 \times 10^{-6} \text{ coulombs}}{3 \times 10^{-6} \text{ f}} = 8 \text{ v}$

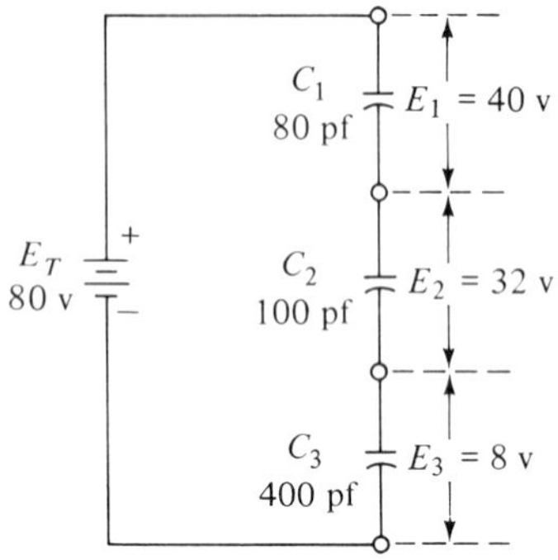

FIGURE 8-10 *Voltage Distribution for Three Series Capacitors*

An example for the distribution of voltage across three capacitors in series is shown in Figure 8–10. The total capacitance is 40 pf [see Figure 8–8(c)]. The charge on each plate is

$$Q = CE = 40 \times 10^{-12} \times 80 = 3200 \times 10^{-12} \text{ coulombs}$$

$$E_1 = \frac{Q}{C_1} = \frac{3200 \times 10^{-12} \text{ coulombs}}{80 \times 10^{-12} \text{ f}} = 40 \text{ v}$$

$$E_2 = \frac{Q}{C_2} = \frac{3200 \times 10^{-12} \text{ coulombs}}{100 \times 10^{-12} \text{ f}} = 32 \text{ v}$$

$$E_3 = \frac{Q}{C_3} = \frac{3200 \times 10^{-12} \text{ coulombs}}{400 \times 10^{-12} \text{ f}} = 8 \text{ v}$$

$$E_T = E_1 + E_2 + E_3 = 40 \text{ v} + 32 \text{ v} + 8 \text{ v} = 80 \text{ v}$$

Since the voltage across each capacitor in a series circuit is inversely proportional to its capacitance, a more direct method can be employed:

$$E = \frac{C_T E_T}{C} \qquad (8\text{--}13)$$

Formula (8–13) can be applied in the following manner.

$$E_1 = \frac{C_T E_T}{C_1}, \qquad E_2 = \frac{C_T E_T}{C_2}, \text{ etc.}$$

For Figure 8–9(a), $C_T = 1 \ \mu\text{f}$.

$$E_1 = \frac{C_T E_T}{C_1} = \frac{1 \ \mu\text{f} \times 12 \text{ v}}{2 \ \mu\text{f}} = \frac{12 \text{ v}}{2} = 6 \text{ v}$$

$$E_2 = E_1 = 6 \text{ v}$$

For Figure 8-9(b), $C_T = 2 \ \mu f$.

$$E_1 = \frac{2 \ \mu f \times 12 \ v}{6 \ \mu f} = \frac{12 \ v}{3} = 4 \ v$$

$$E_2 = \frac{2 \ \mu f \times 12 \ v}{3 \ \mu f} = \frac{2}{3} \times 12 \ v = 8 \ v$$

For Figure 8-10, $C_T = 40 \ pf$.

$$E_1 = \frac{40 \ pf \times 80 \ v}{80 \ pf} = \frac{80 \ v}{2} \times 40 \ v$$

$$E_2 = \frac{40 \ pf \times 80 \ v}{100 \ pf} = 0.4 \times 80 \ v = 32 \ v$$

$$E_3 = \frac{40 \ pf \times 80 \ v}{400 \ pf} = \frac{80 \ v}{10} = 8 \ v$$

8.5 Capacitors in Parallel

The effect of connecting capacitors in parallel is the same effect produced when the area of the plates is increased. Increasing the area of the plates increases the capacitance [see formula (8-6)].

$$C = \frac{kA \times 1.11 \times 10^{-12}}{4\pi d}$$

The relationship is linear when the capacitance is directly proportional to the area of the plates.

Charging and discharging capacitors in parallel follows the same analogy as in Figure 8-3. However, when capacitors are in parallel, the charge (Q) on each plate will not be the same if the capacitance is different. Like parallel resistors, the voltage across each capacitor is the same. According to formula (8-12), the charge will change when the capacitance changes. In Figure 8-11(a), the charge is distributed equally between both capacitors.

$$Q_1 = C_1 E = 2 \times 10^{-6} \times 12 = 24 \times 10^{-6} \ \text{coulombs}$$
$$Q_2 = C_2 E = 2 \times 10^{-6} \times 12 = 24 \times 10^{-6} \ \text{coulombs}$$

The total charge is

$$Q_T = Q_1 + Q_2 = 48 \times 10^{-6} \ \text{coulombs}$$

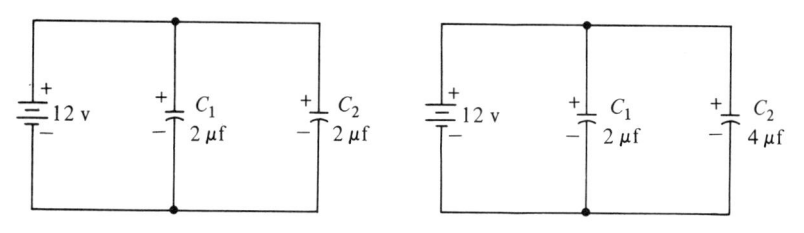

FIGURE 8-11 *Parallel Capacitance*

The total capacitance is then

$$C_T = \frac{Q_T}{E} = \frac{48 \times 10^{-6} \text{ coulombs}}{12 \text{ v}} = 4 \text{ μf}$$

which would indicate that capacitance in parallel adds directly.

$$C_T = C_1 + C_2 + \cdots + C_n \qquad (8\text{-}14)$$
$$C_T = C_1 + C_2 = 2 \text{ μf} + 2 \text{ μf} = 4 \text{ μf}$$

Formula (8-14) is the basic formula used to determine the total capacitance of parallel capacitors. Figure 8-11(b) is another example.

$$C_T = C_1 + C_2 = 2 \text{ μf} + 4 \text{ μf} = 6 \text{ μf}$$

Charge is distributed:

$$Q_1 = C_1 E = 2 \times 10^{-6} \times 12 = 24 \times 10^{-6} \text{ coulombs}$$
$$Q_2 = C_2 E = 4 \times 10^{-6} \times 12 = 48 \times 10^{-6} \text{ coulombs}$$
$$Q_T = Q_1 + Q_2 = 24 \times 10^{-6} + 48 \times 10^{-6} = 72 \times 10^{-6} \text{ coulombs}$$

The total charge for Figure 8-11(b) can also be calculated from the total capacitance and voltage:

$$Q_T = C_T E = 6 \times 10^{-6} \times 12 = 72 \times 10^{-6} \text{ coulombs}$$

As with parallel resistors, the discussion so far would indicate that the charging current is divided between parallel capacitors. This can be shown using formula (8-5). For Figure 8-11(a) assume the charge time is 1 sec (constant).

$$I_1 = \frac{Q_1}{t} = \frac{24 \times 10^{-6} \text{ coulombs}}{1 \text{ sec}} = 24 \text{ μa}$$
$$I_2 = 24 \text{ μa since } Q_1 = Q_2$$

For Figure 8-11(b),

$$I_1 = \frac{Q_1}{t} = \frac{24 \times 10^{-6} \text{ coulombs}}{1 \text{ sec}} = 24 \text{ μa}$$
$$I_2 = \frac{Q_2}{t} = \frac{48 \times 10^{-6} \text{ coulombs}}{1 \text{ sec}} = 48 \text{ μa}$$
$$I_T = I_1 + I_2 = 24 \text{ μa} + 48 \text{ μa} = 72 \text{ μa}$$

also,

$$I_T = \frac{Q_T}{t} = \frac{72 \times 10^{-6} \text{ coulombs}}{1 \text{ sec}} = 72 \text{ μa}$$

So the larger capacitance accumulates a larger charge and a larger share of the charge or discharge current. The analogy is the same for more than two capacitors in parallel. Total capacitance is the sum of the individual capacitances, and the charges are distributed in relation to the capacitance of each capacitor.

8.6 Capacitors in Series/Parallel

When capacitors are connected in series/parallel combinations, the rules governing both series and parallel capacitors apply. The series capacitors are analyzed by the methods described in Section 8-4, while the parallel capacitors are analyzed by the methods described in Section 8-5. The least complicated way to handle combinations is separately.

Charge is distributed equally for each series capacitor, and the same charge is valid for any junction containing parallel capacitors. However, the same charge that was equal for each series capacitor is distributed among the parallel capacitors according to individual capacitance values. A complete analysis follows for the example combination in Figure 8-12.

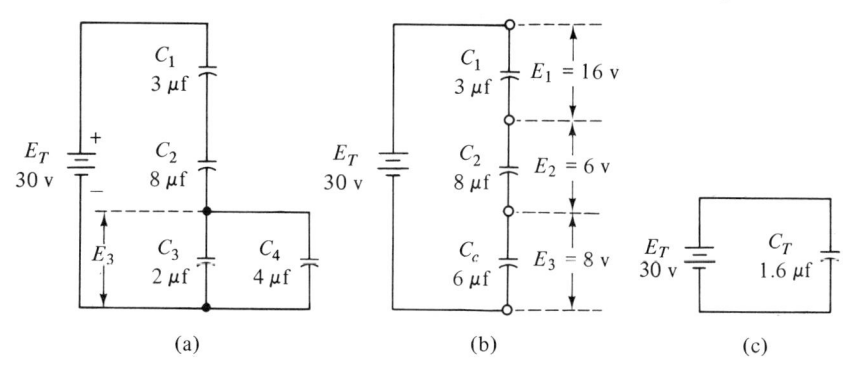

FIGURE 8-12 *Series/Parallel Capacitance*

Example 8-10 Determine the voltage difference across each capacitor and the charge for each capacitor in Figure 8-12(a).

Solution The circuit has to be reduced to one value of capacitance (C_T). The charge Q can be calculated which is the charge for C_1, C_2, and $C_3 - C_4$ together. The voltage drops are then calculated followed by the charge for C_3 and C_4 individually.

C_3 and C_4 are combined in (b).

$$C_c = C_3 + C_4 = 2\ \mu f + 4\ \mu f = 6\ \mu f$$

The total capacitance is shown in (c).

$$C_T = \cfrac{1}{\cfrac{1}{3 \times 10^{-6}} + \cfrac{1}{8 \times 10^{-6}} + \cfrac{1}{6 \times 10^{-6}}}$$
$$= \cfrac{1}{\cfrac{30}{48 \times 10^{-6}}} = \cfrac{48 \times 10^{-6}}{30}$$
$$= 1.6 \times 10^{-6}\ f = 1.6\ \mu f$$

The total charge is calculated:

$$Q_T = CE = 1.6 \times 10^{-6} \times 30 = 48 \times 10^{-6} \text{ coulombs}$$

which is the charge on C_1, C_2, and C_C in diagram (b). It is also the charge that is divided between C_3 and C_4 in diagram (a). The potentials in (b) are calculated:

$$E_1 = \frac{Q_T}{C_1} = \frac{48 \times 10^{-6}}{3 \times 10^{-6}} = 16 \text{ v}$$

$$E_2 = \frac{Q_T}{C_2} = \frac{48 \times 10^{-6}}{8 \times 10^{-6}} = 6 \text{ v}$$

$$E_3 = \frac{Q_T}{C_C} = \frac{48 \times 10^{-6}}{6 \times 10^{-6}} = 8 \text{ v}$$

Now that E_3 is known, and E_3 is the potential across both C_3 and C_4 in diagram (a), the charge distribution for C_3 and C_4 can be calculated:

$$Q_3 = C_3 E_3 = 2 \times 10^{-6} \times 8 = 16 \times 10^{-6} \text{ coulombs}$$
$$Q_4 = C_4 E_4 = 4 \times 10^{-6} \times 8 = 32 \times 10^{-6} \text{ coulombs}$$
$$Q_T = Q_3 + Q_4 = (16 + 32) \times 10^{-6} \text{ coulombs}$$
$$= 48 \times 10^{-6} \text{ coulombs}$$

The voltage potential and charge for each capacitor have been calculated from formulas studied in this chapter. Complex circuit analysis often requires this amount of detail, particularly when high frequency work involving small capacitors causes stray capacitance to become a significant part of the total capacitance.

8.7 Capacitive Reactance

When alternating voltage is applied to a capacitor, current flows by virtue of the charge and discharge of the capacitor. However, the capacitor has a resistance to charging and discharging. The amount of current flowing is dependent upon that resistance to changing current. The resistance to current flow in a capacitive circuit is called **reactance** or, more properly, **capacitive reactance**. The reactance is related to frequency and capacitance. As frequency decreases, reactance increases, and when the frequency is zero or dc, the reactance is high enough to be considered an open circuit. Conversely, as the frequency increases, the reactance decreases, and for high frequencies, a capacitor is often considered a short circuit—an analogy that may be satisfactory for many approximations, even though not entirely true.

As capacitance increases, current flow increases since the charge capacity has increased. The reactance must therefore decrease when the current flow increases, which proves that reactance is inversely proportional to capacitance.

Consider the basic formulas:

$$Q = It \quad Q = CE$$

Substituting for Q, the combined formulas become

$$It = CE \tag{8-15}$$

or

$$I = \frac{CE}{t}$$

For a sinusoidal waveform, formula (8-15) becomes

$$i = C\frac{de}{dt} \tag{8-16}$$

or

$$i = C\frac{\Delta e}{\Delta t}$$

in which i is the alternating current in amp; C, the capacitance in f; de (Δe), the change in voltage in volts, and dt (Δt), the change in time in seconds. Δt is the time it takes e to reach a given value.

In reference to formula (8-16), notice how i is directly proportional to C or Δe. It is inversely proportional to Δt. For example, at higher frequencies the time to reach a value e is smaller, so the i is higher. According to Ohm's law, higher current is the result of lower reactance.

$$i = \frac{e}{R}$$

Figure 8-13 illustrates the change in time as a function of frequency. In (a) the frequency is 250 kHz, and from formula (7-6) in Chapter 7 the following is obtained:

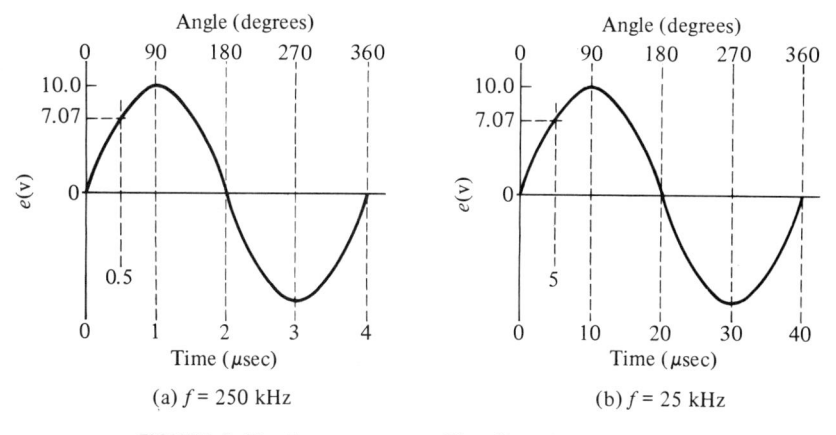

FIGURE 8-13 *Frequency versus Time for Voltage Sine Wave*

$$t = \frac{1}{f} = \frac{1}{250 \text{ kHz}} = 4 \text{ } \mu\text{sec}$$

In this example, the time to reach an rms value of 7.07 v is 0.5 μsec. Assume a capacitance of 0.01 μf. Using formula (8–16) yields

$$i = C\frac{\Delta e}{\Delta t} = 0.01 \times 10^{-6} \frac{7.07}{0.5 \times 10^{-6}} = 141.4 \text{ ma}$$

In (b) the frequency is 25 kHz, and

$$t = \frac{1}{f} = \frac{1}{25 \text{ kHz}} = 40 \text{ } \mu\text{sec}$$

The time it takes to reach an rms value of 7.07 v is 5 μsec, which is 10 times as long as the case for (a). Assume a capacitance of 0.01 μf.

$$i = C\frac{\Delta e}{\Delta t} = 0.01 \times 10^{-6} \frac{7.07}{5 \times 10^{-6}} = 14.14 \text{ ma}$$

Table 8-1 is a summary.

For a constant C and e, frequency and current are directly proportional When the ratio of frequency in (a) is ten times higher than (b), the current is also ten times higher.

Since current increases with frequency while reactance decreases, a formula is needed for calculating capacitive reactance (X_C):

$$X_C = \frac{1}{2\pi f C} \tag{8-17}$$

In the above formula X_C is in ohms, f is in Hz, and C is in farads. The constant 2π in the formula comes from a sine wave. (A one-cycle sine wave equals 2π radians, which is the circumference of a circle.)

To validate the results in Table 8-1 and Figure 8-13, examples for the same frequencies and capacitance are given.

TABLE 8–1

	(a)	(b)
Frequency (f)	250 kHz	25 kHz
Capacitance (C)	0.01 μf	0.01 μf
Voltage (e)	7.07 v	7.07 v
Time (t)	0.5 μsec	5 μsec
Current (i)	141.4 ma	14.14 ma

(a) $$X_C = \frac{1}{2\pi f C} = \frac{1}{2\pi \times 250 \times 10^3 \times 0.01 \times 10^{-6}}$$
$$= \frac{1 \times 10^3}{2\pi \times 2.5} = 63.8 \text{ } \Omega$$

(b) $$X_C = \frac{1}{2\pi fC} = \frac{1}{2\pi \times 25 \times 10^3 \times 0.01 \times 10^{-6}}$$
$$= \frac{1 \times 10^3}{2\pi \times 0.25} = 638 \; \Omega$$

In (a) the reactance is lower by a factor of 10, which agrees with the inverse relationship to current shown in Table 8–1.

Another way to use formula (8–17) for X_C is to divide out the constant 2π.

$$X_C = \frac{1}{2\pi fC} = \frac{0.159}{fC} \qquad (8\text{–}18)$$

The form with 0.159 in the numerator is often easier to work with. As an example of the formula, the first line in Table 8–2 is calculated for two frequencies and two capacitances. The same capacitive reactance is the result:

$$X_C = \frac{0.159}{fC} = \frac{0.159}{10 \times 10^3 \times 0.001 \times 10^{-6}} = 15.9 \; k\Omega$$
$$X_C = \frac{0.159}{fC} = \frac{0.159}{0.1 \times 10^3 \times 0.1 \times 10^{-6}} = 15.9 \; k\Omega$$

TABLE 8–2

$f = 10$ kHz		$C = 0.1 \; \mu f$	
C	X_C	f	
0.001 μf	15,900 Ω	0.1	kHz
0.01 μf	1590 Ω	1	kHz
0.1 μf	159 Ω	10	kHz
1 μf	15.9 Ω	100	kHz
10 μf	1.59 Ω	1	MHz

Table 8–2 demonstrates two points. First, the reactance (X_C) decreases as the capacitance increases (when the frequency is held constant). Second, the reactance decreases as the frequency increases (when the capacitance is held constant). The points in the table are plotted in Figure 8–14. In both graphs (a) and (b), X_C is plotted on the left side. The graphs are scaled in logarithmic functions for the vertical and horizontal axes. In essence, log scales increase by a factor of ten at every division. The graphs illustrate the inverse function of X_C to capacitance and X_C to frequency. The inverse functions are non-linear, even though they show a straight line on a log-log graph.

Formulas (8–17) and (8–18) are rearranged for C or f as follows:

$$C = \frac{1}{2\pi f X_C} \qquad f = \frac{1}{2\pi C X_C} \qquad (8\text{–}17)$$

150   Capacitance   [8.7]

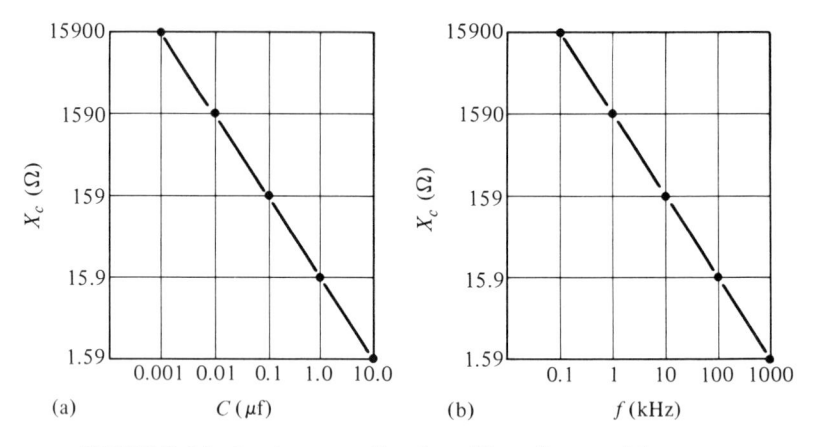

FIGURE 8–14 *Reactance as a Function of Capacitance and Frequency*

$$C = \frac{0.159}{fX_C} \qquad f = \frac{0.159}{CX_C} \qquad (8\text{–}18)$$

Example 8–11 What capacitance is required for a reactance of 400 ohms at 4 MHz?

Solution
$$C = \frac{1}{2\pi f X_C} = \frac{1}{2\pi \times 4 \times 10^6 \times 400}$$
$$= 0.1 \times 10^{-9} = 100 \text{ pf}$$

Example 8–12 What frequency will cause the reactance to be 80 ohms for a capacitor of 0.02 μf?

Solution
$$f = \frac{0.159}{CX_C} = \frac{0.159}{0.02 \times 10^{-6} \times 80}$$
$$= \frac{0.159 \times 10^6}{1.6} \approx 100 \text{ kHz}$$

Reactances in **series** and/or **parallel** are calculated in the same way as resistances. The formulas for series and parallel reactance are the same ones used for resistance. For series reactance,

$$X_{C_T} = X_{C_1} + X_{C_2} + \cdots + X_{C_n} \qquad (8\text{–}19)$$

For parallel reactance,

$$X_{C_T} = \frac{1}{\dfrac{1}{X_{C_1}} + \dfrac{1}{X_{C_2}} + \cdots + \dfrac{1}{X_{C_n}}} \qquad (8\text{–}20)$$

or

$$\frac{1}{X_{C_T}} = \frac{1}{X_{C_1}} + \frac{1}{X_{C_2}} + \cdots + \frac{1}{X_{C_n}} \qquad (8\text{–}21)$$

For two reactances,

$$X_{C_T} = \frac{X_{C_1} X_{C_2}}{X_{C_1} + X_{C_2}} \tag{8-22}$$

When each parallel reactance is the same,

$$X_{C_T} = \frac{X_{C_1}}{2} = \frac{X_{C_2}}{2} \tag{8-23}$$

or

$$X_{C_T} = \frac{X_{C_1}}{3} = \frac{X_{C_2}}{3} = \frac{X_{C_3}}{3} \tag{8-24}$$

As an example of cross-calculations, consider Figure 8–9(a). At 10 kHz, the X_C for one of the 2-μf capacitors is

$$X_C = \frac{1}{2\pi fC} = \frac{0.159}{10 \times 10^3 \times 2 \times 10^{-6}} = 7.95 \, \Omega$$

and the total reactance is

$$X_{C_T} = X_{C_1} + X_{C_2} = 7.95 \, \Omega + 7.95 \, \Omega = 15.9 \, \Omega$$

The total series capacitance was previously calculated at 1 μf. From Table 8–2, the reactance for 1 μf at 10 kHz is 15.9 ohms, which agrees with X_{C_T} above.

Refer to Figure 8–11(a) for an example of parallel reactance at 10 kHz. The reactance for either of the 2-μf capacitors is approximately 8 ohms (as calculated above). The total reactance from formula (8–23) is

$$X_{C_T} = \frac{X_{C_T}}{2} = \frac{8 \, \Omega}{2} = 4 \, \Omega$$

The total capacitance was calculated at 4 μf in Section 8.5. The reactance for 4 μf at 10 kHz is

$$X_C = \frac{1}{2\pi fC} = \frac{0.159}{10 \times 10^3 \times 4 \times 10^{-6}} \approx 4 \, \Omega$$

which agrees with the previous calculations.

Capacitive reactance decreases as frequency increases. The reactances of stray capacitance values of 10 to 20 pf are very high at low frequencies, but can be very low at high frequencies. For example, the reactances of 10 pf for two frequencies are

$$f = 1.59 \text{ kHz} \quad X_C = \frac{0.159}{1.59 \times 10^3 \times 10 \times 10^{-12}} = 10 \text{ M}\Omega$$

$$f = 159 \text{ MHz} \quad X_C = \frac{0.159}{159 \times 10^6 \times 10 \times 10^{-12}} = 100 \, \Omega$$

At frequencies below 1.59 kHz, the reactance is beyond 10 M ohms and is probably not significant in circuit operation. But at higher frequencies (above 159 MHz in the previous example) the reactance is less than 100 ohms. A reactance this low most probably would affect operation of the circuitry. So it becomes important to consider stray capacitance in high frequency applications.

8.8 Phase Angle of Voltage and Current

When a sine wave of voltage is applied to a capacitor, a phase difference or phase angle exists between the voltage and current. The angle is 90° for a capacitive circuit in which the current leads the voltage by 90°. Recall formula (8–16):

$$i = C\frac{\Delta e}{\Delta t}$$

in which the alternating current results from a changing voltage for a given time period. In Figure 8–15(a), voltage (e) and current (i) are shown

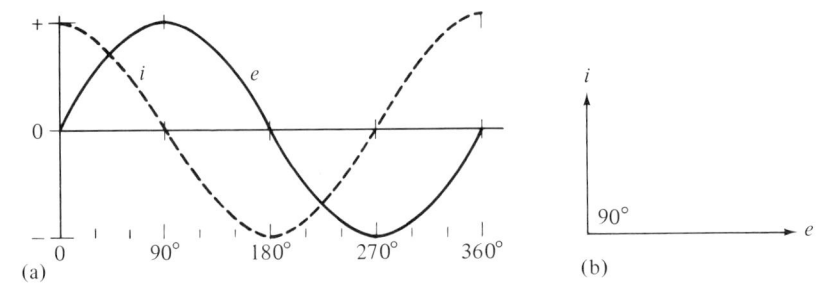

FIGURE 8–15 *Phase Relationship in a Capacitive Circuit*

90° out of phase. The e waveform is a sine wave, and the i waveform is a cosine wave. Notice the steep rise of the voltage waveform during the initial 30°. Maximum i occurs during the initial 30°, and minimum i flows at 90° which is the point of least change in e for a given increment of time. Formula (8–16) shows that the instantaneous value of current (i) is directly proportional to the rate of change in voltage (e).

The current changes polarity (swings negatively), after 90° when e reverses. The current is negative and reaches a maximum negative value when e crosses the reference at 180°; e is very steep at this point. The current changes polarity at 270°, and even though e is a negative value (270° to 360°) it is swinging toward a positive value.

Figure 8–15(b) is the vector representation of e and i. The current is shown on the positive side of the base to illustrate i leading e. The vectors are displaced 90° to represent the phase angle θ.

The first 90° of the sine and cosine waveforms of e and i are plotted in

[8.8] Capacitance 153

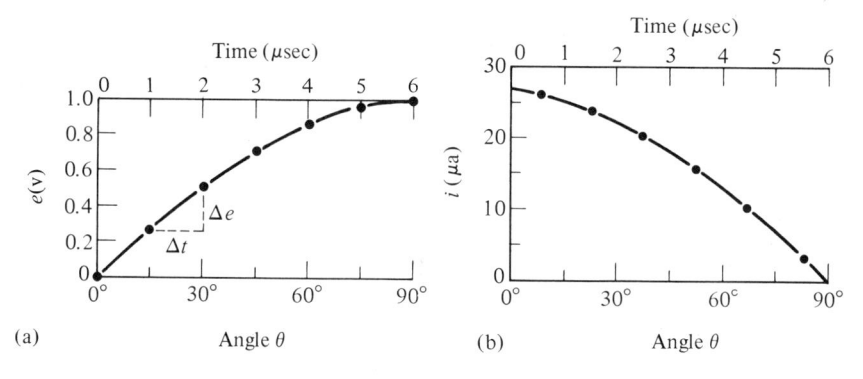

FIGURE 8-16 *Phase Angle versus Amplitude*

Figure 8-16. In (a) e is plotted from 0 to 1.0 v for a time period of 6 μsec. For purposes of illustration, Δe and Δt are plotted in the space $\theta = 15°$ to 30° or $t = 1$ to 2 μsec. Table 8-3 is a tabulation of the variables and results that lead to i for a capacitance of 100 pf. The first 6 μsec of time corresponds to the plot in Figure 8-16(a). The change in time (Δt) is the incremental time for every 15° or 1 μsec. The sin θ for a maximum $e = 1.0$ v is derived from formula (7-1) ($e_i = e_m \sin \theta$) which was partially listed in Table 7-2. The change in voltage (Δe) is the difference in e for every 15° in time. Data for the first 90° of the table are taken from the figure. The current i is plotted in (b). Examples for 30°, 60°, and 90° follow.

TABLE 8-3

θ	Time μsec	Δ θ	Δ Time μsec	Sin θ v	Δe v	i^* μa
15°	1	0–15°	1	0.26	0.26	26.0
30°	2	15°–30°	1	0.50	0.24	24.0
45°	3	30°–45°	1	0.707	0.207	20.7
60°	4	45°–60°	1	0.866	0.159	15.9
75°	5	60°–75°	1	0.966	0.10	10.0
90°	6	75°–90°	1	1.00	0.034	3.4
105°	7	90°–105°	1	0.966	−0.034	−3.4
120°	8	105°–120°	1	0.866	−0.10	−10.0
135°	9	120°–135°	1	0.707	−0.159	−15.9
150°	10	135°–150°	1	0.50	−0.207	−20.7
165°	11	150°–165°	1	0.26	−0.24	−24.0
180°	12	165°–180°	1	0	−0.26	−26.0

*$i = C (\Delta e/\Delta t)$ for $C = 100$ pf

$$t = 30° \quad i = C\frac{\Delta e}{\Delta t} = 100 \times 10^{-12} \times \frac{0.24}{1 \times 10^{-6}}$$
$$= 24 \times 10^{-6} = 24 \ \mu a$$

$$t = 60° \quad i = C\frac{\Delta e}{\Delta t} = 100 \times 10^{-12} \times \frac{0.159}{1 \times 10^{-6}}$$
$$= 15.9 \times 10^{-6} = 15.9 \ \mu a$$

$$t = 90° \quad i = C\frac{\Delta e}{\Delta t} = 100 \times 10 \times \frac{0.034}{1 \times 10^{-6}}$$
$$= 3.4 \times 10^{-6} = 3.4 \ \mu a$$

Between 90° and 180° in Table 8–3, the voltage e has a negative slope, thus Δe carries a minus sign. Notice the relationship to the first 90° as the values repeat in the opposite direction. The current (i) follows a similar pattern. The pattern continues from 180° to 270°, then the voltage changes signs as it begins its positive transition. Other examples of formula (8–16) are given below.

Example 8–13 Find the capacitance for $\Delta e = 32$ mv during 0.5 μsec when $i = 6.4$ ma.

Solution
$$i = C\frac{\Delta e}{\Delta t} \quad C = i\frac{\Delta t}{\Delta e}$$
$$C = 6.4 \times 10^{-3} \frac{0.5 \times 10^{-6}}{32 \times 10^{-3}} = 0.1 \times 10^{-6} = 0.1 \ \mu f$$

Example 8–14 Find the change in voltage for $\Delta t = 2$ msec, $i = 48 \ \mu a$, and $C = 0.02 \ \mu f$.

Solution
$$\Delta e = i\frac{\Delta t}{C} = 48 \times 10^{-6} \frac{2 \times 10^{-3}}{0.02 \times 10^{-6}} = 4.8 \ v$$

QUESTIONS

8–1 The force between two charged plates is 0.135×10^{12} dynes, separated 0.5 cm with a paraffin dielectric. If the charge on one plate is 100 μcoulombs, what is the charge on the other plate?

8–2 What is the force between two plates separated 1 cm with a dielectric constant of 4.5, if each plate is charged 10 μcoulombs?

8–3 What is the capacitance for a stored charge of 80 μcoulombs and a potential difference of 16 v?

Capacitance

8-4 What is the stored charge in a 0.02-μf capacitor across 150 v?

8-5 How much charge is stored in a capacitor for a constant current of 4 ma for 30 msec?

8-6 What is the capacitance of two plates which are 1.5 cm^2 and 0.1 cm apart, and are separated by air?

8-7 For a capacitance of 42.4 pf, the plates are 4 cm^2 and 0.05 cm apart. What is the value of the dielectric constant?

8-8 If the e sine wave in Figure 8-4 were delayed 90°, which way would the current (i) flow in the diagram from 270° to 360°?

8-9 If the e sine wave in Figure 8-4 were earlier by 90°, which way would the current (i) flow in the diagram for 90° to 180°?

8-10 If a sine wave of 200 mv p-p were swinging around a dc reference of -3 v, what would be the maximum negative rms voltage?

8-11 What is the total capacitance of four 0.2-μf capacitors in series?

8-12 What is the capacitance of a 4-μf capacitor and a 16-μf capacitor in series?

8-13 What is the total capacitance of three capacitors in series if $C_1 = 100$ pf, $C_2 = 500$ pf, and $C_3 = 2000$ pf?

8-14 In Figure 8-9(b), if the charge on the capacitors is 6 μcoulombs, what is the voltage across each capacitor?

8-15 In Figure 8-10, assume $C_1 = 1.2$ μf, $C_2 = 0.6$ μf, and $C_3 = 0.1$ μf. What is the voltage across each capacitor?

8-16 For three capacitors in parallel across 5 v, what is the capacitance of each, and what is the total capacitance? The charge on each is $Q_1 = 4$ μcoulombs, $Q_2 = 2.5$ μcoulombs, and $Q_3 = 1$ μcoulomb.

8-17 If the charge time across the parallel capacitors of Question 8-16 is a constant 0.5 sec, what current flows for each capacitor? What is the total current in the circuit?

8-18 A 10-v battery has two capacitors across it in parallel. For a constant current of 30 μsec, C_1 has a charge of 1200×10^{-12} coulombs, and the total charging current is 160 μa. What is the capacitance of each capacitor?

8-19 Refer to Figure 8-12. Assume $E_T = 18$ v, $C_1 = 450$ pf, $C_2 = 900$ pf, $C_3 = 60$ pf, and $C_4 = 90$ pf. What are the voltage across each capacitor and the charge for each capacitor?

8-20 What is the capacitive reactance of a 2-μf capacitor at 1590 Hz?

Capacitance

8-21 What frequency will provide a capacitive reactance of 250 ohms for a 159-pf capacitor?

8-22 What is the total capacitive reactance for two capacitors in parallel? Assume $X_{C_1} = 160$ ohms, and $X_{C_2} = 640$ ohms.

8-23 What is the total capacitive reactance for three capacitors in parallel? Assume $X_{C_1} = 4$ kΩ, $X_{C_2} = 16$ kΩ, and $X_{C_3} = 0.8$ kΩ.

8-24 What is the total capacitive reactance of three capacitors in series? Assume $X_{C_1} = 90$ Ω, $X_{C_2} = 170$ Ω, and $X_{C_3} = 110$ Ω.

8-25 In Figure 8-16(a), what is the current (i) between 1.5 and 2.0 μsec for a capacitance of 400 pf?

8-26 What is the change in sinusoidal voltage across a 0.3-μf capacitor during a period of 5 msec if the current is 6 ma?

9 Inductance

Chapter 2 discussed how changing current in a conductor produces a magnetic field and how changing flux lines induces a voltage along a conductor. The ability of a conductor to induce voltage from a changing current is called **inductance**. This chapter correlates theory from Chapter 2 with terms and values associated with electrical circuits.

9.1 Induction

As varying current through a conductor increases, a magnetic field is generated that expands away from the conductor. As the current decreases, the magnetic field collapses into the conductor. Because the flux from the magnetic field is in motion, a voltage potential is induced within the conductor. A conductor that moves across flux lines of force induces voltage within the conductor (see Chapter 7). In either instance, the conductor is exposed to changes in flux, and these changes in flux produce voltage.

Figure 9-1 illustrates a changing field for the first 180° of a sine wave. The dashed circular lines beneath the sine wave represent the magnetic fields. The shaded circles in the center of the fields represent end views of the conductor at different points in time. Notice how the field increases from 0° to 90°, when the current (i) is maximum. Beyond 90°, the i

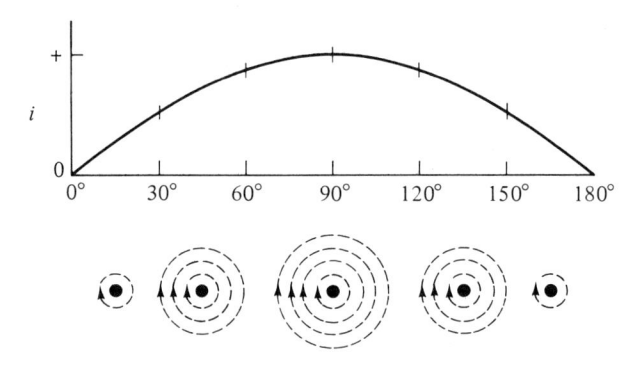

FIGURE 9-1 *Magnetic Fields Produced from Changing Current*

decreases to zero, and the field shows a corresponding decrease as it collapses into the conductor. The flux lines are clockwise in the illustration. A reversal of the flux takes place between 180° and 360° because the current reverses direction. The field would be counterclockwise to correspond with the reversal in flux.

In summary, a changing current produces a magnetic field. The magnetic field increases and decreases with the current and consists of flux lines which also increase and decrease. The flux variations cause an induced voltage within the conductor in the same way that would occur if the conductor moved across a field of flux.

The changing field caused by a changing current extends outward from the conductor and collapses into the conductor as the current varies. If another conductor were close enough for the field to encompass it, the changing field would induce voltage into that conductor. The principle is shown in Figure 9-2, where the primary conductor carries a variable current that produces a variable magnetic field. The secondary conductor

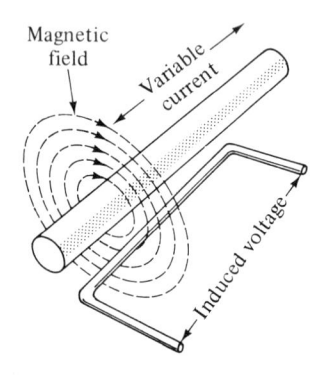

FIGURE 9-2 *Induction Due to Magnetic Field*

is parallel and close enough to be affected by the flux changes. As the flux increases and decreases, a voltage is induced in the secondary conductor. This voltage is in addition to the voltage produced in the primary conductor.

Induction, then, is the process of inducing a voltage within a conductor that has a variable current, or the process of inducing a voltage in an adjacent conductor as a result of flux lines intercepting the adjacent conductor.

Induction, or induced voltage, cannot occur unless the current is changing. The current changes continuously in ac circuits, producing a magnetic field that is capable of inducing voltage. Pulsating dc, or any dc that increases or decreases, will also generate a field that induces voltage. A pure dc level that does not vary cannot produce a magnetic field that will induce voltage. However, a variable field is produced at the instant the dc is switched on or off since the current must either increase or decrease.

9.2 Inductance

Inductance is the ability to induce voltage as a result of a changing current. It is measured in terms of the **henry** (h) after Joseph Henry (1797–1878). The henry is defined as the amount of inductance that will induce one volt when the current changes at a rate of one ampere per second. The formula for inductance is

$$L = \frac{e}{di/dt} \tag{9-1}$$

or

$$L = \frac{e}{\Delta i/\Delta t}$$

in which L is in henrys, e is in volts of induced voltage, and $\Delta i/\Delta t$ is the change in amperes per second. Several examples follow.

Example 9-1 What is the inductance for $\Delta i/\Delta t = 5$ A/sec and an induced voltage of 1 v?

Solution
$$L = \frac{e}{\Delta i/\Delta t} = \frac{1 \text{ v}}{5 \text{ A/sec}} = \frac{1 \text{ v} \times 1 \text{ sec}}{5 \text{ A}}$$
$$= 0.2 \text{ h}$$

Example 9-2 How much inductance produces a voltage of 5 mv when the current changes from 40 ma to 65 ma in 200 msec?

Solution
$$\Delta i = 65 \text{ ma} - 40 \text{ ma} = 25 \text{ ma}$$
$$\frac{\Delta i}{\Delta t} = \frac{25 \times 10^{-3}}{200 \times 10^{-3}} = \frac{0.125 \text{ A}}{\text{sec}} = \frac{125 \text{ ma}}{\text{sec}}$$

$$L = \frac{e}{\Delta i/\Delta t} = \frac{5 \times 10^{-3} \text{ v}}{(125 \times 10^{-3} \text{A})/\text{sec}}$$
$$= \frac{5 \times 10^{-3} \text{ v} \times 1 \text{ sec}}{125 \times 10^{-3} \text{ A}} = \frac{5}{125} = 40 \times 10^{-3}$$
$$= 40 \text{ mh}$$

Example 9-3 If the current changes 0.8 A in 2 sec for an inductance of 8 h, how much voltage is induced?

Solution
$$e = L\frac{\Delta i}{\Delta t} = 8 \times \frac{0.8}{2} = \frac{6.4}{2} = 3.2 \text{ v}$$

Example 9-4 How much does the current have to change during 0.4 sec to produce 2.5 v for an inductance of 2 h?

Solution
$$\Delta i = \Delta t \frac{e}{L} = 0.4 \times \frac{2.5}{2} = \frac{1.0}{2} = 0.5 \text{ A}$$

Thus far, inductance has been discussed for a straight conductor. Inductors are usually in the form of coils. For example, a given length of wire will exhibit more inductance when wound as a coil than as a straight piece of wire. The coil in Figure 9-3 illustrates the increased effectiveness of magnetic fields caused by a coil. In (a) the opposite sides of a single turn are

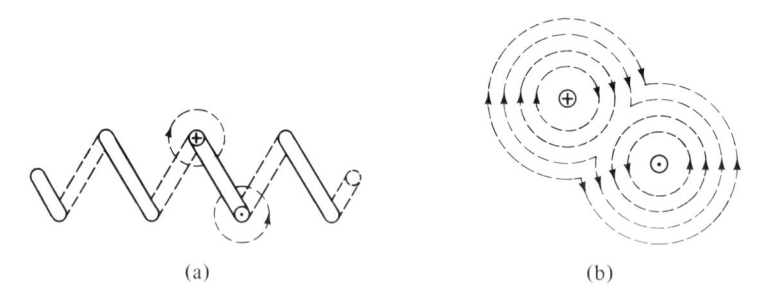

(a)   (b)

FIGURE 9-3 *Magnetic Fields in a Coil*

marked. The circular fields are in opposite directions. In (b) the opposite sides are expanded along with the magnetic fields. Notice that the effects of the two fields are complementary and can become more concentrated where they meet in the center.

As inductance increases, so does the induced voltage [see formula (9-1)]. Inductance is a factor that tends to restrict current flow, somewhat as a resistor does. The nature of inductance, or the resistance to change, is better explained by Lenz's law which can be interpreted in two ways. One way states that the magnetic field produced by a changing current has a direction that tends to oppose the changing current. The second way states that an electromagnetically induced current has a direction such that its magnetic field tends to oppose the current that produced the field.

Lenz's law can be applied to an increasing current. The increasing current generates a magnetic field which in turn produces an induced voltage. But the direction or polarity of the induced voltage opposes the direction of the increasing current. Conversely, a decreasing current reverses the polarity of the voltage—again, opposing a change in the current.

The inductance of a coil can be calculated from its dimensions and shape. For a single layer coil of uniform geometry,

$$L = \frac{4\pi N^2 A \mu}{l} \times 10^{-9} \tag{9-2}$$

L is in henrys, N is the number of turns, A is the cross-sectional area of the core of the coil in square centimeters, l is the length of the coil in centimeters, and μ is the permeability of the core. Air or a vacuum is used as a reference point, so the permeability of an air core is one. Notice that the permeability of the core has a direct effect on the inductance. The permeability of iron, for example, is around 10^3. An iron core concentrates the magnetic lines of force and subsequently increases the inductance. A coil form is shown in Figure 9-4. The area (A) is actually the circular area within the inside diameter of the wires. The drawing has the turns loosely wrapped for illustrative purposes. The core can be air or some material such as iron. Notice also that the length 1 is only for the coil, not the core.

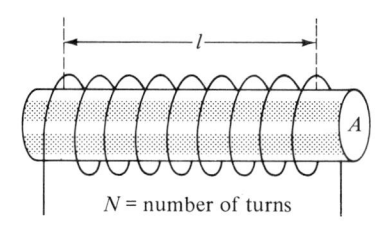

FIGURE 9-4 *Parameters for Calculating Inductance*

Examples of formula (9-2) follow.

Example 9-5 What is the inductance for 300 single-layer turns, 4 cm long with an air core cross-sectional area of 1.5 cm?

Solution
$$L = \frac{4\pi N^2 A \mu}{l} \times 10^{-9}$$
$$= \frac{4\pi \times 300^2 \times 1.5 \times 1}{4} \times 10^{-9}$$
$$= \pi \times 0.135 \times 10^{-3} = 0.424 \times 10^{-3} \text{ h} = 424 \text{ }\mu\text{h}$$

162 Inductance [9.2]

Example 9-6 An iron core has a permeability of 1500 and a cross-sectional area of 4 cm. To achieve an inductance of 1.06 mh, how many turns are required for a distance of 16 cm?

Solution
$$N = \sqrt{\frac{L1 \times 10^9}{4\pi A \mu}} = \sqrt{\frac{1.06 \times 10^{-3} \times 16 \times 10^9}{4\pi \times 4 \times 1.5 \times 10^3}}$$
$$= \sqrt{\frac{1.06 \times 10^3}{\pi \times 1.5}} = \sqrt{\frac{1.06 \times 10^2}{4.7}} = \sqrt{225}$$
$$= 15 \text{ turns}$$

Example 9-7 What is the permeability of a core 3 cm², wrapped with 100 turns of wire for a length of 5 cm, and with an inductance of 60 mh?

Solution
$$\mu = \frac{L1 \times 10^9}{4\pi N^2 A} = \frac{60 \times 10^{-3} \times 5 \times 10^9}{4\pi \times 100^2 \times 3}$$
$$= \frac{300 \times 10^6}{12\pi \times 10^4} = 8 \times 10^2 = 800$$

Example 9-8 What is the inductance for 200 turns of wire on a core that is 6 cm in diameter, has a permeability of 2500, and is wound for 12.5 cm of its length?

Solution Formula (9-2) calls for circular area, thus the diameter (6 cm) is converted to radius (3 cm), which is then converted to area.
$$A = \frac{\pi d^2}{4} = \pi r^2 = \pi 3^2 = 28.25 \text{ cm}^2$$

Inductance is calculated directly.
$$L = \frac{4\pi N^2 A \mu \times 10^{-9}}{1}$$
$$= \frac{4\pi \times 200^2 \times 28.25 \times 2500 \times 10^{-9}}{12.5}$$
$$= 4\pi \times 0.4 \times 0.2825 \times 2 = 2.84 \text{ h}$$

Variable inductances are shown in Figure 9-5. The drawings illustrate methods of varying an inductance to meet specific requirements of a circuit. In (a) the inductance is varied by changing the position of the iron core. An iron core increases the inductance because it concentrates the magnetic flux. Its effect was shown as μ (permeability) in formula (9-2). The core is usually fastened to a threaded rod that can be screwed in or out. The inductance decreases as the core moves out of the coil and increases as it moves into the core. Maximum inductance for the coil occurs when the core is centered within the coil.

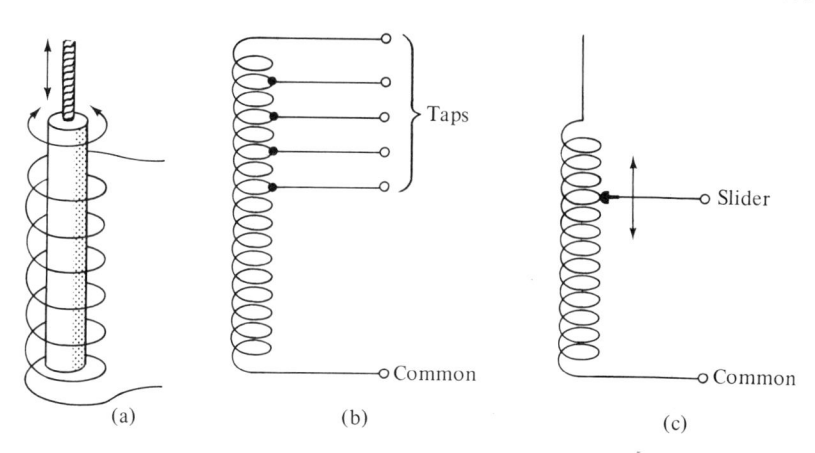

FIGURE 9-5 *Variable Inductances*

A tapped coil is shown in drawing (b). A suitable value of inductance is selected by using the bottom of the coil (common) and one of the taps at the top. Drawing (c) is a slider arrangement. The surface of the coil is free from insulation, and a contact slides across the winding to select the desired value of inductance.

Stray inductance is the inductance of wiring used to connect various components within a circuit. Typical values of stray inductance are in the lower μh range. However, high frequency inductors are also in the μh range, so stray values can have a significant effect on the overall inductance of a circuit. As a result, high frequency radios, televisions, and particularly radar keep the lengths of wire to a bare minimum. Even the wire leads on components can have significant inductance at high frequencies. They too are kept to minimum lengths.

9.3 Transformer Action

Transformer action is the process of induction from one conductor to another conductor or coil, or from one coil to another coil or conductor. The basic approach to induction for two conductors was discussed in Section 9.1 and illustrated in Figure 9-2. Figure 9-6 is a drawing of two coils that are side-by-side and are linked by magnetic lines of flux. Coil L_1 shows Δi across its input terminals to indicate a variable current. The variation in current produces a changing field around the wires of the coil (L_1). If L_2 is close enough, a percentage of the magnetic field cuts across the windings of the second coil (L_2). A voltage is generated across L_1, and because of the coupling by the field, a voltage is induced across L_2. The voltage across L_2 is the output voltage, labeled e_{out} in the drawing.

The voltage induced in L_2 is dependent upon several factors each of

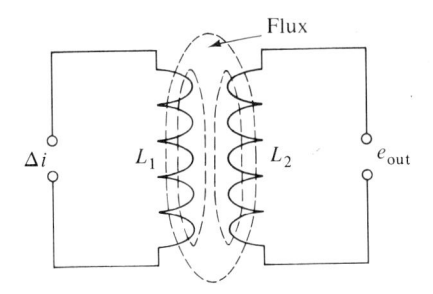

FIGURE 9-6 *Flux Field for Adjacent Coils*

which will be discussed in sequence. First, there is the coupling from L_1 to L_2, called the **coefficient of coupling**. The coefficient of coupling (k) is the percentage of magnetic flux from the primary coil (L_1 for this example) that is coupled to the secondary coil (L_2).

$$k = \frac{\text{Flux Linkage}}{\text{Primary Flux}} = \frac{L_2 \text{ flux}}{L_1 \text{ flux}} \qquad (9\text{-}3)$$

The coefficient of coupling is normally expressed in a decimal fraction (100 percent = 1, 80 percent = 0.8, etc).

Example 9-9 A primary coil creates a magnetic field of 120×10^3 maxwells. A secondary coil is coupled by 108×10^3 maxwells. What is the coefficient of coupling?

Solution
$$k = \frac{L_2 \text{ flux}}{L_1 \text{ flux}} = \frac{108 \times 10^3}{120 \times 10^3} = 0.9$$

Example 9-10 If the coefficient of coupling between two coils is 0.7, and the magnetic field produced by the primary coil is 50×10^3 maxwells, what is the magnitude of the field linking the coils?

Solution
$$L_2 \text{ flux} = kL_1 \text{ flux} = 0.7 \times 50 \times 10^3$$
$$= 35 \times 10^3 \text{ maxwells}$$

Figure 9-7 shows two methods of coupling by winding a primary and secondary on the same core. If both cores are made from some type of iron such as laminated, powdered, or ferrite, the degree of coupling can be nearly as efficient as 100 percent or 1, depending on the construction of the transformer. Efficient coupling with iron cores is achieved from the high concentration of the magnetic flux which causes the fields to be linked in a direct manner. Drawing (a) shows two coils wound serially on a common cylindrical core. The effect of close coupling is achieved by

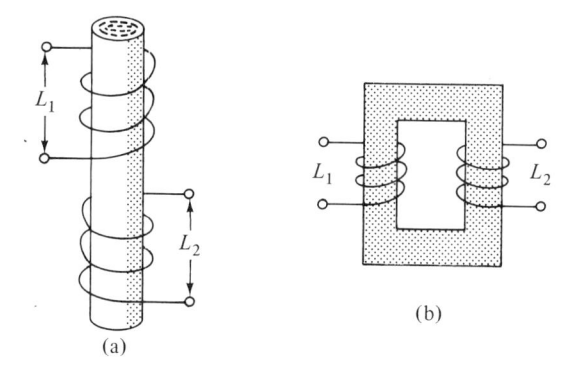

FIGURE 9-7 *Transformer Types*

wrapping one coil over the other. In fact, winding in layers can improve the coupling since more of the windings are closer together.

In drawing (b) a rectangular iron core is used for coupling. Here the windings can be done serially or in layers. Either way coupling is improved because the ends of the core are magnetically coupled, which further concentrates the magnetic lines of flux.

The higher the coupling coefficient is between coils, the higher the voltage that is induced in the secondary coil. A lower coefficient of coupling results in a lower induced voltage in the secondary coil. The effective or **mutual inductance** of the two coils has a similar effect on the amount of voltage induced in the secondary. A higher mutual induction increases the induced voltage. The formula for calculating mutual inductance is

$$L_m = k\sqrt{L_1 L_2} \qquad (9\text{-}4)$$

L_m is the mutual inductance in henrys, k is the coefficient of coupling, and L_1 and L_2 are the inductances of the primary and secondary coils in henrys. The value L_m can be substituted for L in formula (9-1).

$$L_m = \frac{e_s}{\Delta i/\Delta t} \qquad (9\text{-}5)$$

In the above equation e_s is the induced voltage in the secondary. By further substitution between formulas (9-4) and (9-5),

$$k\sqrt{L_1 L_2} = \frac{e_s}{\Delta i/\Delta t}$$

and
$$e_s = k\sqrt{L_1 L_2}\, \Delta i/\Delta t \qquad (9\text{-}6)$$

The above illustrates the effect on induced voltage (e_s) when the factors on the right side of the formula are increased (higher e_s for higher coupling, inductance, and/or current change). Examples of formulas (9-4) and (9-6) are:

Example 9-11 What is the mutual inductance for a coefficient of coupling of 0.95, a primary inductance of 50 mh, and a secondary inductance of 32 mh?

Solution
$$L_m = k\sqrt{L_1 L_2} = 0.95\sqrt{50 \times 10^{-3} \times 32 \times 10^{-3}}$$
$$= 0.95\sqrt{1600 \times 10^{-6}} = 0.95 \times 40 \times 10^{-3}$$
$$= 38 \text{ mh}$$

Example 9-12 If the coefficient of coupling is 0.8, and the primary inductance is 4 h, what secondary inductance is required to achieve a mutual inductance of 2.4 h?

Solution
$$L_2 = \left(\frac{L_m}{k\sqrt{L_1}}\right)^2 = \left(\frac{2.4}{0.8\sqrt{4}}\right)^2 = \left(\frac{2.4}{1.6}\right)^2 = 2.25 \text{ h}$$

Example 9-13 What is the secondary voltage if the coupling is unity, L_1 is 20 μh, L_2 is 45 μh, and the current changes 10 ma per msec?

Solution
$$e_s = k\sqrt{L_1 L_2}\, \Delta i/\Delta t = 1 \times \sqrt{20 \times 10^{-6} \times 45 \times 10^{-6}}$$
$$\times \frac{10 \times 10^{-3}}{1 \times 10^{-3}} = \sqrt{900 \times 10^{-12}} \times 10$$
$$= 30 \times 10^{-6} \times 10 = 300\ \mu\text{v}$$

Up to this point, transformer action has been studied in terms of a changing current in a primary coil that induces a voltage in the same coil. And due to flux linkage with a secondary coil, a voltage is further induced in that coil. The induced secondary voltage has been calculated from values of changing current, the coefficient of coupling, and the inductance of each coil (mutual inductance). Other parameters are sometimes required and are often more convenient for purposes of circuit design and analysis.

An important application of transformers is changing a specific primary voltage to some different value. If an ac voltage is connected to a primary coil, according to Ohm's law the current also alternates which is the requirement for inducing voltage in the secondary. An example is a 10-v ac door buzzer. The buzzer requires 10 v ac. A convenient source is ordinary house power which is 120 v ac. Figure 9-8 shows a primary intended for house power. The secondary supplies 10 v to the buzzer. The vertical lines between the coils indicate an iron core, and the coefficient of coupling is considered unity for the example.

The design of the transformer circuit (Figure 9-8) can be analyzed in terms of the **turns ratio** and **voltage ratio**. The turns ratio is

$$\text{Turns Ratio} = \frac{N_s}{N_p} \quad (9\text{-}7)$$

Inductance

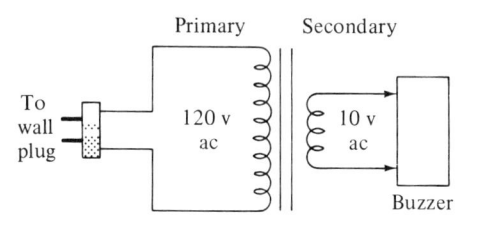

FIGURE 9–8 *Step-Down Transformers*

N_s is the number of turns for the secondary, and N_p is the number of turns for the primary. For example, if the primary has $N_p = 100$, and the secondary has $N_s = 20$, the ratio is calculated:

$$\text{Turns Ratio} = \frac{N_s}{N_p} = \frac{100 \text{ v}}{20 \text{ v}} = 5 \text{ or } 5:1$$

The voltage ratio is

$$\text{Voltage Ratio} = \frac{E_s}{E_p} \tag{9-8}$$

E_s is the voltage across the secondary, and E_p is the voltage across the primary. If the primary voltage is 120 v, and the secondary voltage is 10 v, the ratio is calculated:

$$\text{Voltage Ratio} = \frac{E_s}{E_p} = \frac{120 \text{ v}}{10 \text{ v}} = 12 \text{ or } 12:1$$

The turns ratio is directly proportional to the voltage ratio when the coupling is unity (iron core). Formulas (9–7) and (9–8) are combined as follows.

$$\frac{N_s}{N_p} = \frac{E_s}{E_p} \tag{9-9}$$

Formula (9–9) indicates that the turns ratio can also be reversed, and that the secondary voltage will be higher than the primary voltage.

Example 9–14 The primary of the transformer in Figure 9–8 has 600 turns. How many turns are in the secondary?

Solution Transposing formula (9–9) and solving for N_s gives

$$N_s = \frac{N_p E_s}{E_p} = \frac{600 \times 10 \text{ v}}{120 \text{ v}} = 50 \text{ turns}$$

Example 9–15 The secondary of a transformer has 1000 turns, and the primary has 100 turns. What is the primary voltage if the secondary voltage measures 200 v ac?

168 Inductance [9.3]

Solution The secondary has a higher number of turns than the primary. The primary voltage will be lower than the secondary which means that the voltage is being increased, or stepped-up from primary to secondary.

$$E_p = \frac{E_s N_p}{N_s} = \frac{200 \text{ v} \times 100}{1000} = 20 \text{ v ac}$$

Following the rules of induction, voltage can be induced into more than one secondary winding. The requirement is a common core in which the primary flux encompasses the secondary coils. Figure 9–9(a) illustrates

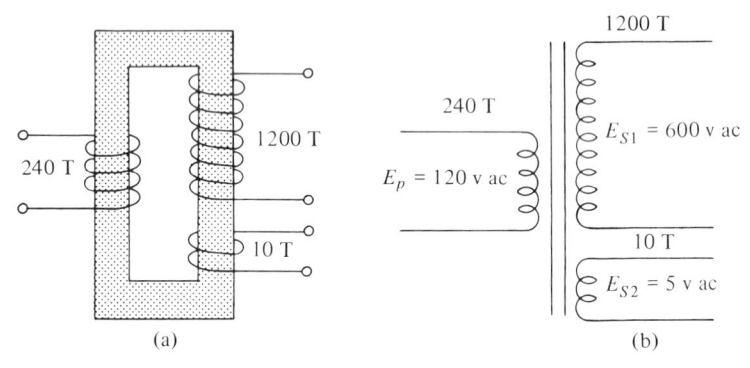

FIGURE 9–9 *Transformer with Two Secondary Windings*

a rectangular iron core with a primary coil on the left side and two secondary coils on the right side. The magnetic lines of flux are concentrated within the core and cut through all of the coils. The same principle holds true for any number of secondary windings. Drawing (b) shows the voltages and turns for each coil. The ratios in terms of secondary voltages are verified in the following manner.

$$E_{s1} = \frac{E_p N_{s1}}{N_p} = \frac{120 \text{ v} \times 1200}{240} = 600 \text{ v ac}$$

$$E_{s2} = \frac{E_p N_{s2}}{N_p} = \frac{120 \text{ v} \times 10}{240} = 5 \text{ v ac}$$

The power consumption of a transformer is normally controlled by the secondary circuitry since the primary would be considered the same as a source voltage and any loading would be that of the secondary. Examples of loading are lamps, vacuum tube filaments, high voltage supplies, door bells, and motors. Assuming that transformer losses are insignificant, power dissipated in the secondary is equal to the power consumed in the primary. Applying Ohm's law gives

$$P_p = P_s$$

or
$$E_p I_p = E_s I_s \tag{9-10}$$

Formula (9–10) can be rearranged and stated in terms of a current ratio,

$$\frac{E_s}{E_p} = \frac{I_p}{I_s} \tag{9-11}$$

which indicates that current is inversely proportional to voltage, or the current ratio is the inverse of the voltage ratio.

Current in the secondary of a transformer is derived from Ohm's law:

$$I_s = \frac{E_s}{R_L} \tag{9-12}$$

R_L is the load resistance connected across the secondary terminals. Assuming that the dc resistance of the secondary is insignificant, the current drain is entirely controlled by R_L. The secondary coil can now be considered a source voltage as far as R_L is concerned. In that case, the secondary voltage drives R_L, and formula (9–12) is a statement of the relationship. In Figure 9–10(a) the secondary current is

$$I_s = \frac{E_s}{R_2} = \frac{24 \text{ v}}{48 \text{ }\Omega} = 0.5 \text{ A}$$

The primary current is found from formula (9–11):

$$I_p = \frac{E_s I_s}{E_p} = \frac{24 \text{ v} \times 0.5 \text{ A}}{240 \text{ v}} = 0.05 \text{ A}$$

Power dissipation of the secondary is

$$E_s I_s = 24 \text{ v} \times 0.5 \text{ A} = 12 \text{ w}$$

and the power consumed by the primary is

$$E_p I_p = 240 \text{ v} \times 0.05 \text{ A} = 12 \text{ w}$$

verifying that primary power is equal to secondary power. Based on the previously stated assumptions (coupling efficiency is 100 percent and secondary resistance is insignificant), Figure 9–10(b) illustrates two points.

(a)

(b)

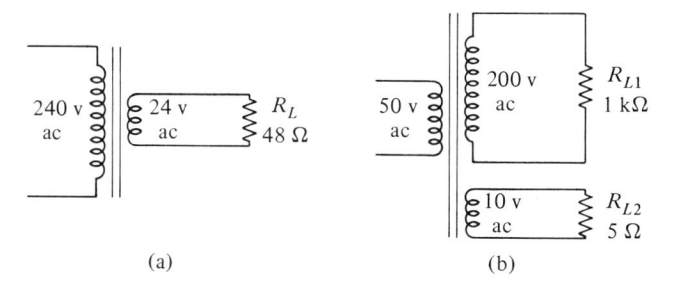

FIGURE 9–10 *Transformer Examples for Calculations*

Total secondary current is inversely proportional to primary current, and **total secondary power** is equal to primary power.

$$I_{s1} = \frac{E_{s1}}{R_{L1}} = \frac{200 \text{ v}}{1 \text{ k}\Omega} = 0.2 \text{ A}$$

$$P_{s1} = E_{s1}I_{s1} = 200 \text{ v} \times 0.2 \text{ A} = 40 \text{ w}$$

$$I_{s2} = \frac{E_{s2}}{R_{L2}} = \frac{10 \text{ v}}{5 \Omega} = 2 \text{ A}$$

$$P_{s2} = E_{s2}I_{s2} = 10 \text{ v} \times 2 \text{ A} = 20 \text{ w}$$

Total secondary current and power are

$$I_{s(\text{tot})} = I_{s1} + I_{s2} = 0.2 \text{ A} + 2 \text{ A} = 2.2 \text{ A}$$

$$P_{s(\text{tot})} = P_{s1} + P_{s2} = 40 \text{ w} + 20 \text{ w} = 60 \text{ w}$$

Total primary current and power are

$$I_{p1} = \frac{E_{s1}I_{s1}}{E_p} = \frac{200 \text{ v} \times 0.2 \text{ A}}{50 \text{ v}} = 0.8 \text{ A}$$

$$I_{p2} = \frac{E_{s2}I_{s2}}{E_p} = \frac{10 \text{ v} \times 2 \text{ A}}{50 \text{ v}} = 0.4 \text{ A}$$

$$I_{p(\text{tot})} = I_{p1} + I_{p2} = 0.8 \text{ A} + 0.4 \text{ A} = 1.2 \text{ A}$$

$$P_p = E_p I_{p(\text{tot})} = 50 \text{ v} \times 1.2 \text{ A} = 60 \text{ w}$$

$$P_p = P_s = 60 \text{ w}$$

An **autotransformer** operates by many of the principles already studied. However, rather than having separate coil windings, an autotransformer has a single winding. The coil is then tapped in various ways (see Figure 9-11) to achieve ratios of turns, current, and voltage. For example, the voltage ratio is directly proportional to the number of turns. Recall that changing primary current induces a voltage across each turn, and that the voltage is determined by the number of turns. Therefore, in Figure 9-11(a) the total primary voltage is subdivided for a proportional secon-

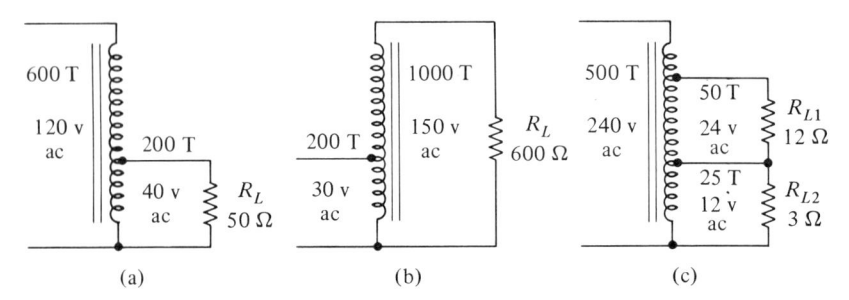

FIGURE 9-11 *Autotransformers*

dary voltage. Notice the coil is tapped one-third of the way, so the secondary voltage is one-third of the primary voltage or

$$E_s = \frac{E_p}{3} = \frac{120 \text{ v}}{3} = 40 \text{ v}$$

Complete analysis of (a) is

$$I_s = \frac{E_s}{R_L} = \frac{40 \text{ v}}{50 \text{ }\Omega} = 0.8 \text{ A}$$
$$P_s = E_s I_s = 40 \text{ v} \times 0.8 \text{ A} = 32 \text{ w}$$
$$I_p = \frac{E_s I_s}{E_p} = \frac{40 \text{ v} \times 0.8 \text{ A}}{120 \text{ v}} = 0.267 \text{ A}$$
$$P_p = E_p I_p = 120 \text{ v} \times 0.267 \text{ A} = 32 \text{ w}$$
$$P_s = P_p = 32 \text{ w}$$

In Figure 9–11(b) the secondary voltage is stepped up from 30 v to 150 v, a factor of five. The turns ratio is also five for the secondary.

$$I_s = \frac{E_s}{R_L} = \frac{150 \text{ v}}{600 \text{ }\Omega} = 0.25 \text{ A}$$
$$P_s = E_s I_s = 150 \text{ v} \times 0.25 \text{ A} = 37.5 \text{ w}$$
$$I_p = \frac{E_s I_s}{E_p} = \frac{150 \text{ v} \times 0.25 \text{ A}}{30 \text{ v}} = 1.25 \text{ A}$$
$$P_p = E_p I_p = 30 \text{ v} \times 1.25 \text{ A} = 37.5 \text{ w}$$

Figure 9–11(c) illustrates two taps in the secondary circuit.

$$I_{s1} = \frac{E_{s1}}{R_{L1}} = \frac{24 \text{ v}}{12 \text{ }\Omega} = 2 \text{ A}$$
$$P_{s1} = E_{s1} I_{s1} = 24 \text{ v} \times 2 \text{ A} = 48 \text{ w}$$
$$I_{s2} = \frac{E_{s2}}{R_{L2}} = \frac{12 \text{ v}}{3 \text{ }\Omega} = 4 \text{ A}$$
$$P_{s2} = E_{s2} I_{s2} = 12 \text{ v} \times 4 \text{ A} = 48 \text{ w}$$
$$P_{s(tot)} = P_{s1} + P_{s2} = 48 \text{ w} + 48 \text{ w} = 96 \text{ w}$$
$$I_{p1} = \frac{E_{s1} I_{s1}}{E_p} = \frac{24 \text{ v} \times 2 \text{ A}}{240 \text{ v}} = 0.2 \text{ A}$$
$$I_{p2} = \frac{E_{s2} I_{s2}}{E_p} = \frac{12 \text{ v} \times 4 \text{ A}}{240 \text{ v}} = 0.2 \text{ A}$$
$$I_{p(tot)} = I_{p1} + I_{p2} = 0.2 \text{ A} + 0.2 \text{ A} = 0.4 \text{ A}$$
$$P_p = E_p I_{p(tot)} = 240 \text{ v} \times 0.4 \text{ A} = 96 \text{ w}$$
$$P_p = P_s = 96 \text{ w}$$

9.4 Transformer Cores

In the last section the secondary power was considered equal to the primary power, and this assumed insignificant transformer losses. If there were no losses, the efficiency of the transformer would be 100 percent because the output power (secondary) equals the input power (primary). When the output power is less than the input power, the efficiency of the transformer is less than 100 percent.

$$\text{Efficiency} = \frac{P_{out}}{P_{in}} \quad (9\text{-}13)$$

For example, if the input power is 75 w, and the output power is 60 w, the efficiency is

$$\text{Eff} = \frac{P_o}{P_i} = \frac{60 \text{ w}}{75 \text{ w}} = 0.8 = 80\%$$

Typical transformers of reasonable quality are approximately 90 percent efficient.

The two principle factors that control the efficiency of a transformer are **eddy currents** and **hysteresis**. An eddy current is produced in the core of a transformer (see Figure 9–12). The variable current that induces a voltage in the winding of a transformer (or coil) also induces a voltage in the core. The induced voltage subsequently causes current to flow in the cross-section of the core since it is also a conductor of electric current. The eddy currents also cause a path of magnetic flux in the core. Unfortunately, flux induced by eddy currents flows in a direction that opposes the direction of flux induced by the coil. The opposing currents require that additional current be supplied to a coil to compensate for losses caused by the eddy currents. In Figure 9–12 the circular direction of current in the coil is opposite the direction of the eddy current. The figure also illustrates the opposing flux fields.

Hysteresis losses are the results of additional power lost at each end of the B and H curve (see Chapter 2). As the current reverses, additional power is required to reverse the magnetic field. Power needed to reverse the

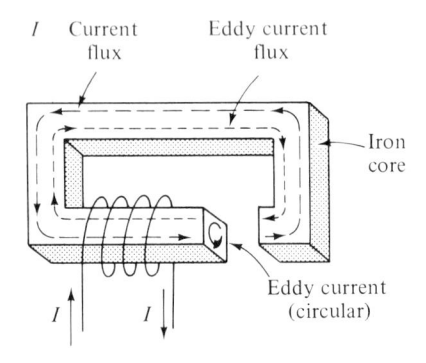

FIGURE 9–12 *Eddy Current Effect in an Iron Core*

field must be supplied by the primary and is not available as secondary power for driving a load.

Several examples of cores are shown in Figure 9–13. In (a) the core is cylindrical. This type is normally used for coils. The coil may be wound tightly around the core, providing the wiring is properly insulated from the core. A tubular insulator may be used that is hollow on the inside; this type

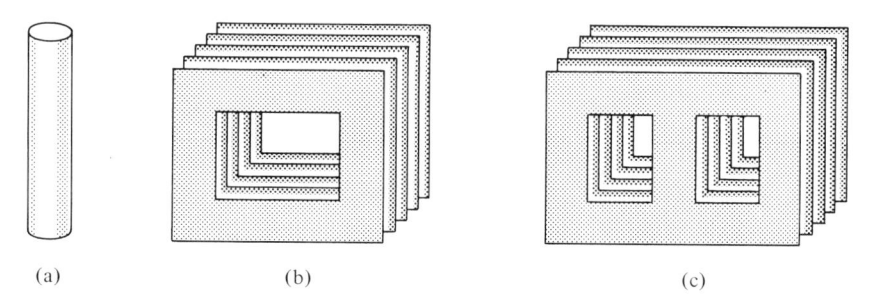

FIGURE 9–13 *Core Types for Transformers*

permits a core to slide in and out, providing a variable inductance. Solid cores are normally made of powdered iron. The grains are individually insulated to reduce the eddy currents. Ferrite cores are also used because they are made of insulating ceramic materials. Both the powdered iron and ferrite cores are capable of high magnetic flux density. They also offer resistance to eddy currents because of their insulating qualities.

Drawings (b) and (c) represent two types of laminated iron cores. The core is made from sheets of iron that are insulated from one another by paper and/or varnish. Laminations increase the cross-sectional resistance, which in turn reduces the eddy currents.

An air core transformer (or coil) does not experience losses from eddy currents or hysteresis. Eddy currents are dependent upon a core capable of conducting current, and hysteresis losses are essentially zero since the magnetic field can be switched with little or no loss in power.

9.5 Series and Parallel Inductance

The total inductance of coils connected in series is the sum of the individual inductances, the same as series resistors.

$$L_T = L_1 + L_2 + \cdots + L_n \qquad (9\text{--}14)$$

When the coils are in series, current changes during a specified time the same amount in all of the coils. Total inductance, from formula (9–1), is directly proportional to the total voltage (e) across the coils. When coils are connected in series, and if the voltage (e) across them is held at a constant value, the change in current (Δt) for a given time will decrease proportionally. Thus, series inductances are additive in proportion to their individual values as though the coils were extended with additional turns. Examples of series inductance follow.

Example 9-16 What is the total inductance for Figure 9-14(a)?

Solution $\quad L_T = L_1 + L_2 = 40 \text{ mh} + 70 \text{ mh} = 110 \text{ mh}$

Example 9-17 What is the inductance of L_2 in Figure 9-14(b)?

Solution $\quad L_2 = L_T - L_1 = 65 \ \mu\text{h} - 20 \ \mu\text{h} = 45 \ \mu\text{h}$

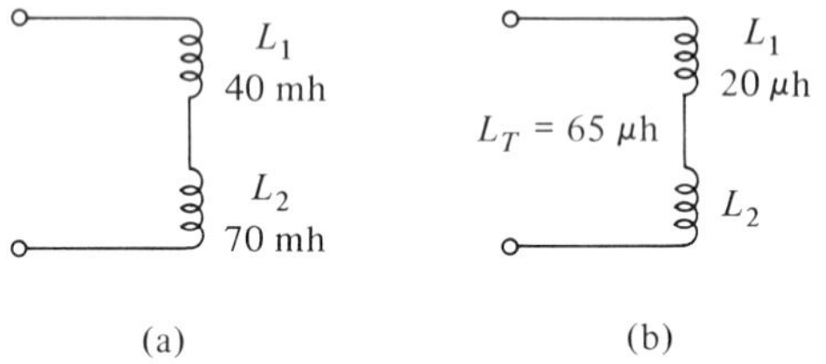

(a)

(b)

FIGURE 9-14 *Series Inductance*

The total inductance of coils connected in parallel is found in the same way as the total resistance of resistors in parallel.

$$L_T = \frac{1}{\frac{1}{L_1} + \frac{1}{L_2} + \cdots + \frac{1}{L_n}} \tag{9-15}$$

$$\frac{1}{L_T} = \frac{1}{L_1} + \frac{1}{L_2} + \cdots + \frac{1}{L_n} \tag{9-16}$$

$$L_T = \frac{L_1 L_2}{L_1 + L_2} \tag{9-17}$$

for $L_1 = L_2$

$$L_T = \frac{L_1}{2} = \frac{L_2}{2} \tag{9-18}$$

Formula (9-18) can be solved for any number (n) of equal inductances where the number n is the denominator.

Example 9-18 Calculate the total inductance for Figure 9-15(a) using the first three parallel formulas.

Solution $\quad L_T = \dfrac{1}{\dfrac{1}{5 \text{ h}} + \dfrac{1}{20 \text{ h}}} = \dfrac{1}{\dfrac{5}{20}} = \dfrac{20}{5} = 4 \text{ h} \tag{9-15}$

$\dfrac{1}{L_T} = \dfrac{1}{5 \text{ h}} + \dfrac{1}{20 \text{ h}} = \dfrac{5}{20} \tag{9-16}$

Inductance

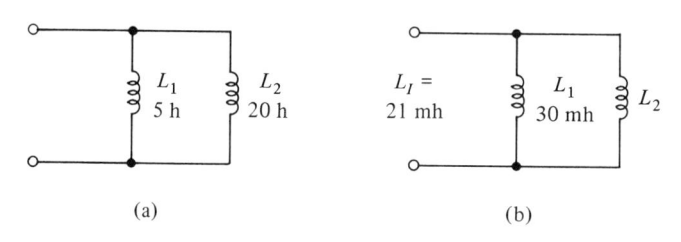

FIGURE 9-15 *Parallel Inductance*

$$L_T = \frac{20}{5} = 4 \text{ h}$$

$$L_T = \frac{5 \text{ h} \times 20 \text{ h}}{5 \text{ h} + 20 \text{ h}} = \frac{100}{25} = 4 \text{ h} \tag{9-17}$$

Example 9-19 What is the inductance of L_2 in Figure 9-15(b)?

Solution Rearranging formula 9-17 results in

$$L_2 = \frac{L_1 L_T}{L_1 - L_T} = \frac{30 \text{ mh} \times 21 \text{ mh}}{30 \text{ mh} - 21 \text{ mh}} = \frac{630}{9} = 70 \text{ mh}$$

When two or more coils are connected in series, the current is the same in each coil. The voltages (e) across each coil are added to make up the total voltage. In Figure 9-16 the input is $\Delta i/\Delta t$, which occurs in both coils. The total voltage across the coils is $e_1 + e_2$. The voltage is proportional to the inductance, which means that e_2 is four times the value of e_1.

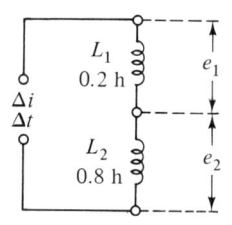

FIGURE 9-16 *Voltage Drops across Series Inductors*

$$e_1 = \frac{L_1}{L_1 + L_2} \times e_T = \frac{0.2 \text{ h}}{0.2 \text{ h} + 0.8 \text{ h}} \times e_T$$
$$= 0.2 \, e_T = 20\% \text{ of } e_T$$

$$e_2 = \frac{L_2}{L_2 + L_1} \times e_T = \frac{0.8 \text{ h}}{0.8 \text{ h} + 0.2 \text{ h}} \times e_T$$
$$= 0.8 \, e_T = 80\% \text{ of } e_T$$

In Figure 9-16, $\Delta i = 2.5$ ma for 0.5 msec.

$$L_T = L_1 + L_2 = 0.2 \text{ h} + 0.8 \text{ h} = 1.0 \text{ h}$$

$$e_T = L\frac{\Delta i}{\Delta t} = 1 \times \frac{2.5 \times 10^{-3}}{0.5 \times 10^{-3}} = 5 \text{ v}$$

$$e_1 = L_1\frac{\Delta i}{\Delta t} = 0.2 \times \frac{2.5 \times 10^{-3}}{0.5 \times 10^{-3}} = 1 \text{ v}$$

$$e_2 = L_2\frac{\Delta i}{\Delta t} = 0.8 \times \frac{2.5 \times 10^{-3}}{0.5 \times 10^{-3}} = 4 \text{ v}$$

$$e_T = e_1 + e_2 = 1 \text{ v} + 4 \text{ v} = 5 \text{ v}$$

e_1 is 20 percent of e_T, and e_2 is 80 percent of e_T. The sum of e_1 and e_2 equals the total e_T.

$$e_1 (\%) = \frac{e_1}{e_T} = \frac{1 \text{ v}}{5 \text{ v}} = 0.2 = 20\%$$

$$e_2 (\%) = \frac{e_2}{e_T} = \frac{4 \text{ v}}{5 \text{ v}} = 0.8 = 80\%$$

Figure 9–17 illustrates parallel coils in which the current changes 3 ma during a period of 0.5 msec. The voltage across L_1 is the same as the voltage

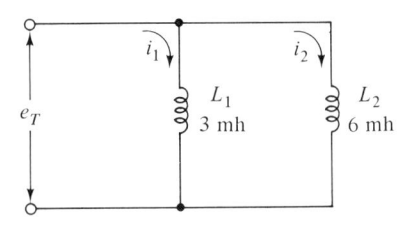

FIGURE 9–17 *Current through Parallel Inductors*

across L_2 according to the principle which also applies to parallel resistors or parallel capacitors. However, in this case the current divides with a portion of the total current going through each coil. To analyze the circuit, the total inductance is calculated.

$$L_T = \frac{L_1 L_2}{L_1 + L_2} = \frac{3 \text{ mh} \times 6 \text{ mh}}{3 \text{ mh} + 6 \text{ mh}} = 2 \text{ mh}$$

Now e_T can be calculated.

$$e_T = L\frac{\Delta i}{\Delta t} = 2 \times 10^{-3} \times \frac{3 \times 10^{-3}}{0.5 \times 10^{-3}}$$
$$= 12 \times 10^{-3} = 12 \text{ mv}$$

The current Δi for each coil over a period of 0.5 msec is

$$\Delta i_1 = \Delta t \frac{e_T}{L_1} = 0.5 \times 10^{-3} \times \frac{12 \times 10^{-3}}{3 \times 10^{-3}}$$
$$= 2 \times 10^{-3} = 2 \text{ ma}$$

[9.5]   Inductance   177

$$\Delta i_2 = \Delta t \frac{e_T}{L_2} = 0.5 \times 10^{-3} \times \frac{12 \times 10^{-3}}{6 \times 10^{-3}}$$
$$= 1 \times 10^{-3} = 1 \text{ ma}$$
$$\Delta i_T = \Delta i_1 + \Delta i_2 = 2 \text{ ma} + 1 \text{ ma} = 3 \text{ ma}$$

L_1 is one-half the inductance of L_2, but the current is twice as much. Values of inductance are similar to values of resistance. Lower inductance means higher ac current. Consider the following examples.

Example 9-20 What are the inductances of L_1 and L_2 in Figure 9-16 if $e_2 = 75$ mv, $e_T = 300$ mv, and the current changes 50 ma during 0.1 msec?

Solution L_2 can be calculated directly.

$$L_2 = \frac{e_2}{\frac{\Delta i}{\Delta t}} = \frac{75 \times 10^{-3}}{\frac{50 \times 10^{-3}}{0.1 \times 10^{-3}}} = \frac{75 \times 10^{-3}}{500}$$
$$= 150 \text{ }\mu\text{h}$$
$$L_T = \frac{e_T}{\frac{\Delta i}{\Delta t}} = \frac{300 \times 10^{-3}}{\frac{50 \times 10^{-3}}{0.1 \times 10^{-3}}} = \frac{300 \times 10^{-3}}{500}$$
$$= 600 \text{ }\mu\text{h}$$
$$L_1 = L_T - L_2 = 600 \text{ }\mu\text{h} - 150 \text{ }\mu\text{h} = 450 \text{ }\mu\text{h}$$

Example 9-21 In Figure 9-18, if the total inductance is 2.4 h, what is the inductance of L_2? (Solve by the current method.)

Solution   The total current can be calculated first.

$$\Delta i_T = \Delta t \frac{e_T}{L_T} = 3 \text{ sec} \times \frac{12 \text{ v}}{2.4 \text{ h}} = 15 \text{ A}$$

The current through L_1 is then

$$\Delta i_1 = \Delta t \frac{e_T}{L_1} = 3 \text{ sec} \times \frac{12 \text{ v}}{3 \text{ h}} = 12 \text{ A}$$
$$\Delta i_2 = \Delta i_T - \Delta i_1 = 15 \text{ A} - 12 \text{ A} = 3 \text{ A}$$

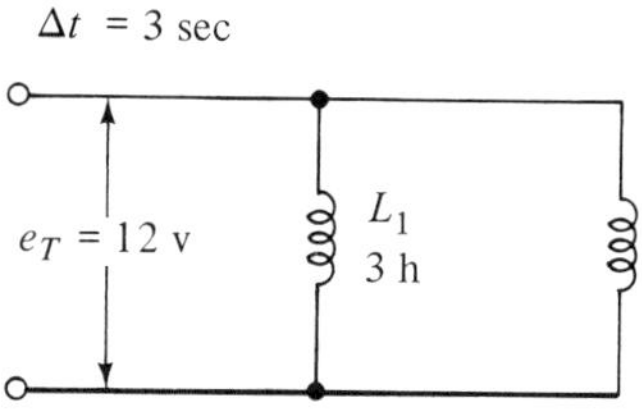

$\Delta t = 3$ sec

FIGURE 9-18 *Determining Unknown Inductance*

Now L_2 is calculated directly:

$$L_2 = \frac{e_T}{\frac{\Delta i_2}{\Delta t}} = \frac{12 \text{ v}}{\frac{3 \text{ A}}{3 \text{ sec}}} = 12 \text{ h}$$

9.6 Inductive Reactance

Inductive reactance is the resistance in ohms that a coil offers an alternating voltage. Reactance is ac dependent, which should not be confused with the dc resistance of the wire that makes up the coil. For example, when a sine wave of voltage is applied to an inductor, an alternating current flows through the windings of the inductor. The alternating current causes a voltage to be induced across the inductor. In general, when the applied voltage increases, the current increases, and then the induced voltage increases.

The induced voltage across a coil is in direct opposition to the applied voltage. In a pure reactance, assuming zero dc resistance (which is theoretically impossible since the wiring has dc resistance), the induced voltage (back emf) is 180° out-of-phase with the applied voltage. Figure 9-19(a) shows an alternating voltage source (e_a) that applies a sinusoidal voltage to a coil (e_L). The graph in (b) illustrates the applied voltage in relationship to the back emf.

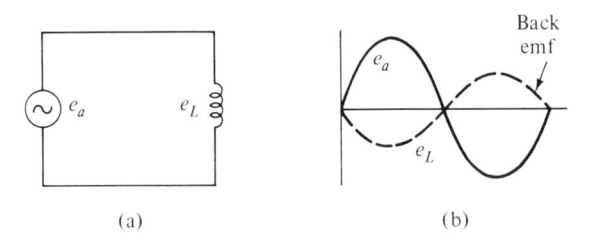

FIGURE 9-19 *Opposing Voltage*

Since the back emf is opposing the applied voltage, a resistance to the current is created within the coil which is called **reactance**. The reactance of an inductive circuit increases with higher frequencies and/or higher inductances. The reactance of a coil is called inductive reactance and is abbreviated X_L. Inductive reactance is calculated in the following manner.

$$X_L = 2\pi f L \tag{9-19}$$

In the above equation X_L is in ohms, f is in Hz, and L is in henrys. As with X_C, the constant 2π represents the circumference of a sine wave which is 2π radians. Frequency and inductance are directly proportional to the reactance. Examples of formula (9-19) follow.

[9.6] Inductance 179

Example 9-22 What is the reactance of a 10-mh coil for a frequency of 100 kHz?

Solution
$$X_L = 2\pi f L = 6.28 \times 100 \times 10^3 \times 10 \times 10^{-3}$$
$$= 6.28 \text{ k}\Omega$$

Example 9-23 If a frequency of 2 kHz provides a reactance of 6.28 ohms, how much is the inductance?

Solution By rearranging formula (9-19), the following is obtained:
$$L = \frac{X_L}{2\pi f} = \frac{6.28 \text{ }\Omega}{2\pi \times 2 \times 10^3 \text{ Hz}} = 0.5 \times 10^{-3} \text{ h}$$
$$= 0.5 \text{ mh}$$

In the first two columns of Table 9-1 inductance and inductive reactance are compared for a constant frequency of 10 kHz. As the inductance increases, the reactance increases. The reactance also increases as the frequency increases as shown in the second and third columns. Figure

TABLE 9-1

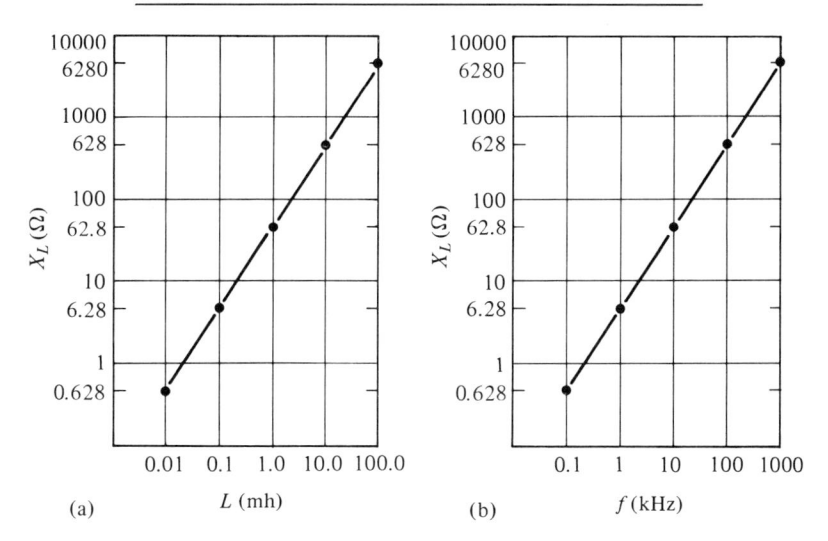

$f = 10$ kHz L	X_L	$L = 1.0$ mh f
10 µh	0.628 Ω	0.1 kHz
100 µh	6.28 Ω	1.0 kHz
1.0 mh	62.8 Ω	10 kHz
10.0 mh	628 Ω	100 kHz
100 mh	6280 Ω	1 MHz

FIGURE 9-20 *Inductive Reactance versus Inductance and Frequency*

9–20 is a plot of the table, and illustrates the relationships between X_L and inductance, and between X_L and frequency. The graphs also show changes in reactance which are proportional to changes in inductance or frequency.

9.7 Series/Parallel Inductive Reactance

The reactances of inductors in series are additive in much the same way as resistances in series are.

$$X_{LT} = X_{L1} + X_{L2} + \cdots + X_{Ln} \tag{9-20}$$

Figure 9–21(a) shows two coils in series. The total reactance is

$$X_{LT} = X_{L1} + X_{L2} = 1.28\ \Omega + 5\ \Omega = 6.28\ \Omega$$

Figure 9–21(b) shows an equivalent inductor in which $X_{LT} = 6.28\ \Omega$. The sum of the inductances is 100 μh, which is found by applying formula (9–14).

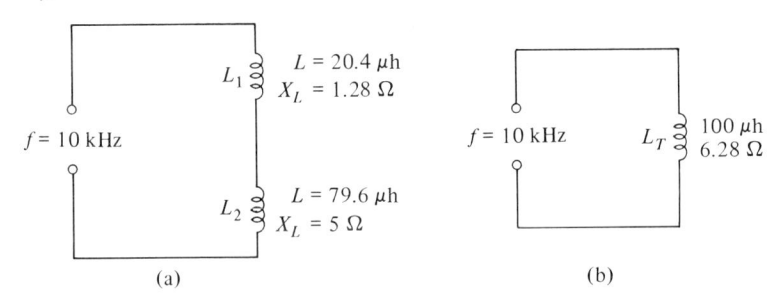

FIGURE 9–21 *Series Reactance*

$$L_T = L_1 + L_2 = 20.4\ \mu h + 79.6\ \mu h = 100\ \mu h$$

To cross-check the results, the total inductive reactance of (b) is calculated for the total inductance of 100 μh.

$$X_{LT} = 2\pi f L = 6.28 \times 10 \times 10^3 \times 100 \times 10^{-6}$$
$$= 6.28\ \Omega$$

The following further illustrates the application of formula (9–20).

Example 9–24 In Figure 9–22 the total inductance is 55 mh. What are the inductance and reactance of L_2?

Solution A logical procedure is to calculate L_3 and then L_2 since L_T is given. Then X_{L2} should be calculated.

$$L_3 = \frac{X_{L3}}{2\pi f} = \frac{150}{2\pi \times 1.59 \times 10^3} = 15 \times 10^{-3}$$
$$= 15\ \text{mh}$$

Inductance

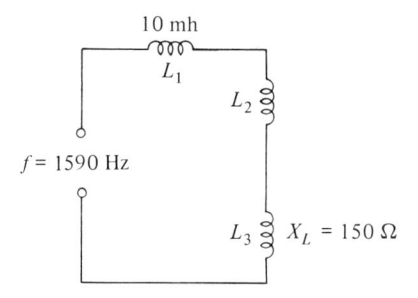

FIGURE 9-22 *Example for Three Inductors in Series*

$$L_2 = L_T - (L_1 + L_3) = 55 \text{ mh} - (10 \text{ mh} + 15 \text{ mh})$$
$$= 30 \text{ mh}$$
$$X_{L2} = 2\pi f L = 2\pi \times 1.59 \times 10^3 \times 30 \times 10^{-3}$$
$$= 300 \text{ }\Omega$$

The circuit can be further analyzed.

$$X_{L1} = 2\pi f L = 2\pi \times 1.59 \times 10^3 \times 10 \times 10^{-3}$$
$$= 100 \text{ }\Omega$$
$$X_{LT} = X_{L1} + X_{L2} + X_{L3} = 100 \text{ }\Omega + 300 \text{ }\Omega + 150 \text{ }\Omega$$
$$= 550 \text{ }\Omega$$

The calculations are then cross-checked.

$$X_{LT} = 2\pi f L_T = 2\pi \times 1.59 \times 10^3 \times 55 \times 10^{-3}$$
$$= 550 \text{ }\Omega$$

Parallel reactances are calculated from the parallel resistance formulas.

$$X_{LT} = \frac{1}{\dfrac{1}{X_{L1}} + \dfrac{1}{X_{L2}} + \cdots + \dfrac{1}{X_{Ln}}} \tag{9-21}$$

$$\frac{1}{X_{LT}} = \frac{1}{X_{L1}} + \frac{1}{X_{L2}} + \cdots + \frac{1}{X_{Ln}} \tag{9-22}$$

$$X_{LT} = \frac{X_{L1} X_{L2}}{X_{L1} + X_{L2}} \tag{9-23}$$

$$X_{LT} = \frac{X_{L1}}{2} = \frac{X_{L2}}{2} \text{ for two equal reactances} \tag{9-24}$$

As usual, the reactance for one series inductor is divided by the total number of equal reactances (see formula 9-24).

Figure 9-23(a) illustrates parallel inductance and reactance and also illustrates the fact that reactance is higher with a higher inductance. The circuit in (b) is the equivalent for (a). The reactance for L_1 and L_2 is calculated in the following manner.

182   Inductance   [9.7]

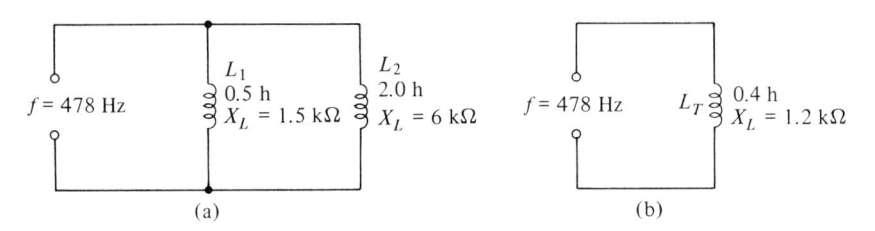

$f = 478$ Hz L_1 0.5 h L_2 2.0 h $f = 478$ Hz L_T 0.4 h
               $X_L = 1.5$ kΩ $X_L = 6$ kΩ $X_L = 1.2$ kΩ

                    (a)   (b)

FIGURE 9-23 *Parallel Reactance*

$$X_{L1} = 2\pi f L_1 = 2\pi \times 478 \times 0.5 = 1.5 \times 10^3 = 1.5 \text{ k}\Omega$$
$$X_{L2} = 2\pi f L_2 = 2\pi \times 478 \times 2.0 = 6 \times 10^3 = 6 \text{ k}\Omega$$

The parallel reactance shown in (b) is calculated below.

$$X_{LT} = \frac{X_{L1} X_{L2}}{X_{L1} + X_{L2}} = \frac{1.5 \times 10^3 \times 6 \times 10^3}{1.5 \times 10^3 + 6 \times 10^3}$$
$$= 1.2 \times 10^3 = 1.2 \text{ k}\Omega$$

As a cross-check, the parallel inductance and the total X_L based on total inductance are

$$L_T = \frac{L_1 L_2}{L_1 + L_2} = \frac{0.5 \times 2.0}{0.5 + 2.0} = 0.4 \text{ h}$$
$$X_{LT} = 2\pi f L_T = 2\pi \times 478 \times 0.4 = 1.2 \times 10^3 = 1.2 \text{ k}\Omega$$

The following is an example of three coils in parallel.

Example 9-25 If the total reactance in Figure 9-24 is 8 k ohms, what are the inductance and reactance of L_2?

Solution The most obvious way to proceed is to calculate X_{L2} using either formula (9-21) or (9-22) and then to calculate the inductance using formula (9-19).

$$\frac{1}{X_{L2}} = \frac{1}{X_{LT}} - \frac{1}{X_{L1}} - \frac{1}{X_{L3}}$$
$$= \frac{1}{8 \times 10^3} - \frac{1}{10 \times 10^3} - \frac{1}{120 \times 10^3}$$
$$= \frac{15}{120 \times 10^3} - \frac{12}{120 \times 10^3} - \frac{1}{120 \times 10^3}$$
$$= \frac{2}{120 \times 10^3}$$
$$X_{L2} = \frac{120 \times 10^3}{2} = 60 \times 10^3 = 60 \text{ k}\Omega$$
$$L_2 = \frac{X_{L2}}{2\pi f} = \frac{60 \times 10^3}{2\pi \times 15.9 \times 10^6}$$
$$= 600 \times 10^{-6} = 600 \text{ }\mu\text{h}$$

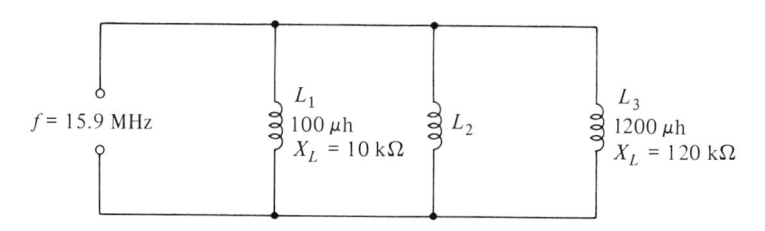

FIGURE 9-24 Example for Three Inductors in Parallel

9.8 Phase Angle of Voltage and Current

The phase angle for a capacitive circuit is 90°, and the current leads the voltage by 90°. When a sine wave of voltage is applied across an inductor, the phase angle is also 90°. Now, however, the voltage leads the current by 90°. Figure 9-25(a) is a plot of the phase angle for an inductive circuit. The current (i) is a normal sine wave. The voltage is maximum at 90° ahead of the current when the current begins. The maximum and minimum values are displaced by 90°.

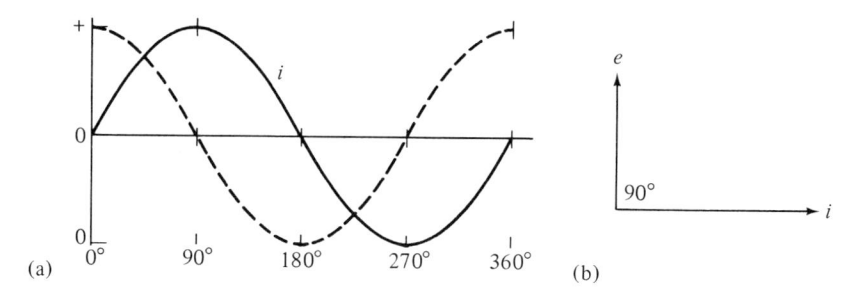

FIGURE 9-25 Phase Relation for Inductive Circuits

To understand more easily the 90° difference, recall that the induced voltage in a coil depends upon a current change through its windings. Therefore, e in Figure 9-25 represents the result of i. A rearrangement of formula (9-1) mathematically describes the phenomenon.

$$e = L\frac{\Delta i}{\Delta t} \qquad (9\text{-}25)$$

As the Δt factor becomes lower (higher rate of change), the induced voltage (e) increases. In the plot, the sine wave of current changes at the maximum rate when it crosses the base reference line at 0°, 180°, and 360°. Notice that the voltage (e) is maximum at these points. The minimum rate of change occurs when the current is in the region of its maximum or minimum value, 90° and 270°. At these points, the voltage is minimum or zero. The phase relationship is shown in vector form in (b). The voltage is shown on the positive side of the base line (current) to indicate that voltage leads current in the relationship.

The first 90° of the sinusoidal waveforms for i and e are plotted in Figure 9–26. A time base of 1 μsec per 15° is used for simplicity. The current is scaled from 0 to 1.0 ma, while the voltage is scaled from 0 to 3 v. According

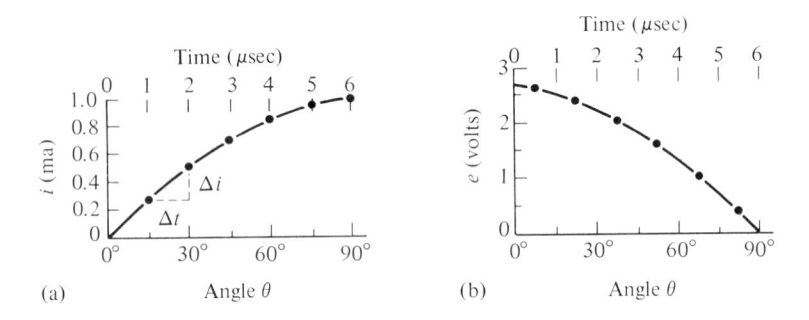

FIGURE 9–26 Current-voltage Plots for L = 10 mh

to formula (9–25), $\Delta i / \Delta t$ causes an induced voltage, so the voltage in (b) is plotted as a function of changing current for 1-μsec increments. The values are presented in Table 9–2 for an inductance of 10 mh. The column under "Time" follows the horizontal axes of the plots in Figure 9–26. The "Δ Time" column shows the region used for calculating $\Delta i / \Delta t$ in degrees, and the time is 1 μsec for each region of 15°. The current amplitude at each 15° point is

$$i_i = i_m \sin \theta \qquad (9\text{–}26)$$

i_i is the instantaneous value of current, i_m is the maximum amplitude of the current, and $\sin \theta$ is the sine of the angle at the instant i_i is desired. In the table, the θ value used for $\sin \theta$ is in the first column. The Δi value is the difference in current for the θ defined under "Δ Time."

Formula (9–26) is derived from formula (7–1) ($e_i = e_m \sin \theta$). The equations are intended for sine wave applications in which the instantaneous amplitude of any point (angle) along a sine wave is $\sin \theta$ times the maximum amplitude. Examples for 30°, 60°, and 90° follow.

$$t = 30° \qquad e = L\frac{\Delta i}{\Delta t} = 10 \times 10^{-3} \times \frac{0.24 \times 10^{-3}}{1 \times 10^{-6}} = 2.4 \text{ v}$$

$$t = 60° \qquad e = L\frac{\Delta i}{\Delta t} = 10 \times 10^{-3} \times \frac{0.159 \times 10^{-3}}{1 \times 10^{-6}} = 1.59 \text{ v}$$

$$t = 90° \qquad e = L\frac{\Delta i}{\Delta t} = 10 \times 10^{-3} \times \frac{0.034 \times 10^{-3}}{1 \times 10^{-6}} = 0.34 \text{ v}$$

Between 90° and 180° in Table 9-2, the current has a negative slope as it approaches the zero reference line. Changes in current, then, are minus values which produce negative values of voltage. Note that the Δi and e values are the reverse of those between 0° and 90°. The pattern is true sinusoidal and repeats from 180° to 360°. From 180° to 270° the Δi and e values are minus, and from 270° to 360° the values are positive.

TABLE 9-2

θ	Time μsec	Δ Time θ	μsec	Sin θ ma	Δi ma	e* v
15°	1	0–15°	1	0.26	0.26	2.60
30°	2	15°–30°	1	0.50	0.24	2.40
45°	3	30°–45°	1	0.707	0.207	2.07
60°	4	45°–60°	1	0.866	0.159	1.59
75°	5	60°–75°	1	0.966	0.10	1.00
90°	6	75°–90°	1	1.00	0.034	0.34
105°	7	90°–105°	1	0.966	−0.034	−0.34
120°	8	105°–120°	1	0.866	−0.10	−1.00
135°	9	120°–135°	1	0.707	−0.159	−1.59
150°	10	135°–150°	1	0.50	−0.207	−2.07
165°	11	150°–165°	1	0.26	−0.24	−2.40
180°	12	165°–180°	1	0	−0.26	−2.60

*$e = L\,(\Delta i/\Delta t)$ for $L = 10$ mh

QUESTIONS

9-1   What causes a voltage to be induced in a conductor?

9-2   If 160 ma per 2 msec in a coil induces 40 v, what is the inductance?

9-3   How much voltage will a 0.6-h coil induce for a change in current of 3 A in 0.25 sec?

9-4   If a coil has an inductance of 27.5 h, how long does a current of 400 ma have to change to induce 0.22 v?

9-5   What is the inductance for a single-layer coil which is 20 cm long, has 80 turns, a cross-sectional area of 1.59 cm, and a core with a permeability of 1000?

9-6   To wind a single-layer coil for 50 μh, how long must an air core tube be when it has 120 turns and a cross-sectional area of 5 cm?

186 Inductance

9-7 What cross-sectional area is required of an iron core for a single-layer coil of 15.7 mh which is 8 cm long, has a core permeability of 2000, and has 50 turns?

9-8 If the coefficient of coupling between two coils is 95 percent, what is the flux field of the primary in maxwells if the secondary field is 76×10^3 maxwells?

9-9 What is the mutual inductance of two coils on a common iron core if the primary inductance is 20 mh, and the secondary inductance is 5 mh?

9-10 What is the secondary voltage of two coils if their mutual inductance is 0.3 h, and the primary current is 600 ma per sec?

9-11 What is the coefficient of coupling between two coils, 3 h and 6 h, if the primary current is 40 A in 8 sec and produces a secondary voltage of 17 v?

9-12 In Figure 9-8, if the secondary winding has 12 turns, how many turns does the primary winding have?

9-13 The primary of a transformer has 500 turns connected to 240 v ac. How many turns are required for each secondary winding for secondary voltages of 120 v ac and 24 v ac?

9-14 In Figure 9-8, what is the primary current if the secondary current is 30 A?

9-15 Assuming zero losses, what is the power dissipated in the primary and secondary for Question 9-14?

9-16 In Figure 9-10(b) assume that $R_{L1} = 5$ k ohms, and that $R_{L2} = 20$ ohms. What are the current and power dissipation for each winding?

9-17 In Figure 9-27, what are the number of turns and voltage in the secondary winding?

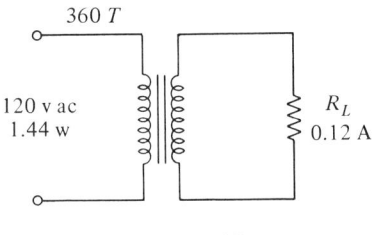

FIGURE 9-27

Inductance

9–18 The entire primary of an autotransformer, 800 turns, is connected to 60 v ac. How many turns out of the total 800 are required for a 6-v ac tap?

9–19 An autotransformer draws 5 A in the primary circuit and draws 0.5 A through a 2-k ohm load. What is the primary voltage?

9–20 What is the input power to a transformer that is 95 percent efficient if the output power is 20 w?

9–21 What is the advantage of using laminated cores in transformers rather than solid iron cores?

9–22 What is the total inductance for three coils in series if $L_1 = 100$ μh, $L_2 = 300$ μh, and $L_3 = 1200$ μh?

9–23 If the total inductance of three coils in series is 7 h, and L_1 and L_2 are both 1.5 h, what is the inductance of L_3?

9–24 What is the total inductance of four parallel coils if the inductance of each is 50 mh?

9–25 In Figure 9–15(b) what is the inductance of L_2 if L_1 is 6 mh, and L_T is 4.8 mh?

9–26 What is the total inductance of three coils in parallel if $L_1 = 90$ mh, $L_2 = 180$ mh, and $L_3 = 40$ mh?

9–27 Refer to Figure 9–14(a). If 0.22 v ac are connected across the two coils, what is the voltage across each individual coil?

9–28 Refer to Figure 9–16. Assume that $e_1 = 6$ v, $e_2 = 4$ v, and the input is 0.5 A per sec. What are the inductances of L_1 and L_2?

9–29 In Figure 9–17, if the input is 96 ma for 2 msec, what is the current through each coil?

9–30 In Figure 9–18, if L_1 is 120 μh, and L_T is 96 μh, what is the current in L_2?

9–31 What is the reactance of a 2-h coil at a frequency of 500 Hz?

9–32 At what frequency will the reactance of a 100-μh coil be 10 ohms?

9–33 In Figure 9–21(a), what are the reactances of L_1, L_2, and L_T if the frequency is 0.5 MHz?

9–34 A sine wave voltage, $f = 79.6$ Hz, is applied to two coils in parallel. The total reactance is 300 ohms, and $L_2 = 1.8$ h. What are the inductance and reactance of L_1?

9-35 What is the total reactance of three coils in parallel when $X_{L1} = 20$ ohms, $X_{L2} = 80$ ohms, and $X_{L3} = 4$ ohms?

9-36 The total reactance of four coils in parallel is 3.2 k ohms. What is the reactance of L_1 if $L_2 = 6$ k ohms, $L_3 = 20$ k ohms, and $L_4 = 30$ k ohms?

9-37 Refer to Table 9-2. What is the voltage (e) at 165° if $\Delta t = 40$ μsec?

9-38 Referring to Figure 9-26 and Table 9-2, what is the voltage (e) for a change from 30° to 35°?

10 Impedance

Current leads voltage by 90° in an ac capacitive circuit (see chapter 8). For an ac inductive circuit, voltage leads current by 90° (see chapter 9). In addition, the phase difference through a resistive circuit is zero. Therefore, when reactance and resistance are combined, the phase difference is somewhere between $\pm 90°$ and 0°, depending on the balance between reactance and resistance and whether the reactance is inductive ($+$) or capacitive ($-$). The vector addition (combined reactance and resistance) in an ac circuit is called **impedance.** Since reactance is frequency dependent, impedance must also be frequency dependent.

This chapter covers impedance and its effects on ac circuits. Analysis is oriented around vector quantities, which describe the ac voltages and currents. Because frequency variations with reactance have been covered in earlier chapters, problems in this chapter will, for the most part, concern given reactances that are assumed solved for specific frequencies.

10.1 Series Resistance and Capacitive Reactance

If resistance is considered alone, the current and voltage waveforms are of the same phase. If capacitive reactance is considered alone, the current leads the voltage by 90°. A series combination of resistance and capacitive reactance results in a phase angle that is something less than 90°. The total reactance, then, is something less than the straight sum of the resistance and reactance.

Figure 10–1(a) is a circuit with a resistor and a capacitor in series. The resistance and reactance are each 10 ohms. In (b) the resistance is shown on the X axis because the phase is zero. X_C is shown negatively displaced 90°

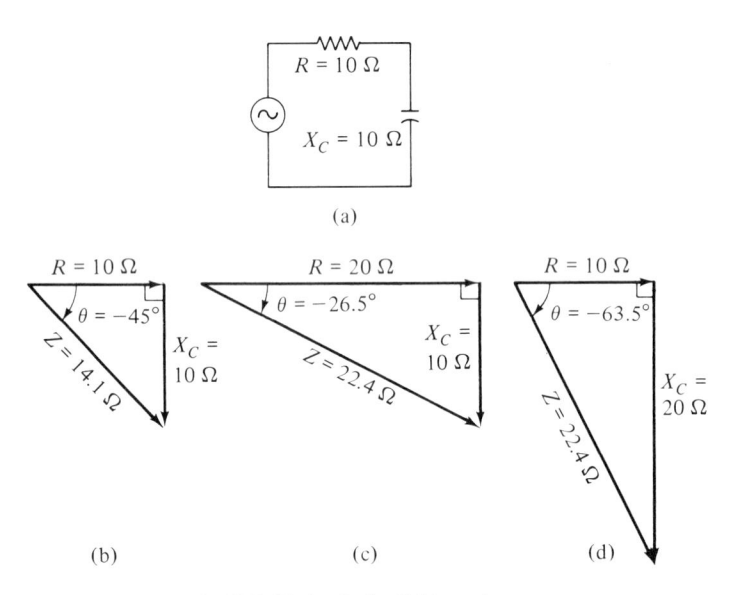

FIGURE 10–1 *Series RC Impedance*

because the voltage across a capacitor lags the current in the circuit. The impedance of the combination is the length of the hypotenuse since the R and X_C vectors form a right triangle.

$$Z = \sqrt{R^2 + X_C^2} \qquad (10\text{--}1)$$

The impedance of (b) is calculated:

$$Z = \sqrt{R^2 + X_C^2} = \sqrt{10^2 + 10^2} = \sqrt{200} = 14.1\ \Omega$$

The phase angle (θ) between the voltage and current is obviously 45° since the two vectors that form the legs of the triangle are the same length. Mathematically, the phase angle is

$$\tan \theta = \frac{Y \text{ axis}}{X \text{ axis}} = \frac{-X_C}{R} \qquad (10\text{--}2)$$

and for Figure 10–1(b),

$$\tan \theta = \frac{-X_C}{R} = \frac{-10\ \Omega}{10\ \Omega} = -1$$
$$\theta = -45°$$

In drawing (c) R is twice as long as X_C. The impedance is
$$Z = \sqrt{R^2 + X_C^2} = \sqrt{20^2 + 10^2} = \sqrt{500} = 22.4\ \Omega$$
and the phase angle is
$$\tan \theta = \frac{-X_C}{R} = \frac{-10\ \Omega}{20\ \Omega} = -0.5$$
$$\theta = -26.5°$$

In drawing (d), X_C is twice as long as R. The impedance is
$$Z = \sqrt{R^2 + X_C^2} = \sqrt{10^2 + 20^2} = \sqrt{500} = 22.4\ \Omega$$
and the phase angle is
$$\tan \theta = \frac{-X_C}{R} = \frac{-20\ \Omega}{10\ \Omega} = -2$$
$$\theta = -63.5°$$

Notice that (c) and (d) have the same impedance, even though the phase angles change. The hypotenuse is the same length for both. The vector diagram in (d) is more reactive, meaning that the current leads the voltage by a greater angle than in (c).

An ac voltage-current analysis is shown in Figure 10-2. The series circuit in (a) has a maximum sine wave current (i) of 1 A. The IR drop across

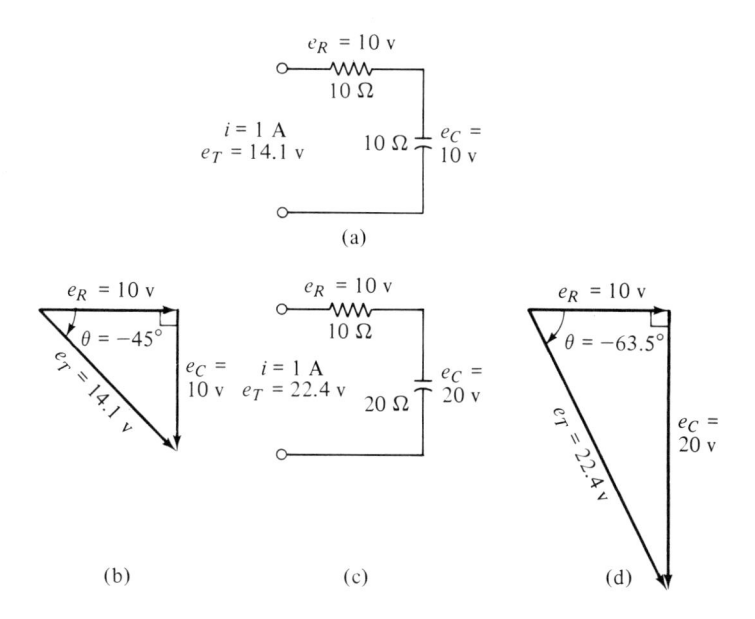

FIGURE 10-2 *Series RC Voltage*

each component is 10 v since R and X_C each equal 10 ohms. Because of the 90° phase shift, one is at a maximum value while the other is at a minimum value. Therefore, the total voltage (e_T) is not 20 v, but is the vector sum of the two voltages. A vector diagram is shown in (b). The voltage across the resistance is shown as the base line because the phase between current and voltage is the same. The voltage across the capacitor is shown in the negative direction since voltage lags current across a capacitor.

As in Figure 10–1, the hypotenuse represents the net quantity, which in this case is voltage.

$$e_T = \sqrt{e_R^2 + e_C^2} \qquad (10\text{--}3)$$

For Figure 10–2(a) and (b),

$$e_T = \sqrt{10^2 + 10^2} = \sqrt{200} = 14.1 \text{ v}$$

The phase angle is determined in the same way it was in formula (10–2).

$$\tan \theta = \frac{Y \text{ axis}}{X \text{ axis}} = \frac{e_C}{e_R} \qquad (10\text{--}4)$$

For Figure 10–2(a) and (b),

$$\tan \theta = \frac{e_C}{e_R} = \frac{-10 \text{ v}}{10} = -1$$

$$\theta = -45°$$

In Figure 10–2(c) and (d), the capacitive reactance is doubled. The analysis is similar to that used in Figure 10–1(d), except for the variations in the basic formulas (10–1) and (10–2).

$$e_T = \sqrt{e_R^2 + e_C^2} = \sqrt{10^2 + 20^2} = \sqrt{500} = 22.4 \text{ v} \qquad (10\text{--}3)$$

$$\tan \theta = \frac{e_R}{e_C} = \frac{-10 \text{ v}}{20 \text{ v}} = -0.5 \qquad (10\text{--}4)$$

$$\theta = -63.5°$$

The following are examples of the application of formulas (10–1) through (10–4).

Example 10–1 For R and C in series, what are the impedance and phase angle if $R = 200$ ohms, and $X_C = 140$ ohms?

Solution
$$Z = \sqrt{R^2 + X_C^2} = \sqrt{200^2 + 140^2} = 244 \text{ }\Omega$$

$$\tan \theta = \frac{-X_C}{R} = \frac{-140 \text{ }\Omega}{200 \text{ }\Omega} = -0.7$$

$$\theta = -35°$$

Example 10–2 For R and C in series, what are the capacitive reactance and impedance for a phase angle of $-78.7°$ when $R = 100$ ohms?

[10.1] Impedance

Solution By rearranging formula (10–2), the following is obtained.
$$X_C = R \tan \theta = 100 \, \Omega \times 5 = 500 \, \Omega$$
$$Z = \sqrt{R^2 + X_C^2} = \sqrt{100^2 + 500^2} = \sqrt{26 \times 10^4}$$
$$= 510 \, \Omega$$

Example 10–3 What is the total ac voltage across a series resistor and capacitor, and what is the phase angle, if $R = 1.5$ k ohms, $X_C = 0.3$ k ohms, and $i = 20$ ma?

Solution
$$e_R = iR = 20 \times 10^{-3} \times 1.5 \times 10^3 = 30 \text{ v}$$
$$e_C = iX_C = 20 \times 10^{-3} \times 0.3 \times 10^3 = 6 \text{ v}$$
$$e_T = \sqrt{e_R^2 + e_C^2} = \sqrt{30^2 + 6^2} = \sqrt{936} = 30.6 \text{ v}$$
$$\tan \theta = \frac{-e_C}{e_R} = \frac{-6 \text{ v}}{30 \text{ v}} = -0.2$$
$$\theta = -11.3°$$

Example 10–4 What are the voltage across the resistor and the total voltage for a series resistor and capacitor when the phase angle is $-36.8°$, and e_C is 3 v?

Solution Using formula (10–4) gives
$$e_R = \frac{e_C}{\tan \theta} = \frac{3 \text{ v}}{\tan 36.8°} = \frac{3 \text{ v}}{0.75} = 4 \text{ v}$$
$$e_T = \sqrt{e_R^2 + e_C^2} = \sqrt{4^2 + 3^2} = \sqrt{25} = 5 \text{ v}$$

When more than one resistor and/or capacitor are in a series circuit, they should be reduced to a single resistance and reactance for ease of analysis. Frequency and current remain the same throughout a series circuit, and the phase relationship is still the vector addition of the total magnitude of R and X_C.

An example of two resistors and three capacitors in series is shown in Figure 10–3(a). The vector diagram is shown in (b). The vector com-

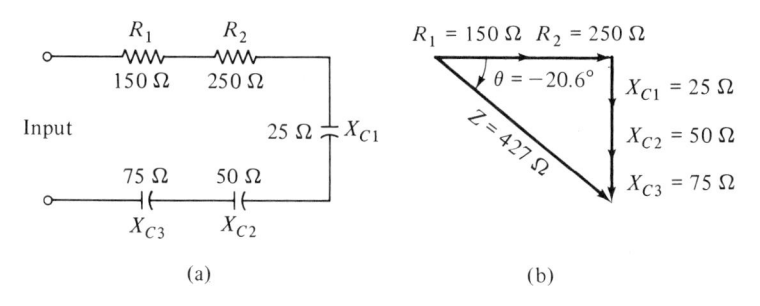

FIGURE 10–3 Series Impedance for More Than One R and C

ponents add directly since resistance and capacitive reactance each add directly when in series. The totals for resistance and reactance are

$$R_T = R_1 + R_2 = 15\ \Omega + 250\ \Omega = 400\ \Omega$$
$$X_{CT} = X_{C1} + X_{C2} + X_{C3} = 25\ \Omega + 50\ \Omega + 75\ \Omega = 150\ \Omega$$

The impedance and phase angle are calculated in the conventional way now that the components are reduced to a single R and a single X_C.

$$Z = \sqrt{R^2 + X_C^2} = \sqrt{400^2 + 150^2} = \sqrt{182{,}500} = 427\ \Omega$$
$$\tan\theta = \frac{-X_C}{R} = \frac{-150\ \Omega}{400\ \Omega} = -0.375$$
$$\theta = -20.6°$$

In a series circuit, involving R and X_C, the voltage is directly proportional to the resistance or reactance, and the total voltage across the components is directly proportional to the impedance. A comparison of Figures 10–1 and 10–2 illustrates the relationship. For example, X_C doubled from Figure 10–2(a) to 10–2(c), and the voltage also doubled.

10.2 Parallel Resistance and Capacitive Reactance

The phase difference is 90° for a parallel combination as it is for series resistance and capacitance. In the series circuit, the current was the same for each component. In that case, e_R and e_C were added vectorially, and so were R and X_C. In parallel combinations, the voltage is the same across each branch. The currents are therefore added vectorially since there can be different amounts in each branch.

An example is shown in Figure 10–4(a) The resistance and reactance are each 100 ohms. Since the voltage, $e = 10$ v, is the same for each, the current for each branch is

$$i = \frac{e}{X} = \frac{10\ \text{v}}{100\ \Omega} = 100\ \text{ma}$$

The vector is diagrammed in (b). The horizontal reference axis is the resistive current since the current is in-phase with the applied voltage. The vector diagram shows the capacitive current moving in a positive direction because it leads the voltage. The formula for total parallel current is

$$i_T = \sqrt{i_R^2 + i_C^2} \tag{10–5}$$

Applying the formula to Figure 10–4(a) and (b) results in

$$i_T = \sqrt{i_R^2 + i_C^2} = \sqrt{(100 \times 10^{-3})^2 + (100 \times 10^{-3})^2}$$
$$= 141 \times 10^{-3} = 141\ \text{ma}$$

The formula for the phase angle is

$$\tan\theta = \frac{i_C}{i_R} \tag{10–6}$$

[10.2] Impedance

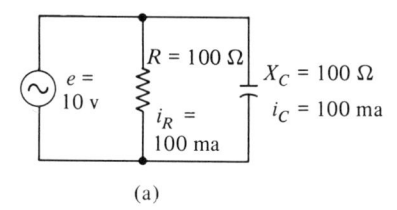

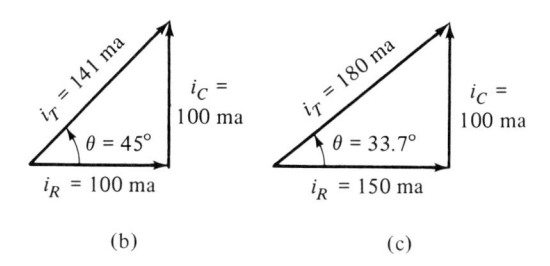

FIGURE 10-4 *Parallel* RC *Current*

And the phase angle for the example is

$$\tan \theta = \frac{i_C}{i_R} = \frac{100 \text{ ma}}{100 \text{ ma}} = 1$$

$\theta = 45°$ (positive for the positive direction of the angle)

Impedance is

$$Z = \frac{e}{i_T} \qquad (10\text{-}7)$$

An example using formula (10-7) is

$$Z = \frac{e}{i_T} = \frac{10 \text{ v}}{141 \text{ ma}} = 70.7 \; \Omega$$

Figure 10-4(c) is a vector example for $R = 100$ ohms, and $X_C = 66.7$ ohms.

$$i_R = \frac{e}{R} = \frac{10 \text{ v}}{100 \; \Omega} = 100 \text{ ma}$$

$$i_C = \frac{e}{X_C} = \frac{10 \text{ v}}{66.7 \; \Omega} = 150 \text{ ma}$$

$$i_T = \sqrt{i_R^2 + i_C^2} = \sqrt{(150 \times 10^{-3})^2 + (100 \times 10^{-3})^2}$$
$$= 180 \times 10^{-3} = 180 \text{ ma}$$

$$\tan \theta = \frac{i_C}{i_R} = \frac{100 \text{ ma}}{150 \text{ ma}} = 0.667$$
$$\theta = 33.7°$$
$$Z = \frac{e}{i_T} = \frac{10 \text{ v}}{180 \text{ ma}} = 55.6 \text{ }\Omega$$

Example 10-5 Solve the circuit in Figure 10-5 for i_C, i_R, R, and Z.

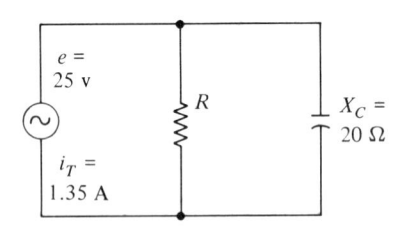

FIGURE 10-5 *Parallel RC Circuit*

Circuit values: $e = 25$ v, $i_T = 1.35$ A, $X_C = 20$ Ω

Solution i_C is calculated directly.

$$i_C = \frac{e}{X_C} = \frac{25 \text{ v}}{20 \text{ }\Omega} = 1.25 \text{ A}$$

Rearranging formula (10-5) results in

$$i_R = \sqrt{i_T^2 - i_C^2} = \sqrt{1.35^2 - 1.25^2} = \sqrt{0.25} = 0.5 \text{ A}$$
$$R = \frac{e}{i_R} = \frac{25 \text{ v}}{0.5 \text{ A}} = 50 \text{ }\Omega$$
$$Z = \frac{e}{i_T} = \frac{25 \text{ v}}{1.35 \text{ A}} = 18.5 \text{ }\Omega$$

When the resistance and capacitive reactance of a parallel circuit are known, a formula for calculating impedance directly is

$$Z = \frac{RX_C}{\sqrt{R^2 + X_C^2}} \tag{10-8}$$

Examples for Figure 10-4(b) and (c) are:

(b) $$Z = \frac{RX_C}{\sqrt{R^2 + X_C^2}} = \frac{100 \times 100}{\sqrt{100^2 + 100^2}} = \frac{10{,}000}{\sqrt{20{,}000}}$$
$$= \frac{10{,}000}{141} = 70.7 \text{ }\Omega$$

(c) $$Z = \frac{RX_C}{\sqrt{R^2 + X_C^2}} = \frac{66.7 \times 100}{\sqrt{66.7^2 + 100^2}} = \frac{6670}{\sqrt{14{,}430}}$$
$$= \frac{6670}{120} = 55.6 \text{ }\Omega$$

In circuits with parallel resistors and capacitors multiple components

[10.3] Impedance

are easier to analyze if reduced to a single resistor and a single capacitor. Similar rules apply to parallel and series circuits, except that the vector lengths in parallel circuits add for currents instead of for reactance, resistance, or voltage. Figure 10-6 is an example of parallel combinations.

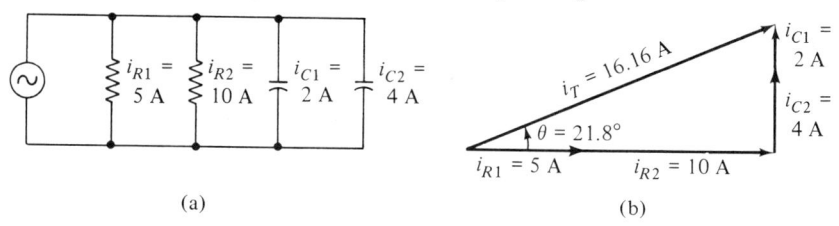

FIGURE 10-6 *Parallel Currents for Two Resistors and Capacitors*

Since the voltage and current are in-phase for the resistors, their currents are added directly on the X axis of the vector diagram. The capacitances have a 90° phase shift, but the phase shift is the same for each of the capacitors. Therefore, the capacitive currents may also be added directly, but on the Y axis. The analysis would be the same even if the equivalent resistance (single resistor) and equivalent reactance (single capacitor) were calculated. The total current is calculated:

$$i_T = \sqrt{i_R^2 + i_C^2} = \sqrt{15^2 + 6^2} = \sqrt{261} = 16.16 \text{ A}$$

and the phase angle is

$$\tan \theta = \frac{i_C}{i_R} = \frac{6 \text{ A}}{15 \text{ A}} = 0.4$$
$$\theta = 21.8°$$

10.3 RC Time Constant

In the two preceding sections, combinations of resistance and capacitance were analyzed for sine wave voltages and currents. Formulas (10-1) through (10-8) are valid only for sinusoidal waveforms. Pulsating dc has an entirely different effect in resistance/capacitance (*RC*) circuits. The time to charge or discharge a capacitor is instantaneous without series resistance (see Chapter 8). The charge current studied in Section 8.1 resulted from a difference in potential across the plates of a capacitor. Electron flow was from one plate to the other (through the circuit) until one was charged negatively, and the other was charged positively. If resistance is added in series with the capacitance, the charge or discharge is time dependent. Formulas (8-4) and (8-5) can be combined to show the relationship to time.

$$C = \frac{Q}{E} \quad (8\text{-}4)$$

$$Q = It \quad (8\text{-}5)$$

Substituting C for Q in formula (8–5) and IR for E in formula (8–4) results in

$$C = \frac{It}{IR}$$

By dividing out I and rearranging the equation, the following is obtained:

$$t = RC \qquad (10\text{–}9)$$

Time (t) is in seconds, resistance (R) is in ohms, and capacitance (C) is in farads. The formula calculates the time it takes for a capacitor to charge or discharge for a period of one time constant. One time constant is the period it takes for a capacitor to either charge to 63 percent of its maximum value, or to discharge 63 percent down from its maximum value (down to 37 percent of maximum).

Figure 10–7(a) represents a charge plot for a period of five time constants. In one time constant, the curve reaches the 63 percent point. At

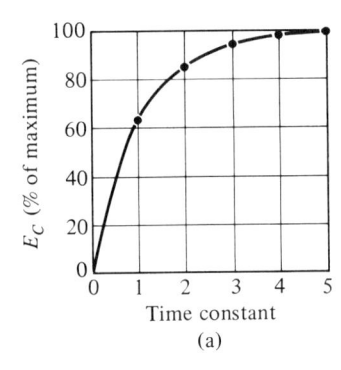

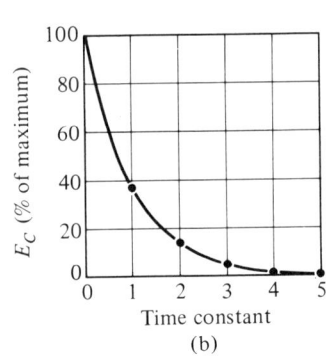

FIGURE 10–7 RC *Time Constant Curves*

two time constants, the curve reaches 63 percent of the remaining amplitude.

$$E_C = (63\% \times 37) + 63\% \approx 86\%$$

The curve continues at a rate of 63 percent of the remaining amplitude. Five time constants is better than 99 percent and is usually considered fully charged.

Figure 10–7(b) represents a discharge plot for five time constants. The first time constant causes a discharge down to 37 percent of the maximum.

$$E_C = 100\% - 63\% = 37\%$$

Each succeeding time constant discharges to 37 percent of the remaining amplitude. Once again, five time constants is usually considered fully discharged since the amplitude is below 1 percent of maximum.

[10.3]   Impedance   199

Notice that the rate of charge is fastest during the first time constant. As the capacitor takes on charge, the potential difference between it and the source decreases, which decreases the charging current. As a result, the charge rate becomes less and less. The discharge curve follows the same pattern. Initially, the discharge current is high, and the capacitor discharges at a high rate. As the capacitor approaches equilibrium, the rate slows accordingly.

In Figure 10–8(a), a charging circuit is shown for a 10-v source. With SW1 closed (SW2 open), the battery charges the capacitor through the 10-k

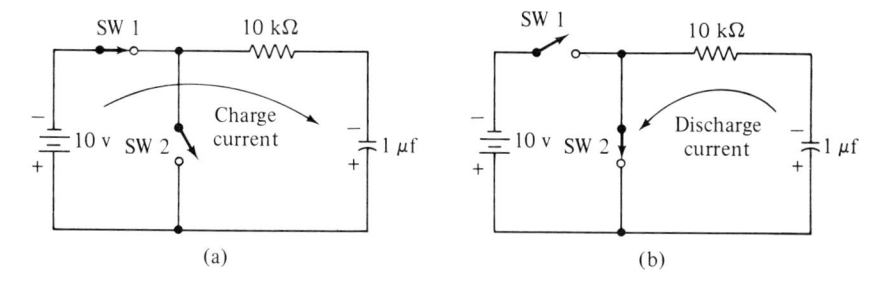

FIGURE 10–8 RC *Charge and Discharge Circuits*

ohm resistor. The time it takes to charge to one time constant (63 percent of 10 v, or 6.3 v) is

$$t = RC = 10 \times 10^3 \times 1 \times 10^{-6} = 10 \times 10^{-3} = 10 \text{ msec}$$

Referring back to Figure 10–7(a), each time constant would be equal to 10 msec. In 10 msec E_C would be 6.3 v; in 20 msec, E_C would be 8.6 v; etc. The charge section of Table 10–1 illustrates the progression from 1 to 5 time constants.

TABLE 10–1

TC	TIME msec	CHARGE %	E_C	DISCHARGE %	E_C
1	10	63	6.3 v	37	3.7 v
2	20	86	8.6 v	14	1.4 v
3	30	95	9.5 v	5	0.5 v
4	40	98	9.8 v	2	0.2 v
5	50	99	9.9 v	1	0.1 v

In Figure 10–8(b), a discharge circuit is shown. SW1 is open, and SW2 is closed, which provides a discharge path so the capacitor can neutralize its charge. The time constant of 10 msec is the same as for circuit (a) since the same values of resistance and capacitance are used. Each time constant is equal to 10 msec in the discharge graph of Figure 10–7(b). The discharge section of Table 10–1 shows the progression through each time constant.

Several examples follow for calculating time constants.

Example 10–6 How much is a time constant for a 2-M ohm resistor and a 0.25-μf capacitor?

Solution $t = RC = 2 \times 10^6 \times 0.25 \times 10^{-6} = 0.5$ sec

Example 10–7 How much resistance is needed in series with 2 μf to achieve a time constant of 66 msec?

Solution $$R = \frac{t}{C} = \frac{66 \times 10^{-3}}{2 \times 10^{-6}} = 33 \times 10^3 = 33 \text{ k}\Omega$$

Example 10–8 If the dc source voltage is 20 v, what is the potential across a 0.01-μf capacitor in 4.5 msec if $R = 150$ kΩ?

Solution $t = RC = 150 \times 10^3 \times 0.01 \times 10^{-6} = 1.5 \times 10^{-3}$
$= 1.5$ msec

One time constant is 1.5 msec, so 4.5 msec is 3 time constants. Table 10–1 shows 3 time constants to be 95 percent. Therefore, 95 percent of the source is the potential across the capacitor.

$$E_C = 0.95 \times 20 \text{ v} = 19 \text{ v}$$

Example 10–9 The voltage across a 400-pf capacitor discharges to 7 v in 400 μsec. If the series resistance is 0.5 M ohms, what is the source voltage?

Solution $t = RC = 0.5 \times 10^6 \times 400 \times 10^{-12} = 200 \times 10^{-6}$
$= 200$ μsec

$$TC = \frac{\text{elapsed time}}{t} = \frac{400 \text{ }\mu\text{sec}}{200 \text{ }\mu\text{sec}} = 2$$

Two time constants indicate that the E_C will be 14 percent of the source potential (E_S) (see Table 10–1).

$$E_S = \frac{E_C}{\%} = \frac{7 \text{ v}}{0.14} \times 50 \text{ v}$$

10.4 Series Resistance and Inductive Reactance

When combined with resistance, inductive reactance acts very much as capacitive reactance does, except for the phase difference. Voltage leads the current by 90° in an inductive circuit, and like a capacitive circuit,

[10.4] Impedance

phase is less than 90° when combined with resistance. Drawings and examples in this section are presented as parallels to those of Section 10.1 so that a direct comparison can be made.

In Figure 10–9(a), a resistor and inductor are connected in series. The vector diagram, (b), shows R on the X axis due to zero phase shift. X_L is

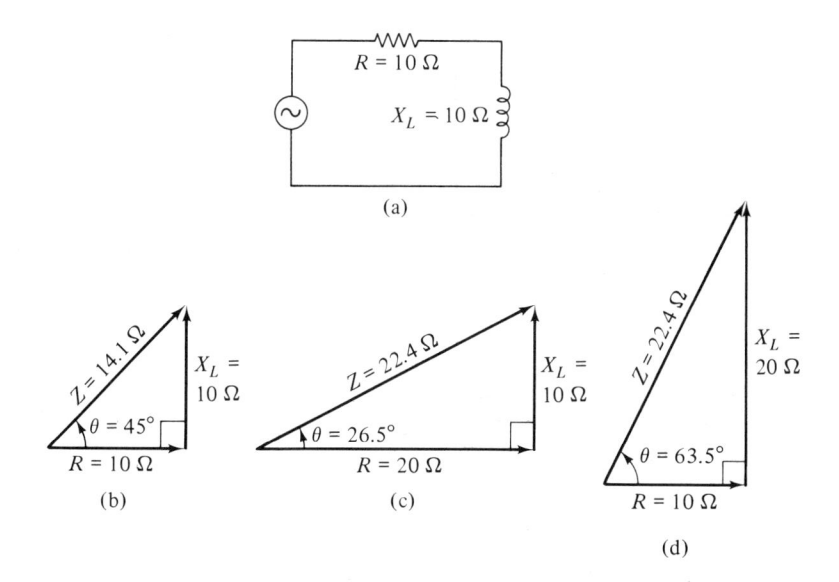

FIGURE 10–9 *Series LR Impedance*

on the Y axis in a positive direction because the voltage leads the current. In Sections 10–1 and 10–2, capacitive reactance was negative because capacitive voltage lags the current. The formula for impedance is

$$Z = \sqrt{R^2 + X_L^2} \qquad (10\text{–}10)$$

Calculating for (b) of the figure gives:

$$Z = \sqrt{R^2 + X_L^2} = \sqrt{10^2 + 10^2} = \sqrt{200} = 14.1 \ \Omega$$

The formula for the phase angle is similar to formula (10–2).

$$\tan \theta = \frac{Y \text{ axis}}{X \text{ axis}} = \frac{X_L}{R} \qquad (10\text{–}11)$$

However, notice that the Y axis is positive (instead of negative) to represent a positive angle. The phase angle in 10–9(b) is

$$\tan \theta = \frac{X_L}{R} = \frac{10 \ \Omega}{10 \ \Omega} = 1$$
$$\theta = 45°$$

Drawings (c) and (d) are parallels of those in Figure 10–1.

(c) $Z = \sqrt{R^2 + X_L^2} = \sqrt{20^2 + 10^2} = \sqrt{500} = 22.4 \, \Omega$

$\tan \theta = \dfrac{X_L}{R} = \dfrac{10 \, \Omega}{20 \, \Omega} = 0.5$

$\theta = 26.5°$

(d) $Z = \sqrt{R^2 + X_L^2} = \sqrt{10^2 + 20^2} = \sqrt{500} = 22.4 \, \Omega$

$\tan \theta = \dfrac{X_L}{R} = \dfrac{20 \, \Omega}{10 \, \Omega} = 2$

$\theta = 63.5°$

A voltage analysis for a different example is shown in Figure 10–10. The current is 0.1 A, so the individual voltages are 30 v for the resistor and 4.2 v for the inductor according to Ohm's law. The vector diagram shows

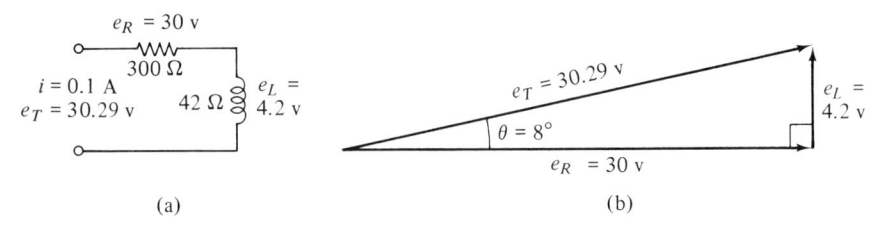

FIGURE 10–10 Series LR Voltage

e_R (no phase shift) on the X axis, and the e_L positive on the Y axis (voltage leads current). The formulas for total voltage and the phase angle are

$$e_T = \sqrt{e_R^2 + e_L^2} \qquad (10\text{--}12)$$

$$\tan \theta = \dfrac{e_L}{e_R} \qquad (10\text{--}13)$$

Formulas (10–12) and (10–13) are applied to Figure 10–10 in the following manner.

$e_T = \sqrt{e_R^2 + e_L^2} = \sqrt{30^2 + 4.2^2} = \sqrt{917.6} = 30.29 \text{ v}$

$\tan \theta = \dfrac{e_L}{e_R} = \dfrac{4.2}{30} = 0.14$

$\theta = 8°$

Figure 10–11 is an example of several resistors and inductors in series. Series resistance and inductive reactance each add directly. The addition is shown vectorially in (b). The process is similar to that used in Figure 10–3, but here the angle is positive instead of negative. Total resistance and reactance are

$R_T = R_1 + R_2 = 75 \, \Omega + 125 \, \Omega = 200 \, \Omega$
$X_{LT} = X_{L1} + X_{L2} + X_{L3} = 75 \, \Omega + 275 \, \Omega + 150 \, \Omega = 500 \, \Omega$

[10.4] Impedance 203

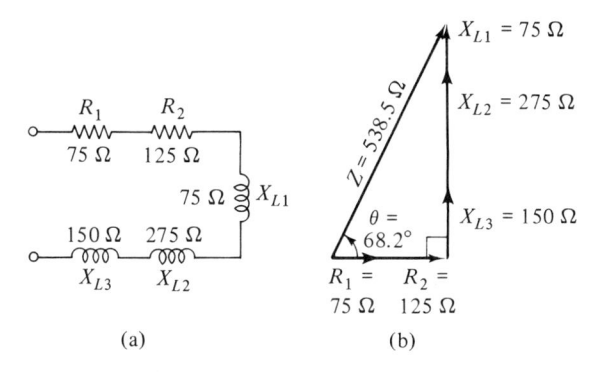

FIGURE 10-11 *Series Impedance for More Than One L and R*

The impedance and phase angle are calculated from formulas (10–10) and (10–11), using total resistance and reactance.

$$Z = \sqrt{R^2 + X_L^2} = \sqrt{200^2 + 500^2} = \sqrt{29 \times 10^4} = 538.5 \, \Omega$$

$$\tan \theta = \frac{X_L}{R} = \frac{500 \, \Omega}{200 \, \Omega} = 2.5$$

$$\theta = 68.2°$$

Voltages are handled in the same way as impedance is in Figure 10–11. The voltages for resistors add directly since the phase is the same through each resistor. The voltages for inductors also add directly since the phase is the same through each inductor. Also, a 90° phase shift exists between the resistor voltage and the inductor voltage.

Examples using formulas (10–10) to (10–13) follow.

Example 10-10 What are the impedance and phase angle for $R = 2 \, \text{k}$ ohms, and $X_L = 1.2 \, \text{k}$ ohms when in series?

Solution
$$Z = \sqrt{R^2 + X_L^2} = \sqrt{(2 \times 10^3)^2 + (1.5 \times 10^3)^2}$$
$$= \sqrt{6.25 \times 10^6} = 2.5 \, \text{k}\Omega$$
$$\tan \theta = \frac{X_L}{R} = \frac{1.5 \times 10^3}{2 \times 10^3} = 0.75$$
$$\theta = 36.8°$$

Example 10-11 What are the values of a resistance and inductance in series if the phase angle is 45°, and the impedance is 56.57 ohms?

Solution The tangent of the angle θ is 1 when $\theta = 45°$. If $\theta = 45°$, the sides of the vector diagram are the same, which means $R = X_L$.

$$\tan \theta = \frac{X_L}{R}$$

$$\tan 45° = 1 = \frac{X_L}{R}$$

Since

$$Z = \sqrt{R^2 + X_L^2}$$

then

$$R^2 + X_L^2 = Z^2$$

Since $R^2 = X_L^2$ (phase angle of 45°)

$$2R^2 = Z^2$$

or

$$R^2 = \frac{Z^2}{2}$$

or

$$R = \sqrt{\frac{Z^2}{2}}$$

For an impedance of 56.57 ohms,

$$R = \sqrt{\frac{Z^2}{2}} = \sqrt{\frac{56.57^2}{2}} = \sqrt{\frac{3200}{2}} = \sqrt{1600} = 40\ \Omega$$

$$R = X_L = 40\ \Omega$$

Example 10-12 What are the total voltage and the phase angle when R and L are in series, $e_R = 5$ v, and $e_L = 20$ v?

Solution

$$e_T = \sqrt{e_R^2 + e_L^2} = \sqrt{5^2 + 20^2} = \sqrt{425} = 20.6\ \text{v}$$

$$\tan \theta = \frac{e_L}{e_R} = \frac{20\ \text{v}}{5\ \text{v}} = 4$$

$$\theta = 76°$$

Example 10-13 Two resistors and two inductors are connected in series. The voltages across three of the components are: $e_{R1} = 0.5$ v, $e_{R2} = 1.5$ v, and $e_{L1} = 0.8$ v. For a phase angle of 31°, what is the voltage across L_2, and what is the total voltage?

Solution The tangent for an angle of 31° is 0.6, so tangent 31° = $0.6 = e_{LT}/e_{RT}$.

$$e_{LT} = 0.6\ e_{RT}$$

$$e_{RT} = e_{R1} + e_{R2} = 0.5\ \text{v} + 1.5\ \text{v} = 2\ \text{v}$$

$$e_{LT} = 0.6 \times 2 \text{ v} = 1.2 \text{ v}$$
$$e_{L2} = e_{LT} - e_{Li} = 1.2 \text{ v} - 0.8 \text{ v} = 0.4 \text{ v}$$
$$e_T = \sqrt{e_{RT}^2 + e_{LT}^2} = \sqrt{2^2 + 1.2^2} = \sqrt{5.44} = 2.33 \text{ v}$$

10.5 Parallel Resistance and Inductive Reactance

In LR series circuits, current and resistance are the common factors. The current has the same value and phase through each resistor and inductor. The current and voltage have the same phase for the resistor. In parallel circuits, voltage is the common factor and has the same value and phase across each branch. As in parallel resistor-capacitor circuits, the current is 90° out of phase with the applied voltage in the inductive reactance branch. The total current is somewhere between 0° and 90°, depending upon the vector addition.

A parallel circuit is shown in Figure 10–12(a). The voltage, 10 v, is the same across each branch. According to Ohm's law, the branch currents

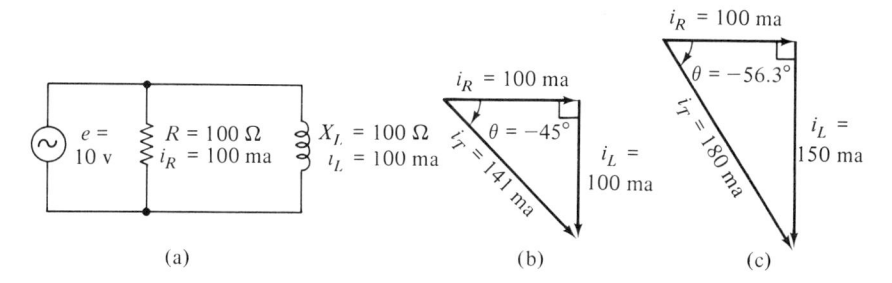

FIGURE 10–12 *Parallel LR Current*

are each 100 ma, since R and X_L each are 100 ohms. The vector diagram is in (b). The current through the resistor is shown as the reference X axis since the phase is the same between the current and voltage. The inductive current is negative since the current lags the voltage. The formula for total current is

$$i_T = \sqrt{i_R^2 + i_L^2} \qquad (10\text{–}14)$$

and the total current for Figure 10–12(a) and (b) is

$$i_T = \sqrt{i_R^2 + i_L^2} = \sqrt{(100 \times 10^{-3})^2 + (100 \times 10^{-3})^2}$$
$$= \sqrt{20 \times 10^{-3}} = 141 \text{ ma}$$

The formula for the negative phase angle is

$$\tan \theta = \frac{-i_L}{i_R} \qquad (10\text{–}15)$$

Appling the formula to Figure 10–12(a) and (b) yields

$$\tan \theta = \frac{-i_L}{i_R} = \frac{-100 \text{ ma}}{100 \text{ ma}} = -1$$

$$\theta = -45°$$

The impedance is calculated from formula (10–7).

$$Z = \frac{e}{i_T} = \frac{10 \text{ v}}{141 \text{ ma}} = 70.7 \text{ }\Omega$$

The vector diagram in Figure 10–12(c) illustrates a parallel circuit for $e = 10$ v, $i_R = 100$ ma, and $i_L = 150$ ma. Using formulas (10–14) and (10–15) gives

$$i_T = \sqrt{i_R^2 + i_L^2} = \sqrt{(100 \times 10^{-3})^2 + (150 \times 10^{-3})^2}$$
$$= \sqrt{3.25 \times 10^{-2}} = 180 \text{ ma}$$

$$\tan \theta = \frac{-i_L}{i_R} = \frac{-150 \text{ ma}}{100 \text{ ma}} = -1.5$$

$$\theta = -56.3°$$

The impedance is

$$Z = \frac{e}{i_T} = \frac{10 \text{ v}}{180 \text{ ma}} = 55.6 \text{ }\Omega$$

A formula for calculating the impedance for parallel resistance and inductance directly is

$$Z = \frac{RX_L}{\sqrt{R^2 + X_L^2}} \qquad (10\text{–}16)$$

Verification of the formula for Figure 10–12(a) and (b) is

$$Z = \frac{RX_L}{\sqrt{R^2 + X_L^2}} = \frac{100 \times 100}{\sqrt{100^2 + 100^2}} = \frac{1 \times 10^4}{\sqrt{2 \times 10^4}}$$
$$= \frac{1 \times 10^4}{1.41 \times 10^2} = 70.7 \text{ }\Omega$$

For Figure 10–12(c),

$$X_L = \frac{e}{i_L} = \frac{10 \text{ v}}{150 \text{ ma}} = 66.7 \text{ }\Omega$$

$$Z = \frac{RX_L}{\sqrt{R^2 + X_L^2}} = \frac{100 \times 66.7}{\sqrt{100^2 + 66.7^2}} = \frac{0.667 \times 10^4}{\sqrt{1.443 \times 10^4}}$$
$$= \frac{0.667 \times 10^4}{1.2 \times 10^2} = 55.6 \text{ }\Omega$$

Figure 10–13 illustrates multiple components in parallel. As with other circuits involving multiple components, analysis is simplified when the circuit is reduced to an equivalent of a single resistor and inductor. In

[10.5] Impedance

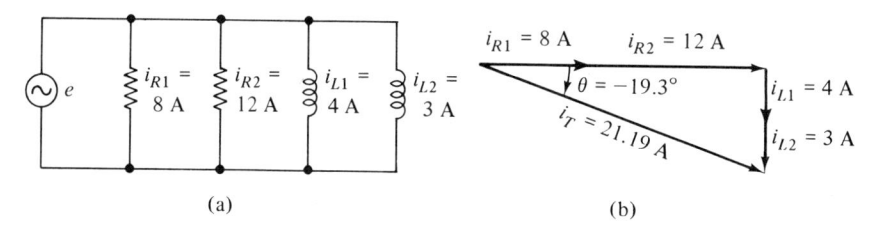

FIGURE 10-13 *Parallel Current for Two Inductors and Resistors*

the figure, the branch currents are added directly for the resistors and then directly for the inductors. The total current is the vector sum. In drawing (b), the resistor currents are

$$i_R = i_{R1} + i_{R2} = 8 \text{ A} + 12 \text{ A} = 20 \text{ A}$$

The inductor currents are

$$i_L = i_{L1} + i_{L2} = 4 \text{ A} + 3 \text{ A} = 7 \text{ A}$$

The vector sum for total current is

$$i_T = \sqrt{i_R^2 + i_L^2} = \sqrt{20^2 + 7^2} = \sqrt{449} = 21.19 \text{ A}$$

The phase angle is

$$\tan \theta = \frac{-i_L}{i_R} = \frac{-7 \text{ A}}{20 \text{ A}} = -0.35$$
$$\theta = -19.3°$$

The following are examples for parallel resistors and inductors.

Example 10-14 For a parallel circuit, if $e = 5$ v, $i_R = 3$ A, and $i_L = 9$ A, what are i_T, θ, and Z?

Solution
$$i_T = \sqrt{i_R^2 + i_L^2} = \sqrt{3^2 + 9^2} = \sqrt{90} = 9.49 \text{ A}$$
$$\tan \theta = \frac{-i_L}{i_R} = \frac{-9 \text{ A}}{3 \text{ A}} = -3$$
$$\theta = -71.6°$$
$$Z = \frac{e}{i_T} = \frac{5 \text{ v}}{9.49 \text{ A}} = 0.53 \text{ Ω}$$

Example 10-15 In Figure 10-13, if $e = 96$ v, what are the total impedance, the resistance of each resistor, and the reactance of each inductor?

Solution Since i_T is already calculated as 21.19 A,

$$Z = \frac{e}{i_T} = \frac{96 \text{ v}}{21.19 \text{ A}} = 4.53 \text{ Ω}$$

The resistance and reactance of the individual branches are figured in the following manner:

$$R_1 = \frac{e}{i_{R1}} = \frac{96 \text{ v}}{8 \text{ A}} = 12 \text{ }\Omega$$

$$R_2 = \frac{e}{i_{R2}} = \frac{96 \text{ v}}{12 \text{ A}} = 8 \text{ }\Omega$$

$$X_{L1} = \frac{e}{i_{L1}} = \frac{96 \text{ v}}{4 \text{ A}} = 24 \text{ }\Omega$$

$$X_{L2} = \frac{e}{i_{L2}} = \frac{96 \text{ v}}{3 \text{ A}} = 32 \text{ }\Omega$$

Example 10-16 The impedance of a parallel resistor and inductor is 5.6 k ohms. If the applied voltage is 40 v, and the current through the resistor is 6.4 ma, what is the current through the inductor?

Solution The total current is

$$i_T = \frac{e}{Z} = \frac{40 \text{ v}}{5.6 \text{ k}\Omega} = 7.16 \text{ ma}$$

Rearranging formula (10-14) results in

$$i_L = \sqrt{i_T^2 - i_R^2} = \sqrt{(7.16 \text{ ma})^2 - (6.4 \text{ ma})^2}$$
$$= \sqrt{51.2 \times 10^{-6} - 40.96 \times 10^{-6}}$$
$$= \sqrt{10.24 \times 10^{-6}} = 3.2 \times 10^{-3} = 3.2 \text{ ma}$$

10.6 *LR* Time Constant

Resistance and inductive reactance have been analyzed in the last two sections for sinusoidal waveforms. Chapter 9 deals with Lenz's law which states that a changing current is opposed by the inductance of an inductor. When resistance is placed in series with an inductor, the current changes at an exponential rate. The relationship to time for the change in current is explained by rearranging formula (9-1)

$$e = L\frac{\Delta i}{\Delta t}$$

which is equivalent to

$$L = \frac{E\Delta t}{\Delta I}$$

Dividing both sides by *R* results in

$$\frac{L}{R} = \frac{Et}{IR}$$

$$\frac{L}{R} = \frac{Et}{E} = t$$

so that one time constant is

$$t = \frac{L}{R} \tag{10-17}$$

t is time in seconds, L is inductance in henrys, and R is resistance in ohms. Figure 10–14 illustrates the percentage of maximum (steady-state) current

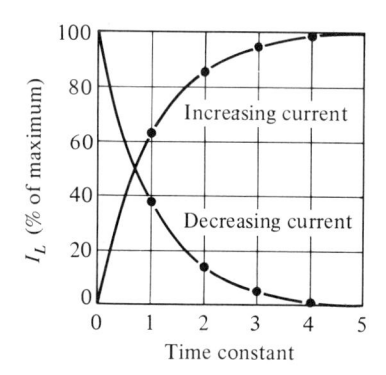

FIGURE 10–14 LR *Time Constant Curves*

versus time constants. Notice the similarity of these curves to *RC* time constant curves. The graph assumes full current at 5 time constants, which is 99 percent of maximum, or minimum current at 5 time constants, which is 1 percent of maximum. The curves increase and decrease at the same rates as the *RC* curves of Figure 10–7. Formula 10–17 is the time for 1 time constant in which the current has increased to 63 percent of its maximum value, or decreased to 37 percent of its maximum value.

Figure 10–15 shows a switch-operated circuit for a series resistor and inductor with a dc source of 10 v. The time constant is

$$t = \frac{L}{R} = \frac{2\,\text{h}}{1\,\text{k}\Omega} = 2\,\text{msec}$$

In Table 10–2, the time for each time constant is 2 msec. For 5 time con-

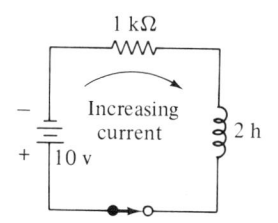

FIGURE 10–15 LR *Time Constant Circuit*

stants, the total time is 10 msec. The maximum (steady-rate) current for Figure 10-15 is

$$I = \frac{E}{R} = \frac{10 \text{ v}}{1 \text{ k}\Omega} = 10 \text{ ma}$$

The dc resistance of the inductor is assumed to be low enough to ignore. Table 10-2 shows the current for each time constant, relates the current

TABLE 10-2

TC	TIME msec	INCREASE %	I_L	DECREASE %	I_L
1	2	63	6.3 ma	37	3.7 ma
2	4	86	8.6 ma	14	1.4 ma
3	6	95	9.5 ma	5	0.5 ma
4	8	98	9.8 ma	2	0.2 ma
5	10	99	9.9 ma	1	0.1 ma

to time ($t = 2$ msec for each time constant) and percentage of the maximum current. Increasing current occurs when the switch is closed, and decreasing current occurs when the switch is open. The RC circuit used a shorting switch to discharge the capacitor. An inductor does not need a shorting switch because it does not store energy as a capacitor does.

Formula (10-17) shows the resistance as inversely proportional to time; that is, higher resistance values shorten the time constant. As the resistance becomes proportionally higher than the inductance, the circuit becomes more resistive and less inductive. Thus, there is less opposition to changes in current, and the current can increase to its full value in less time. Several examples include:

Example 10-17 In a series RL circuit, what value of resistance is required for a time constant of 30 nsec (30×10^{-9} sec) if $L = 360$ μh?

Solution
$$R = \frac{L}{t} = \frac{360 \times 10^{-6}}{30 \times 10^{-9}} = 12 \times 10^3 = 12 \text{ k ohms}$$

Example 10-18 In a series LR circuit, if $R = 100$ ohms, $L = 3$ mh, and the source is 1.5 v dc, what value is the current after 90 μsec (assume the dc resistance of L is zero)?

Solution
$$t = \frac{L}{R} = \frac{3 \text{ mh}}{0.1 \text{ k}\Omega} = 30 \text{ }\mu\text{sec}$$

The number of time constants is

$$TC = \frac{90 \; \mu sec}{30 \; \mu sec} = 3$$

which is 95 percent of maximum or total current. The maximum current is

$$I_T = \frac{E}{R} = \frac{1.5 \text{ v}}{0.1 \text{ k}\Omega} = 15 \text{ ma}$$

The current after 90 μsec is 95 percent of I_T.

$$I = 0.95 \times I_T = 0.95 \times 15 \text{ ma} = 14.25 \text{ ma}$$

Example 10-19 How long does it take for the current in an inductor to decrease to one percent of its total value if $L = 120$ mh, $E = 10$ v, and $I_T = 0.25$ A (assuming zero resistance in the inductor)?

Solution

$$R = \frac{E}{I_T} = \frac{10 \text{ v}}{0.25 \text{ A}} = 40 \; \Omega$$

$$t = \frac{L}{R} = \frac{120 \text{ mh}}{40 \; \Omega} = 3 \text{ msec}$$

1 percent of total current is 5 time constants (for reference see Table 10-2). Therefore, the time (t_T) necessary for the current to decrease to 1 percent is

$$t_T = TC \times t = 5 \times 3 \text{ msec} = 15 \text{ msec}$$

10.7 Series Inductive and Capacitive Reactance

When inductance and capacitance are placed in series, their net reactance is the difference between them. For X_L, voltage leads the current by 90°. For X_C, voltage lags the current by 90°. When current and voltage are combined, the voltage has a phase difference of 180°, which results in cancellation. This would happen for example when one is positive while the other is negative. The relationship is illustrated in Figure 10-16, where

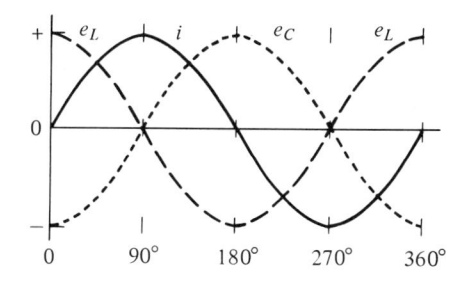

FIGURE 10-16 *Phase Relationships in Series LC Circuits*

the current (i) is the reference sine wave, and e_L and e_C are cosine waves that are displaced 180°. Notice that the cosine waves are symmetrical in their canceling effects. If e_L had the same amplitude as e_C, the net voltage for the two components would be zero. If the two voltages were of different amplitudes, the net voltage would be the difference since cancellation occurs only to the point of their difference.

Figure 10-17 is an example of series reactance and the resultant equivalent reactance. In (a), X_L and X_C are 60 and 90 ohms respectively. The

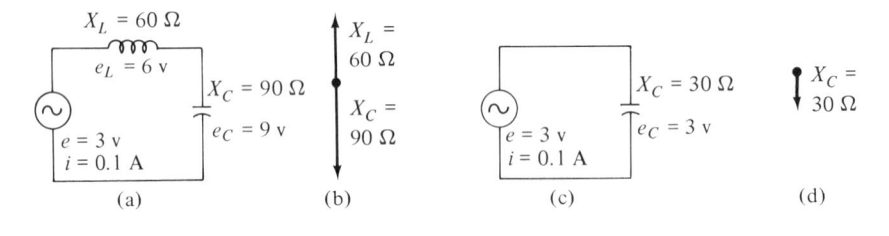

FIGURE 10-17 *Net Reactance in a Series LC Circuit*

current (i) is 0.1 A and has the same value through the entire series circuit. Using Ohm's law results in

$$e_L = iX_L = 0.1 \text{ A} \times 60 \text{ }\Omega = 6 \text{ v}$$
$$e_C = iX_C = 0.1 \text{ A} \times 90 \text{ }\Omega = 9 \text{ v}$$

The net reactance (Z) is found by the formula

$$Z = X_C - X_L \quad \text{or} \quad X_L - X_C \quad \quad (10\text{-}18)$$

In (b) the net reactance is $X_C - X_L = 90 \text{ }\Omega - 60 \text{ }\Omega = 30\Omega$, and because X_C is predominant, the difference is capacitive, X_C. The equivalent circuit is shown in (c) with the same current. According to Ohm's law,

$$e_C = iX_C = 0.1 \text{ A} \times 30 \text{ }\Omega = 3 \text{ v}$$

The net reactance, $X_C = 30$ Ω, is diagrammed in (d). Notice that X_L was shown positively to agree with X_L in Figure 10-9, and that X_C was shown negatively to agree with X_C in Figure 10-1.

In Figure 10-17 the voltage across each component is greater than the applied voltage. It is important to realize that the phase difference causes a net voltage that, in the example, happens to be less than either value. A further example is:

Example 10-20 What are the X_C, e_C, e_L, and the applied voltage (e_T) if X_L is 300 ohms, the net reactance is 200 ohms, and the series current is 1 A?

Solution $X_C = X_L - X_{net} = 300\ \Omega - 200\ \Omega = 100\ \Omega$

Notice that the net reactance of 200 ohms is inductive since X_L is greater than X_C.

$$e_C = iX_C = 1\ A \times 100\ \Omega = 100\ v$$
$$e_L = iX_L = 1\ A \times 300\ \Omega = 300\ v$$
$$e_T = iX_{net} = 1\ A \times 200\ \Omega = 200\ v$$

or

$$e_T = e_L - e_C = 300\ v - 100\ v = 200\ v$$

In this example, the net voltage (e_T) is more than e_C and less than e_L.

10.8 Parallel Inductive and Capacitive Reactance

When inductance and capacitance are placed in parallel, their net current is the difference between the branch currents. The voltage is the same across each parallel component. Since the current for the inductor lags the voltage by 90°, and the current for the capacitor leads the voltage by 90°, the combined phase difference is 180°. The phase difference of 180° in current for a parallel circuit results in a net current in the same way the phase difference results in a net voltage for the series *LC* circuit.

Figure 10–18 is similar to Figure 10–16, except that the voltage (*e*) is shown as the reference. Capacitive and inductive currents are cosine

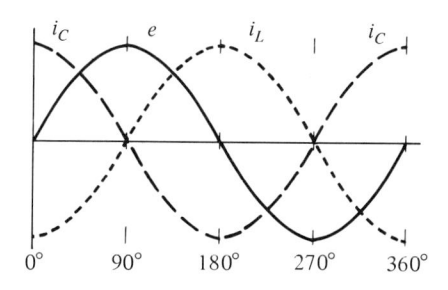

FIGURE 10-18 *Phase Relationships in Parallel* LC *Circuits*

waves, in which i_C leads the voltage, and i_L lags the voltage. The current waveforms are displaced 180°, demonstrating the effects of cancellation.

Figure 10–19 illustrates a parallel inductor and capacitor. The applied voltage (*e*) is 15 v. By Ohm's law the branch currents are

$$i_L = \frac{e}{X_L} = \frac{15\ v}{30\ \Omega} = 0.5\ A$$
$$i_C = \frac{e}{X_C} = \frac{15\ v}{50\ \Omega} = 0.3\ A$$

214　　　　　　　　　　　　　Impedance　　　　　　　　　　　　　[10.8]

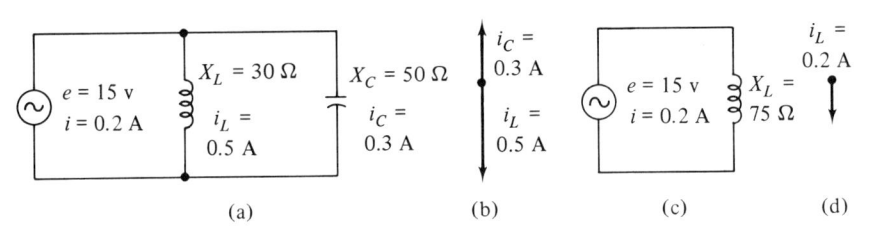

(a) (b) (c) (d)

FIGURE 10-19 *Net Reactance in a Parallel LC Circuit*

The vector diagram in (b) has i_C positive (0.3 A), and i_L negative (0.5 A). This agrees with the polarities of Figures 10-4 and 10-12. The net difference for a parallel circuit is calculated:

$$i_T = i_L - i_C \quad \text{or} \quad i_C - i_L \tag{10-19}$$

Applying this formula to Figure (10-19) gives

$$i_T = i_L - i_C = 0.5 \text{ A} - 0.3 \text{ A} = 0.2 \text{ A}$$

The equivalent circuit and vector are drawn in (c) and (d). Since $i_T = 0.2$ A, and $e = 15$ v, the net impedance (Z) is

$$Z = \frac{e}{i_T} = \frac{15 \text{ v}}{0.2 \text{ A}} = 75 \text{ }\Omega$$

which is inductive here because i_L is greater than i_C, and current lags the voltage by 90°. A formula for calculating the impedance is

$$Z = \frac{X_L X_C}{X_L - X_C} \quad \text{or} \quad \frac{X_C X_L}{X_C - X_L} \tag{10-20}$$

Verification of Figure 10-19 is

$$Z = \frac{X_C X_L}{X_C - X_L} = \frac{50 \times 30}{50 - 30} = \frac{1500}{20} = 75 \text{ }\Omega$$

The impedance for a parallel combination is inductive or capacitive when one reactance is greater than the other. An example for a predominating X_C follows.

Example 10-21 For parallel components, assume the applied voltage is 0.5 v, the net current is 2 ma, and the capacitive current is 2.5 ma. What are i_L, X_L, X_C, and the total Z?

Solution

$$i_L = i_C - i_T = 2.5 \text{ ma} - 2.0 \text{ ma} = 0.5 \text{ ma}$$

$$X_L = \frac{e}{i_L} = \frac{0.5 \text{ v}}{0.5 \text{ ma}} = 1 \text{ k}\Omega$$

$$X_C = \frac{e}{i_C} = \frac{0.5 \text{ v}}{2.5 \text{ v}} = 0.2 \text{ k}\Omega$$

$$Z = \frac{e}{i_T} = \frac{0.5 \text{ v}}{2 \text{ ma}} = 0.25 \text{ k}\Omega$$

or

$$Z = \frac{X_L X_C}{X_L - X_C} = \frac{1 \times 10^3 \times 0.2 \times 10^3}{1 \times 10^3 - 0.2 \times 10^3}$$
$$= \frac{0.2 \times 10^6}{0.8 \times 10^3} = 0.25 \text{ k}\Omega$$

Notice that the impedance is capacitive due to the higher current attributed to the capacitive branch of the circuit, causing the current to lead the voltage by 90°.

10.9 Series Resistance and Reactance

This section discusses the effects of series resistance, capacitive reactance, and inductive reactance. Since series capacitive and inductive reactances cancel to produce a net reactance, a serial circuit readily reduces to resistance and X_C or X_L. The impedance is then found by applying formula (10-1) or (10-10).

$$Z = \sqrt{R^2 + X_C^2} \qquad (10\text{-}1)$$
$$Z = \sqrt{R^2 + X_L^2} \qquad (10\text{-}10)$$

The phase angle is found from formula (10-2) or (10-11).

$$\tan \theta = \frac{-X_C}{R} \qquad (10\text{-}2)$$
$$\tan \theta = \frac{X_L}{R} \qquad (10\text{-}11)$$

Once the net reactance is determined, the methods for calculating impedance, phase angles, and voltages are the same as the method explained in Section 10.1 for X_C or e_C, and in Section 10.4 for X_L or e_L.

An example problem is illustrated in Figure 10-20. The circuit in (a) has R, X_C, and X_L in series. The vector representing the quantities is in (b). The resistance is the base line. The Y axis has $X_L = 5$ ohms going in the positive direction, and $X_C = 2$ ohms in the negative direction. The net reactance is

$$X_{\text{net}} = X_L - X_C = 5 \,\Omega - 2 \,\Omega = 3 \,\Omega$$

The impedance is then found with formula (10-10).

$$Z = \sqrt{R^2 + X_L^2} = \sqrt{4^2 + 3^2} = \sqrt{25} = 5$$

And formula (10-11) provides the phase angle.

$$\tan \theta = \frac{X_L}{R} = \frac{3 \,\Omega}{4 \,\Omega} = 0.75$$
$$\theta = 36.8°$$

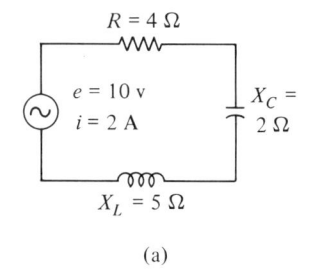

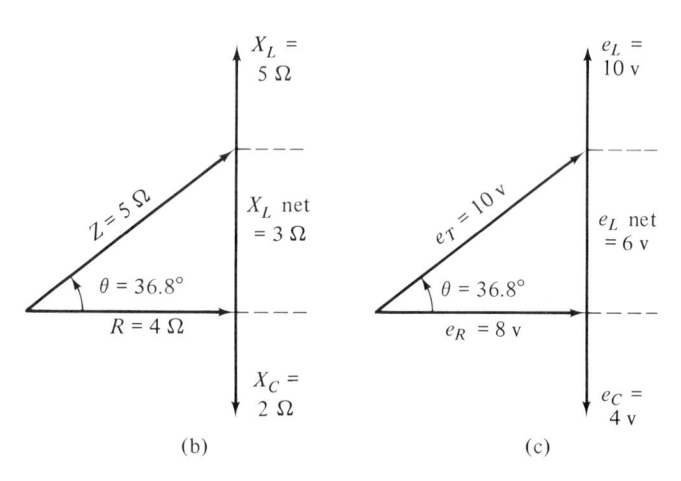

FIGURE 10–20 *Series LRC Circuit*

The vector diagram in (c) is for the voltages. Using Ohm's law gives
$$e_R = iR = 2\text{ A} \times 4\,\Omega = 8\text{ v}$$
$$e_C = iX_C = 2\text{ A} \times 2\,\Omega = 4\text{ v}$$
$$e_L = iX_L = 2\text{ A} \times 5\,\Omega = 10\text{ v}$$

The net voltage is
$$e_{\text{net}} = e_L - e_C = 10\text{ v} - 4\text{ v} = 6\text{ v}$$

Total voltage is obtained from formula (10–12).
$$e_T = \sqrt{e_R^2 + e_L^2} = \sqrt{8^2 + 6^2} = \sqrt{100} = 10\text{ v}$$

The phase angle is calculated for the net values of voltage using formula (10–13).
$$\tan\theta = \frac{e_L}{e_R} = \frac{6\text{ v}}{8\text{ v}} = 0.75$$
$$\theta = 36.8°$$

In Figure 10–20, if X_C had been the dominating reactance, the vectors would have been drawn with a negative phase angle to agree with the rules for X_C and e_C. If there had been more components the vectors could have been plotted serially, or the circuits could be reduced to single components (one resistor, one capacitor, and one inductor). Then, single vectors represent each component. Reduction of serial resistances, reactances, or voltages is achieved by summing the individual components.

Example 10-22 For $e_T = 10.1$ v across a series circuit, what are the impedance and the voltage across the net resistance and reactances (X_{LT} and X_{CT}) if $R_1 = 30$ ohms, $R_2 = 12$ ohms, $X_{L1} = 5$ ohms, $X_{L2} = 20$ ohms, $X_{C1} = 45$ ohms, and $X_{C2} = 8$ ohms? What is the phase angle?

Solution
$$R = R_1 + R_2 = 30\,\Omega + 12\,\Omega = 42\,\Omega$$
$$X_{LT} = X_{L1} + X_{L2} = 5\,\Omega + 20\,\Omega = 25\,\Omega$$
$$X_{CT} = X_{C1} + X_{C2} = 45\,\Omega + 8\,\Omega = 53\,\Omega$$
$$X_{net} = X_{CT} - X_{LT} = 53\,\Omega - 25\,\Omega = 28\,\Omega$$
$$= X_C \text{ (net)}$$
$$Z = \sqrt{R^2 + X_C^2} = \sqrt{42^2 + 28^2} = \sqrt{2548}$$
$$= 50.5\,\Omega$$
$$i_T = \frac{e_T}{Z} = \frac{10.1 \text{ v}}{50.5\,\Omega} = 0.2 \text{ A}$$
$$e_R = i_T R = 0.2 \text{ A} \times 42\,\Omega = 8.4 \text{ v}$$
$$e_{LT} = i_T X_{LT} = 0.2 \text{ A} \times 25\,\Omega = 5 \text{ v}$$
$$e_{CT} = i_T X_{CT} = 0.2 \text{ A} \times 53\,\Omega = 10.6 \text{ v}$$
$$\tan\theta = \frac{-X_C}{R} = \frac{-28\,\Omega}{42\,\Omega} = -0.67$$
$$\theta = -33.7°$$

10.10 Parallel Resistance and Reactance

In parallel circuits, the voltage is the same across each branch. Since the currents vary in amount and phase, the currents are used to calculate total values including the phase angle. The currents for X_C and X_L are 180° apart and cancel. Therefore, the circuits are normally reduced to a single resistance and reactance (X_C or X_L). The total current is then found using formula (10-5) or (10-14).

$$i_T = \sqrt{i_R^2 + i_C^2} \qquad (10\text{--}5)$$
$$i_T = \sqrt{i_R^2 + i_L^2} \qquad (10\text{--}14)$$

The phase angle is found from formula (10-6) or (10-15),

$$\tan\theta = \frac{i_C}{i_R} \qquad (10\text{--}6)$$
$$\tan\theta = \frac{-i_L'}{i_R} \qquad (10\text{--}15)$$

The impedance in general form for i_C or i_L is

$$Z = \frac{e}{i_T} \tag{10-7}$$

Impedance from resistance or reactance is found from formulas (10-8) and (10-16).

A parallel circuit is illustrated in Figure 10-21(a). Since $e = 10$ v, the branch currents are determined in the following manner.

$$i_R = \frac{e}{R} = \frac{10 \text{ v}}{100 \text{ }\Omega} = 100 \text{ ma}$$

$$i_L = \frac{e}{X_L} = \frac{10 \text{ v}}{125 \text{ }\Omega} = 80 \text{ ma}$$

$$i_C = \frac{e}{X_C} = \frac{10 \text{ v}}{500 \text{ }\Omega} = 20 \text{ ma}$$

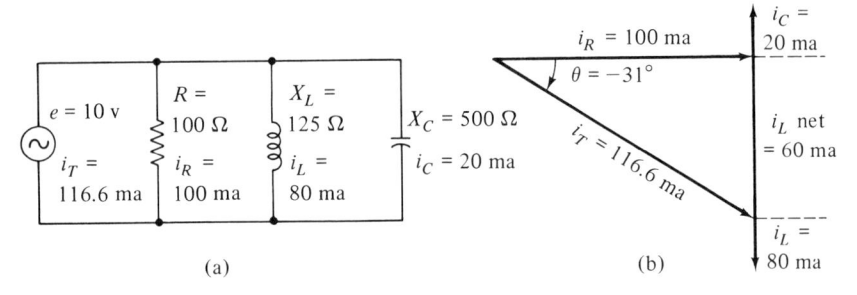

FIGURE 10-21 *Parallel LRC Circuit*

The net current is

$$i_{net} = i_L - i_C = 80 \text{ ma} - 20 \text{ ma} = 60 \text{ ma} = i_L \text{ (net)}$$

Notice that the net current is inductive. The vector diagram in (b) has i_C positive 20 ma, and i_L negative 80 ma. A right-triangle is drawn from the point of difference which is 60 ma. The current and voltage are in-phase for the resistor, which again accounts for i_R as the X axis. Because the net current is inductive, total current is calculated from formula (10-14).

$$i_T = \sqrt{i_R^2 + i_L^2} = \sqrt{(100 \times 10^{-3})^2 + (66 \times 10^{-3})^2}$$
$$= \sqrt{13,600 \times 10^{-6}} = 116.6 \text{ ma}$$

The phase angle is found from formula (10-15), using the net inductive current.

$$\tan \theta = \frac{-i_L}{i_R} = \frac{-60 \text{ ma}}{100 \text{ ma}} = -0.6$$
$$\theta = -31°$$

[10.10] Impedance

The impedance of the circuit is calculated from formula (10–7).

$$Z = \frac{e}{i_T} = \frac{10 \text{ v}}{116.6 \text{ ma}} = 86 \, \Omega$$

Analysis of parallel circuits with more than one R, L, or C is approached in a manner similar to that used for earlier combinations. The circuit is reduced to an equivalent circuit containing a single resistor, inductor, and capacitor. Currents are reduced for similar components by straight addition.

$$i_{RT} = i_{R1} + i_{R2} + \cdots + i_{Rn}$$
$$i_{LT} = i_{L1} + i_{L2} + \cdots + i_{Ln}$$
$$i_{CT} = i_{C1} + i_{C2} + \cdots + i_{Cn}$$

Reactances can also be grouped individually (X_C's and X_L's), and their parallel equivalencies can be determined as discussed in Chapters 8 and 9.

The currents, of course, can be plotted by summing them in a vector diagram, and by keeping the resistive, inductive, and capacitive currents separate. The individual totals of the reactive components are netted (their difference) and combined with the total resistance. Then one of the right-triangle formulas, (10–5) or (10–14), is applied to determine the total current in the circuit.

Example 10–23 What are the total current, impedance, and phase angle of parallel components if $e_T = 40$ v, $R = 20$ ohms, $X_L = 80$ ohms, $X_{C1} = 15$ ohms, and $X_{C2} = 30$ ohms?

Solution

$$i_R = \frac{e_T}{R} = \frac{40 \text{ v}}{20 \, \Omega} = 2 \text{ A}$$

$$i_L = \frac{e_T}{X_L} = \frac{40 \text{ v}}{80 \, \Omega} = 0.5 \text{ A}$$

$$X_C = \frac{X_{C1} X_{C2}}{X_{C1} + X_{C2}} = \frac{15 \, \Omega \times 30 \, \Omega}{15 \, \Omega + 30 \, \Omega} = 10 \, \Omega$$

$$i_C = \frac{e_T}{X_C} = \frac{40 \text{ v}}{10 \, \Omega} = 4 \text{ A}$$

$$i_C \text{ (net)} = i_C - i_L = 4 \text{ A} - 0.5 \text{ A} = 3.5 \text{ A}$$

$$i_T = \sqrt{i_R^2 + i_C^2} = \sqrt{2^2 + 3.5^2} = \sqrt{15.25}$$
$$= 3.91 \text{ A}$$

$$Z = \frac{e_T}{i_T} = \frac{40 \text{ v}}{3.91 \text{ A}} = 10.2 \, \Omega$$

$$\tan \theta = \frac{i_{C(\text{net})}}{i_R} = \frac{3.5 \text{ A}}{2 \text{ A}} = 1.75$$

$$\theta = 60.2°$$

10.11 Series/Parallel Resistance and Reactance

So far, series and parallel circuits have been analyzed by reducing X_C and X_L to single reactances. Reduction was relatively simple because the phase angles were 180°, so it was only a matter of finding the difference between X_C and X_L. The net reactance was X_C or X_L which was then combined with resistance and analyzed by means of vector diagrams or formulas.

When resistance and reactances are mixed in series/parallel branches, analysis becomes more complex. The phase angles, for instance, are not necessarily 180°. The angles for the various branches may be a mixture of positive and negative values between 0° and 180°. When resistance is interspersed with reactance in one or more parallel branches, the analysis is complex because the phase angle of each branch is not an even 90°.

Several methods are discussed in this section involving the less complicated series/parallel circuits. More involved models are studied mathematically using complex variables in Chapter 12.

Figure 10–22(a) illustrates a circuit in which the series line consists of resistance, and the parallel branches are reactance. The circuit is reduced to a simpler version in (b).

$$R_T = R_1 + R_2 = 300 \, \Omega + 950 \, \Omega = 1250 \, \Omega = R$$
$$X_1 = X_{C1} - X_{L1} = 600 \, \Omega - 100 \, \Omega = 500 \, \Omega = X_C$$
$$X_2 = X_{L2} - X_{C2} = 350 \, \Omega - 50 \, \Omega = 300 \, \Omega = X_L$$

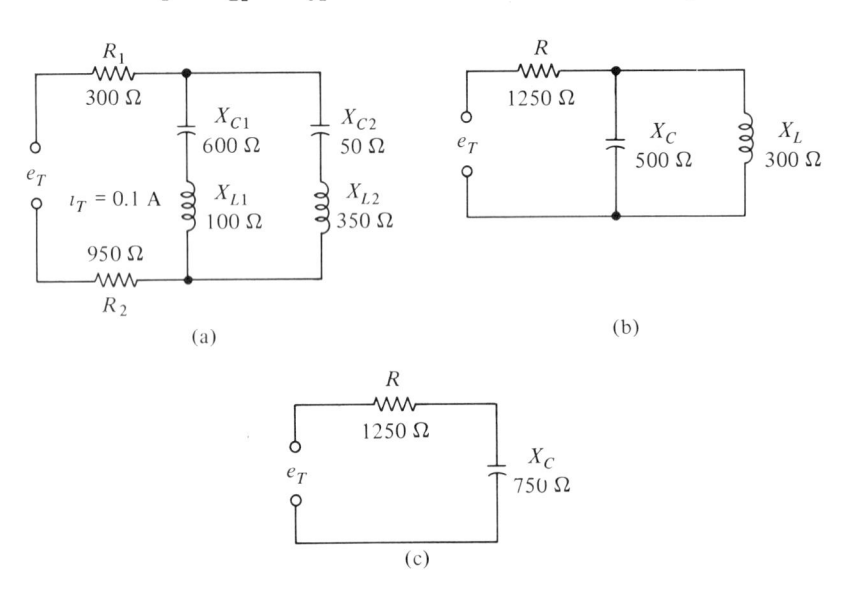

FIGURE 10–22 LC *Reactance in Parallel Branches*

[10.11]   Impedance   221

The circuit is even further reduced in (c).

$$Z = \frac{X_C X_L}{X_C - X_L} = \frac{500 \times 300}{500 - 300} = \frac{150{,}000}{200}$$
$$= 750 \, \Omega = X_C$$

The total impedance for Figure 10–22 is derived from the equivalent components in (c).

$$Z = \sqrt{R^2 + X_C^2} = \sqrt{1250^2 + 750^2}$$
$$= \sqrt{2.125 \times 10^6} = 1460 \, \Omega$$

The phase angle is

$$\tan \theta = \frac{-X_C}{R} = \frac{-750 \, \Omega}{1250 \, \Omega} = -0.6$$
$$\theta = -31°$$

The input voltage, e_T, is calculated from $Z = 1460$ ohms and 0.1 A (given).

$$e_T = iZ = 0.1 \text{ A} \times 1460 \, \Omega = 146 \text{ v}$$

Voltage distribution in circuit (c) is

$$e_R = iR = 0.1 \text{ A} \times 1250 \, \Omega = 125 \text{ v}$$
$$e_C = iX_C = 0.1 \text{ A} \times 750 \, \Omega = 75 \text{ v}$$

The total voltage can be verified by using formula (10–3).

$$e_T = \sqrt{e_R^2 + e_C^2} = \sqrt{125^2 + 75^2}$$
$$= \sqrt{2.125 \times 10^4} = 146 \text{ v}$$

In circuit (b) the branch currents for the reactive components are calculated with $e = 75$ v, which was calculated from (c).

$$i_C = \frac{e}{X_C} = \frac{75 \text{ v}}{500 \, \Omega} = 0.15 \text{ A}$$

$$i_L = \frac{e}{X_L} = \frac{75 \text{ v}}{300 \, \Omega} = 0.25 \text{ A}$$

The total current in the branch is the same as the source current.

$$i_T = i_L - i_C = 0.25 \text{ A} - 0.15 \text{ A} = 0.1 \text{ A}$$

Two combinations of impedance which can be calculated directly are shown in Figure 10–23. For (a) the formula is

$$Z = X_C \sqrt{\frac{R^2 + X_L^2}{R^2 + (X_C - X_L)^2}} \qquad (10\text{–}21)$$

The impedance of Figure 10–23(a) is calculated:

$$Z = 5\sqrt{\frac{4^2 + 2^2}{4^2 + (5-2)^2}} = 5\sqrt{\frac{20}{25}}$$
$$= 5\sqrt{0.8} = 5 \times 0.896 = 4.48 \, \Omega$$

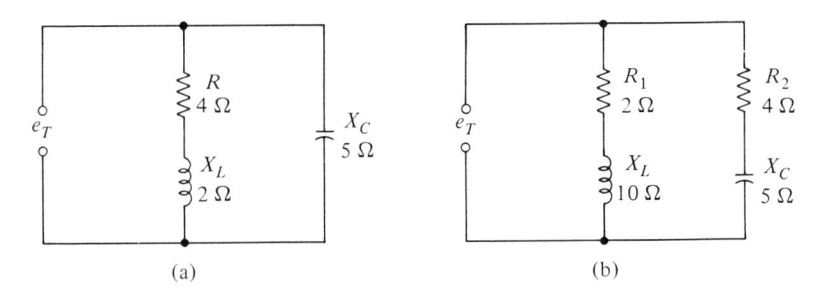

FIGURE 10-23 *Special Cases of R with Reactance in Parallel Branches*

The formula for the combination in (b) is

$$Z = \sqrt{\frac{(R_1^2 + X_L^2)(R_2^2 + X_C^2)}{(R_1 + R_2)^2 + (X_C - X_L)^2}} \quad (10\text{–}22)$$

Total impedance of the circuit in Figure 10–23(b) is calculated:

$$Z = \sqrt{\frac{(2^2 + 10^2)(4^2 + 5^2)}{(2 + 4)^2 + (10 - 5)^2}}$$

$$= \sqrt{\frac{104 \times 41}{36 + 25}} = \sqrt{\frac{4264}{61}} = \sqrt{69.83} = 8.36 \; \Omega$$

The two circuits in Figure 10–23 and formulas (10–21) and (10–22) are special cases. Frequently a circuit can be arranged to match one of the illustrations. For example, if the X_C branch of (a) contained an X_L component, the difference would fit if the X_C component were higher. Or the X_L branch would be valid if a difference in X_L and X_C netted to X_L. The same methods apply to the illustration in (b). The general rule is: reduce a circuit to equivalent values and fit the values to existing formulas.

QUESTIONS

10-1 What is the impedance of a series resistor and capacitor if $R = 75$ ohms, and $X_C = 30$ ohms?

10-2 What is the impedance of a series circuit if the resistance is 1.6 k ohms, and the capacitive reactance is 2.4 k ohms?

10-3 What is the phase angle of Question 10–1?

10-4 What is the phase angle of Question 10–2?

10-5 If the total impedance of a series circuit is 22.36 ohms, and the resistive component is 20 ohms, what is the capacitive reactance?

Impedance

10–6 If the voltage across a capacitor is 15 mv, and the voltage across a capacitor in series is 6 mv, what is the total voltage across both capacitors?

10–7 In Figure 10–3, what is the voltage across the input terminals if the series current is 0.05 A?

10–8 In Figure 10–4, if $R = 200$ ohms, and $X_C = 500$ ohms, what are the total current and total impedance of the circuit?

10–9 What is the phase angle of Question 10–8?

10–10 In Figure 10–5, what value is X_C if $R = 10$ ohms, and $i_T = 2.69$ A?

10–11 If the phase angle of a parallel RC circuit is 35°, and the resistive current is 20 μa, what is the reactive current?

10–12 What are the total current and phase angle of two resistors and two capacitors (all in parallel) when $i_{R1} = 0.2$ A, $i_{R2} = 0.4$ A, $i_{C1} = 0.7$ A, and $i_{C2} = 1.1$ A?

10–13 What is the time constant of a series circuit if the resistance is 0.4 M ohms, and the capacitance is 0.01 μf?

10–14 What value of capacitance will discharge to 37 percent in 240 μsec through 120 ohms?

10–15 A 24-v battery is switched across a series resistor and 300-pf capacitor. How much resistance is required to charge the capacitor to 22.8 v in 9 msec?

10–16 What are the total impedance and phase angle of a series resistance and inductor if $R = 400$ ohms, and $X_L = 260$ ohms?

10–17 For Question 10–16, what are the source voltage (e_T) and phase angle if the voltage across the resistor is 16 v?

10–18 What is the resistance of a resistor that is in series with two inductors with reactances of 15 ohms and 35 ohms if the total impedance is 54.6 ohms?

10–19 In Figure 10–12(a), what are the total current, total impedance, and phase angle if $R = 90$ ohms, and $X_L = 120$ ohms?

10–20 In Figure 10–13(a), what is the reactance of L_2 if $R_1 = 10$ ohms, $i_{R2} = 2.5$ A, $i_{L1} = 4$ A, $e = 20$ v, and $i_T = 10.06$ A?

10–21 What is the phase angle for Question 10–20?

10–22 How long does it take the current to increase to 99 percent of its maximum value in a 75-μh coil with a series resistance of 5 k ohms?

10–23 If one LR time constant is 4 μsec, and the series resistance is 82 k ohms, what is the inductance?

10–24 For a series connected inductor and capacitor, what is the net reactance if $X_L = 1500$ ohms, and $X_C = 700$ ohms?

10–25 A sinusoidal voltage is connected across a coil and capacitor that are in series. If the total current is 0.5 A, the voltage across the capacitor is 75 v, and X_L is 100 ohms, what are the circuit impedance and input voltage?

10–26 In Figure 10–19(a), what are the total current and circuit impedance if $X_L = 1.5$ k ohms, and $X_C = 1.0$ k ohms?

10–27 What is the impedance of a parallel combination of $X_L = 420$ ohms, and $X_C = 300$ ohms?

10–28 What are the total impedance and phase angle of series component if $R = 40$ ohms, $X_C = 10$ ohms, and $X_L = 70$ ohms?

10–29 In Figure 10–20(a), what is the source voltage if $R = 3$ ohms, and $i_T = 0.4$ A?

10–30 What is the phase angle for Question 10–29?

10–31 In Figure 10–2(a), what are the total current and impedance when $R = 500$ ohms, $X_L = 400$ ohms, and $X_C = 1$ k ohm?

10–32 For parallel components, what is the capacitive current if $i_T = 8.544$ A, $i_R = 8$ A, $i_L = 10$ A, and the phase angle is positive?

10–33 Refer to Figure 10–22(a). What is the equivalent series circuit if $R_1 = 20$ ohms, $R_2 = 45$ ohms, $X_{C1} = 75$ ohms, $X_{L1} = 100$ ohms, $X_{C2} = 200$ ohms, and $X_{L2} = 80$ ohms?

10–34 What is the total impedance of the circuit in Figure 10–23(a) if $R = 30$ ohms, $X_L = 40$ ohms, and $X_C = 15$ ohms?

10–35 What is the total impedance of the circuit in Figure 10–23(b) if $R_1 = 20$ ohms, $R_2 = 10$ ohms, $X_L = 5$ ohms, and $X_C = 12$ ohms?

11 Frequency Considerations

Electronic circuits usually handle signals of various amplitudes and frequencies. Inductance and capacitance both affect wave shape, phase, and amplitude of signals. These effects vary according to the frequency. For example, the reactance of an inductor increases as the frequency increases. The reactance of a capacitor decreases as the frequency increases.

When inductors and capacitors are combined in a circuit, their phase angles are opposite (180° apart). In a series circuit, if $X_L = X_C$, the net reactance is nearly zero, so that the net current is high. In a parallel circuit ($X_L = X_C$), the net current is nearly zero, which means the reactance is high. By the selection of components, signals can be attenuated or magnified to satisfy any particular frequency requirement.

11.1 Series Resonance

Whenever X_L is equal to X_C for a given frequency, the circuit is said to be **resonant**. The resonant frequency of an LC combination occurs when $X_L = X_C$. Figure 11-1 shows the linear increase in X_L as frequency increases.

$$X_L = 2\pi f L \tag{11-1}$$

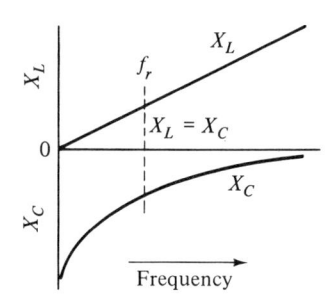

FIGURE 11-1 *Reactance Versus Frequency*

Figure 11-1 also shows the exponential decrease in X_C as frequency increases.

$$X_C = \frac{1}{2\pi f C} \tag{11-2}$$

Depending upon the values of L and C, there is one particular frequency at which the reactances are equal and opposite. That point is the resonant frequency f_r.

At the resonant frequency of a series combination of LC, a sharp increase in the voltage occurs across the inductor and the capacitor. The voltages tend to cancel since they are 180° out-of-phase, but, individually, they can attain amplitudes many times greater than the ac source voltage. Figure 11-2 illustrates the increased amplitude at the resonant frequency. Frequency is shown increasing (the sine waves become closer together) from left to right. At the resonant point, a sharp increase in amplitude occurs. In a series circuit, the increased voltage is the direct result of increased current.

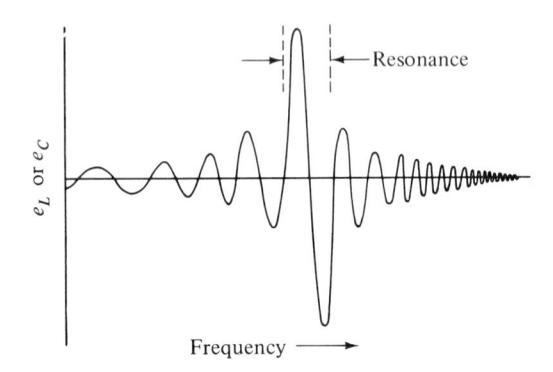

FIGURE 11-2 *Increased Voltage at Resonance*

The resonant frequency of a circuit is derived from formulas (11-1) and (11-2) since the reactances are equal at resonance.

$$2\pi f L = \frac{1}{2\pi f C}$$

Multiplying both sides of the above formula by f and then dividing both sides by $2\pi L$ gives

$$f^2 = \frac{1}{(2\pi)^2 LC}$$

The square root of both sides is

$$f_r = \frac{1}{2\pi\sqrt{LC}} \tag{11-3}$$

This is the formula for finding the resonant frequency. The resonant frequency of the circuit in Figure 11-3 is

$$f_r = \frac{1}{2\pi\sqrt{LC}} = \frac{1}{2\pi\sqrt{636 \times 10^{-6} \times 159 \times 10^{-12}}}$$
$$= \frac{1}{2\pi\sqrt{0.101 \times 10^{-12}}} = \frac{1}{2\pi \times 0.318 \times 10^{-6}}$$
$$= \frac{1 \times 10^6}{2} = 0.5 \text{ MHz}$$

Using formulas (11-1) and (11-2), the reactances of L and C can be calculated at resonance.

$$X_L = 2\pi f L = 2\pi \times 0.5 \times 10^6 \times 636 \times 10^{-6}$$
$$= \pi \times 636 = 2 \times 10^3 = 2 \text{ k}\Omega$$

$$X_C = \frac{1}{2\pi f C} = \frac{0.159}{0.5 \times 10^6 \times 159 \times 10^{-12}}$$
$$= \frac{1}{0.5 \times 10^{-3}} = 2 \times 10^3 = 2 \text{ k}\Omega$$

Notice that $X_L = X_C = 2$ k ohms at the resonant frequency of 0.5 MHz. Table 11-1 is a summary of calculations for Figure 11-3. At the resonant frequency, 0.5 MHz, both X_L and X_C are 2 k ohms. Since X_L leads by 90°, and X_C lags by 90°, the two cancel. The internal ac resistance of the coil becomes significant at this point. The net reactance is zero, but the total impedance Z_T is 20 ohms (represented by R_C), which is purely resistive. The total series current in the circuit is

$$i_T = \frac{e}{Z_T} = \frac{500 \text{ mv}}{20 \text{ }\Omega} = 25 \text{ ma}$$

TABLE 11-1

Frequency MHz	X_L kΩ	X_C kΩ	Z_T kΩ	i_T ma	e_L mv	e_C mv
0.1	0.4	10.0	9.6	0.052	21	520
0.3	1.2	3.33	1.1	0.454	544	1,500
0.5	2.0	2.0	0.02*	25	50,000	50,000
0.7	2.8	1.43	1.37	0.365	1,020	510
0.9	3.6	1.11	2.49	0.201	724	223

*Resistance of the coil R_C

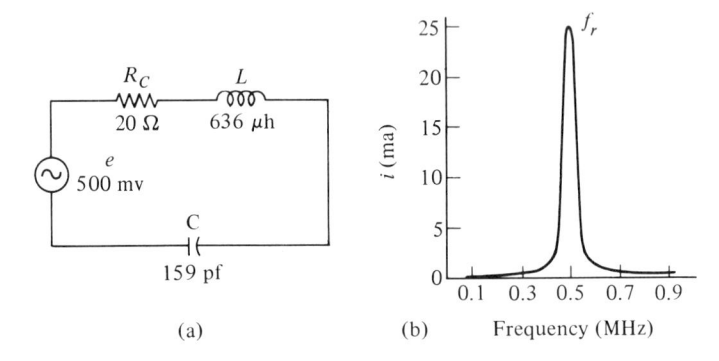

FIGURE 11-3 *Impedance at Series Resonance*

The voltage across the inductor is

$$e_L = i_T X_L = 25 \times 10^{-3} \times 2 \times 10^3 = 50 \text{ v} = 50{,}000 \text{ mv}$$

and the voltage across the capacitor is

$$e_C = i_T X_C = 25 \times 10^{-3} \times 2 \times 10^3 = 50 \text{ v} = 50{,}000 \text{ mv}$$

Notice that the ac voltages across L and C far exceed the source voltage e. The calculations are performed in the same way for the other frequencies in the table. On either side of resonance, the net reactance (Z_T) increases. The resistive part is ignored when its value is small compared to the net reactance. Below resonance, the difference between X_L and X_C increases as the reactances move in opposite directions; the net reactance is capacitive. Above resonance, the same phenomenon takes place, but the net reactance is inductive. Figure 11-1 illustrated the point.

As the total impedance (Z_T) of the circuit increases, the total current (i_T) decreases. The ac voltage across the reactive elements decreases with the current. Actually, the change in Z_T and i_T is quite sharp on either side of f_r.

[11.1]   Frequency Considerations   229

The total current is illustrated in Figure 11-3(b). It peaks at 25 ma and drops to less than 1 ma on each side of resonance.

The voltage drops around a series resonant *LC* circuit are shown in Figure 11-4. The values are taken from Figure 11-3(a) and Table 11-1. Each reactance has a resonant voltage drop of 50,000 mv. The voltages are out-of-phase, so the net voltage of *L* and *C* is zero when measured across both of the reactances. The source voltage of 500 mv, then, is dropped across R_C, the internal resistance of the coil.

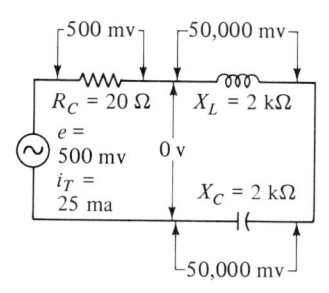

FIGURE 11-4 *Voltage Amplification at Series Resonance*

The following examples illustrate calculations for Figure 11-3.

Example 11-1 For a frequency of 0.1 MHz, verify the calculations in Table 11-1.

Solution
$$X_L = 2\pi f L = 2\pi \times 0.1 \times 10^6 \times 636 \times 10^{-6}$$
$$= 400 = 0.4 \text{ k}\Omega$$

$$X_C = \frac{1}{2\pi f C} = \frac{0.159}{0.1 \times 10^6 \times 159 \times 10^{-12}}$$
$$= \frac{1 \times 10^{-3}}{0.1 \times 10^{-6}} = 10 \times 10^3 = 10 \text{ k}\Omega$$

$$Z_T = X_C - X_L = 10 \text{ k}\Omega - 0.4 \text{ k}\Omega = 9.6 \text{ k}\Omega$$

$$i_T = \frac{e}{Z_T} = \frac{500 \times 10^{-3}}{9.6 \times 10^3} = 52 \times 10^{-6} = 0.052 \text{ ma}$$

$$e_L = i_T X_L = 0.052 \times 10^{-3} \times 0.4 \times 10^3$$
$$= 0.021 = 21 \text{ mv}$$

$$e_C = i_T X_C = 0.052 \times 10^{-3} \times 10 \times 10^3$$
$$= 0.52 = 520 \text{ mv}$$

Example 11-2 Verify the calculations of Table 11-1 for 0.9 MHz.

Solution
$$X_L = 2\pi f L = 2\pi \times 0.9 \times 10^6 \times 636 \times 10^{-6}$$
$$= 3.6 \times 10^3 = 3.6 \text{ k}\Omega$$

$$X_C = \frac{1}{2\pi fC} = \frac{0.159}{0.9 \times 10^6 \times 159 \times 10^{-12}}$$
$$= \frac{1 \times 10^{-3}}{0.9 \times 10^{-6}} = 1.11 \times 10^3 = 1.11 \text{ k}\Omega$$
$$Z_T = X_L - X_C = 3.6 \text{ k}\Omega - 1.11 \text{ k}\Omega = 2.49 \text{ k}\Omega$$
$$i_T = \frac{e}{Z_T} = \frac{500 \times 10^{-3}}{2.49 \times 10^3} = 201 \times 10^{-6} = 0.201 \text{ ma}$$
$$e_L = i_T X_L = 0.201 \times 10^{-3} \times 3.6 \times 10^3$$
$$= 0.724 = 724 \text{ mv}$$
$$e_C = i_T X_C = 0.201 \times 10^{-3} \times 1.11 \times 10^3$$
$$= 0.223 = 223 \text{ mv}$$

Example 11-3 If the resonant frequency of a series LC circuit is 1.59 MHz, and $C = 50$ pf, what is the value of L?

Solution Rearranging formula (11-3) gives

$$L = \frac{1}{(2\pi)^2 f^2 C} = \frac{(0.159)^2}{(1.59 \times 10^6)^2 \times 50 \times 10^{-12}}$$
$$= \frac{(0.1)^2 \times 10^{-12}}{50 \times 10^{-12}} = \frac{0.01}{50} = 200 \times 10^{-6}$$
$$= 200 \text{ } \mu\text{h}$$

Formulas for calculating L or C at resonance are shown here for convenience.

$$L = \frac{1}{(2\pi)^2 f_r^2 C} \qquad (11\text{--}4)$$

$$C = \frac{1}{(2\pi)^2 f_r^2 L} \qquad (11\text{--}5)$$

In summary, the characteristics of a series circuit at its resonant frequency are:

> Maximum current, i_T
> Minimum impedance, Z_T
> Maximum voltage across L or C

11.2 Parallel Resonance

A parallel resonant LC circuit is analyzed in terms of branch currents and impedance. For example, the branch currents are out of phase by 180° since X_L leads by 90°, and X_C lags by 90°. Like a series LC circuit, a parallel LC circuit is resonant when X_L is equal to X_C. At resonance, then, the reactive branch currents cancel so that the main line current is minimal. The main line current would be zero if it were not for the small internal ac resistance of the inductor.

The total impedance of a parallel resonant circuit is maximum at the resonant frequency because the current is minimal. By Ohm's law,

$$Z_T = \frac{e}{i_T} \tag{11-6}$$

The curve in Figure 11-5(a) illustrates a sharp drop in current at $f_r = 0.5$ MHz. In (b), a corresponding peak in the impedance is evident at the same resonant frequency.

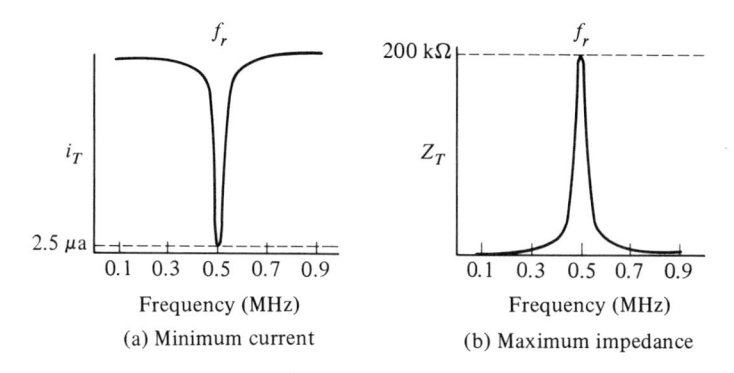

FIGURE 11-5 *Parallel Resonant Curves*

The circuit in Figure 11-6 is designed to resonate at 0.5 MHz. The same values of L and C were used in Figure 11-3, which also resonated at 0.5 MHz.

$$f_r = \frac{1}{2\pi\sqrt{LC}} = \frac{1}{2\pi\sqrt{636 \; \mu h \times 159 \; pf}} = 0.5 \; \text{MHz}$$

X_L and X_C are each 2 k ohms, the same as the series circuit of Figure 11-3. Calculations for Figure 11-6 are itemized in Table 11-2. At the resonant frequency, 0.5 MHz, i_L and i_C are each 250 μa. Since the source voltage (e) is 500 mv,

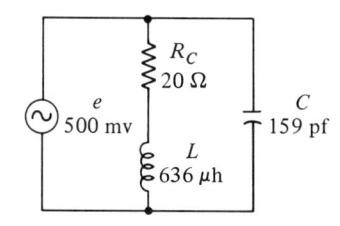

FIGURE 11-6 *Impedance at Parallel Resonance*

TABLE 11-2

Frequency MHz	X_L kΩ	X_C kΩ	i_L μa	i_C μa	i_T μa	Z_T kΩ
0.1	0.4	10.0	1,250	50	1,250	0.4
0.3	1.2	3.33	417	150	267	1.87
0.5	2.0	2.0	250	250	2.5*	200
0.7	2.8	1.43	178	350	172	2.9
0.9	3.6	1.11	139	450	311	1.61

*i_L and i_C cancel so that R_C (20 Ω) becomes significant. $i_T = (R_C i_L)/X_L = 20/2000 \times 250$ μa $= 2.5$ μa

$$i_L = \frac{e}{X_L} = \frac{500 \text{ mv}}{2 \text{ k}\Omega} = 250 \text{ μa}$$

$$i_C = \frac{e}{X_C} = \frac{500 \text{ mv}}{2 \text{ k}\Omega} = 250 \text{ μa}$$

However, the internal coil resistance R_C becomes significant at resonance. The internal coil resistance is in the branch with X_L so that the net current at resonance is a ratio of R_C and X_L. Ignoring the phase angle, an approximation of the total current is

$$i_T = \frac{R_C}{X_L} \times i_L = \frac{20 \text{ } \Omega \times 250 \text{ μa}}{2 \text{ k}\Omega} = 2.5 \text{ μa}$$

As shown in Figure 11-7, i_L is 2.5 μa less than the resonant current of 250 μa. Notice that the branch currents are opposing so that the net current i_T is 2.5 μa.

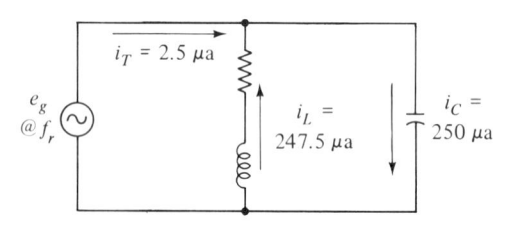

FIGURE 11-7 *Net Current (i_T) at Parallel Resonance*

At resonance, the total impedance is calculated from formula (11-6) (see Table 11-2).

$$Z_T = \frac{e}{i_T} = \frac{500 \times 10^{-3}}{2.5 \times 10^{-6}} = 200 \times 10^3 = 200 \text{ k}\Omega$$

A comparison of Table 11-2 to Table 11-1 will show that for given values of L and C, X_L and X_C are the same for each frequency listed. This is not surprising since the same formulas are used for X_L and X_C [formulas (11-1) and (11-2)]. Also, formula (11-3) is used for determining resonance of either the series or parallel *LC* circuit.

In summary, the characteristics of a parallel circuit at its resonant frequency are:

Minimum current, i_T

Maximum impedance, Z_T

11.3 Q Effects

The letter *Q* is the symbol for *quality* or *figure of merit* for a resonant circuit. This figure of merit is what determines the sharpness of a response curve on each side of resonance. A higher value signifies a sharper response curve. The *Q* of a resonant circuit is determined by the ratio of the inductive reactance to the internal resistance of the coil.

$$Q = \frac{X_L}{R_C} \qquad (11\text{-}7)$$

In the above formula, X_L and R_C are in ohms at f_r. Since R_C is actually the ac resistance of the inductor, it is several times more than its dc resistance.

The *Q* of an *LC* circuit is calculated from the inductor because the ac resistance of a capacitor is normally much lower due to the small losses associated with capacitance. Therefore, capacitors have a *Q* which is much higher than that of an inductor. The *Q* of a circuit is shown in Figure 11-8.

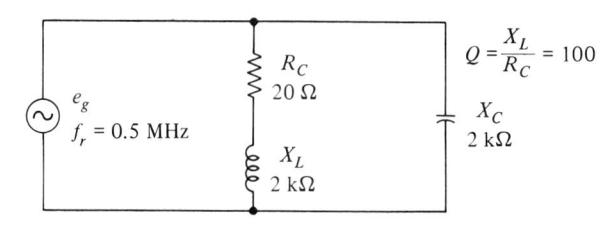

FIGURE 11-8 *Q of a Parallel Resonant Circuit*

The values are the same as those in the series and parallel examples used for the Tables 11-1 and 11-2.

$$Q = \frac{X_L}{R_C} = \frac{2000\ \Omega}{20\ \Omega} = 100$$

Example 11-4 What is the ac resistance of a 426-μh coil if its *Q* is 160 at a resonant frequency of 300 kHz?

Solution To determine *Q* from formula (11-7), X_L must be determined.

Using formula (11-1) yields
$$X_L = 2\pi fL = 2\pi \times 0.3 \times 10^6 \times 426 \times 10^{-6} = 800\ \Omega$$
Rearranging formula (11-7) gives
$$R_C = \frac{X_L}{Q} = \frac{800\ \Omega}{160} = 5\ \Omega$$

A higher L/C ratio will produce a higher Q value as long as the ac resistance of the coil increases at a lesser rate. Typical values of Q are 100 to 200. In general, maximum Q occurs when X_L is from 1 to 2 k ohms which is therefore a good starting point for X_L when calculating values of L and C for a specific f_r.

Example 11-5 Determine L and C values for a resonant frequency of of 100 MHz.

Solution For ease of calculation, a value of 1 k ohm is selected for X_L. Therefore,
$$L = \frac{X_L}{2\pi f} = \frac{1 \times 10^3}{2\pi \times 100 \times 10^6} = \frac{0.159 \times 10^3}{100 \times 10^6}$$
$$= 1.59 \times 10^{-6} = 1.59\ \mu h$$

Using formula (11-5) gives
$$C = \frac{1}{(2\pi)^2 f_r^2 L} = \frac{1}{(2\pi)^2 \times (100 \times 10^6)^2 \times 1.59 \times 10^{-6}}$$
$$= \frac{(0.159)^2}{1 \times 10^{16} \times 1.59 \times 10^{-6}}$$
$$= \frac{0.159}{10 \times 10^{10}} = 1.59 \times 10^{-12} = 1.59\ \text{pf}$$

Verification can be made from formula (11-3).
$$f_r = \frac{1}{2\pi\sqrt{LC}} = \frac{0.159}{\sqrt{1.59 \times 10^{-6} \times 1.59 \times 10^{-12}}}$$
$$= \frac{0.159}{\sqrt{2.53 \times 10^{-18}}} = \frac{0.159}{1.59 \times 10^{-9}}$$
$$= 0.1 \times 10^9 = 100\ \text{MHz}$$

In a **series resonant** circuit, the voltage across the inductor or capacitor is a function of the Q value.

$$f_r\ \text{(series)} \qquad e_L = e_C = Qe_g \tag{11-8}$$

Rearranging formula (11-8) gives
$$Q = \frac{e_L}{e_g} = \frac{e_C}{e_g} \tag{11-9}$$

[11.3] Frequency Considerations 235

As an example, the Q of Figure 11–3 has already been calculated as 100. Table 11–1, the summary for that figure, can be verified for $f_r = 0.5$ MHz. From formula (11–8) the following is obtained:

$$e_L = e_C = Qe_g = 100 \times 500 \text{ mv} = 50{,}000 \text{ mv}$$

Example 11–6 What is the Q of a series resonant circuit if the input voltage (e_g) is 2.4 v ac, and the ac voltage across the capacitor is 600 v?

Solution Using formula (11–9) yields

$$Q = \frac{e_C}{e_g} = \frac{600 \text{ v}}{2.4 \text{ v}} = 250$$

In a **parallel resonant** circuit, the total impedance and current are functions of Q.

$$f_r \text{ (parallel)} \quad Z_T = QX_L \quad \text{or} \quad Z_T = QX_C \quad \textbf{(11–10)}$$

$$i_T = \frac{i_L}{Q} \quad \text{or} \quad i_T = \frac{i_C}{Q} \quad \textbf{(11–11)}$$

Refer to the f_r calculations in Table 11–2 for examples. Q has already been established as 100. To verify Z_T for $X_L = 2$ k ohms,

$$Z_T = QX_L = 100 \times 2 \times 10^3 = 200 \times 10^3 = 200 \text{ k}\Omega$$

The total current is verified:

$$i_T = \frac{i_L}{100} = \frac{250 \text{ } \mu a}{100} = 2.5 \text{ } \mu a$$

Example 11–7 In a parallel circuit, what are Q, e_L, i_T and f_r, if $X_L = 628$ ohms, $Z_T = 125.6$ k ohms, $L = 2.5$ mh, and $e_g = 1.2$ v?

Solution From formula (11–10), the following is obtained:

$$Q = \frac{Z_T}{X_L} = \frac{125.6 \times 10^3}{628} = 200$$

Formula (11–8) yields

$$e_L = Qe_g = 200 \times 1.2 \text{ v} = 240 \text{ v}$$

Determining i_L and using formula (11–11) gives

$$i_L = \frac{e_g}{X_L} = \frac{1.2 \text{ v}}{628 \text{ } \Omega} = 1.91 \text{ ma}$$

$$i_T = \frac{i_L}{Q} = \frac{1.91 \times 10^{-3}}{200} = 9.55 \times 10^{-6} = 9.55 \text{ } \mu a$$

And rearranging formula (11-1) gives
$$f_r = \frac{X_L}{2\pi L} = \frac{628}{2\pi \times 2.5 \times 10^{-3}} = \frac{100}{2.5 \times 10^{-3}}$$
$$= 40 \times 10^3 = 40 \text{ kHz}$$

11.4 Bandwidth

The bandwidth of a circuit defines the width or band of frequencies around the resonant frequency point of a tuned circuit. The width is taken at the 70.7-percent point; that is, those frequencies with an amplitude of 70.7 percent or more of the maximum are counted as the bandwidth (abbreviated BW).

The response curve in Figure 11-9 illustrates the principle of bandwidth. The resonant point is 13 MHz. The amplitude decreases on either side of

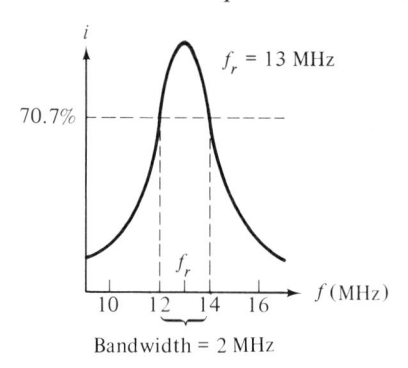

FIGURE 11-9 *Bandwidth at Resonance*

f_r, and at 70.7 percent the frequencies are 12 MHz and 14 MHz. The bandwidth, then, is the difference, which is calculated from the formula

$$\text{BW} = f_h - f_l \quad (11\text{-}12)$$

When the above formula is applied to Figure 11.9, the bandwidth is

$$\text{BW} = f_h - f_l = 14 \text{ MHz} - 12 \text{ MHz} = 2 \text{ MHz}$$

In a series circuit, bandwidth is determined from 70.7 percent of the maximum current. For a parallel circuit, bandwidth is determined from 70.7 percent of the maximum impedance.

A higher Q provides a sharper response curve. Figure 11-10 illustrates variations in sharpness as a function of Q. When $Q = 100$, the curve is narrow, and when $Q = 10$, the curve broadens considerably. Bandwidth is therefore related to Q as follows:

$$\text{BW} = \frac{f_r}{Q} \quad (11\text{-}13)$$

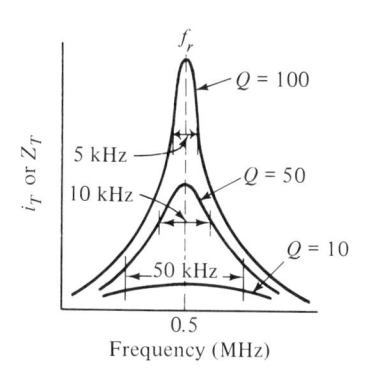

FIGURE 11-10 Q Effect on Bandwidth

This formula is valid for series circuits. It is valid for parallel circuits only when the value of Q is large enough so that the phase angle has little effect. Since $Q = X_L/R_c$, the internal resistance must be less than a tenth to be considered insignificant in terms of the phase angle, which means that Q must be greater than 10.

Example 11-8 Verify the bandwidth of each curve in Figure 11-10.

Solution Using formula (11-13) gives

$$\text{BW} = \frac{f_r}{Q} = \frac{0.5 \times 10^6 \text{ Hz}}{100} = 5 \times 10^3 = 5 \text{ kHz}$$

$$\text{BW} = \frac{f_r}{Q} = \frac{0.5 \times 10^6 \text{ Hz}}{50} = 10 \times 10^3 = 10 \text{ kHz}$$

$$\text{BW} = \frac{f_r}{Q} = \frac{0.5 \times 10^6 \text{ Hz}}{10} = 50 \times 10^3 = 50 \text{ kHz}$$

Example 11-9 What is the value of Q for the series resonant circuit plotted in Figure 11-9?

Solution Rearranging formula (11-13) yields

$$Q = \frac{f_r}{\text{BW}} = \frac{13 \times 10^6 \text{ Hz}}{2 \times 10^6 \text{ Hz}} = 6.5$$

The resonant frequency and bandwidth of a tuned circuit become important in such things as radio and television. For example, a variable tuned circuit at the input of a radio resonates at the frequency of the selected radio station. In a series tuned circuit, the radio circuitry is driven from the output of the resonant inductor or capacitor. The voltage, then, is maximum and amplifies the radio signal. The bandwidth determines how much signal on each side of f_r is allowed to enter the radio. Television operates in a similar way.

Selective tuning is diagrammed in Figure 11-11. The receiving antenna illustrated is the common loop type found in many household radios. The series tuned circuit is adjusted by varying C to resonate at the frequency of

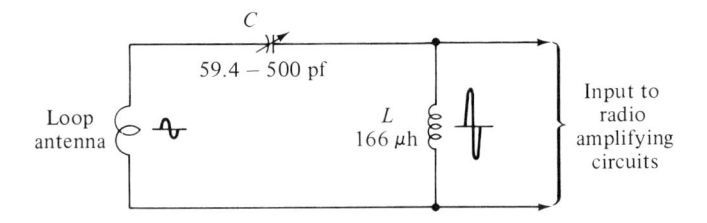

FIGURE 11-11 *AM Radio Input Circuit*

the radio station. For example, the AM broadcast band is from 550 kHz to 1600 kHz. The capacitor and inductor in the figure are designed to resonate within that range. Taking the extremes of the capacitor results in

$$f_r = \frac{1}{2\pi\sqrt{LC}} = \frac{0.159}{\sqrt{0.166 \times 10^{-3} \times 59.4 \times 10^{-12}}}$$
$$= \frac{0.159}{\sqrt{98.5 \times 10^{-16}}} = \frac{0.159}{9.95 \times 10^{-8}} = 1.6 \times 10^6$$
$$= 1600 \text{ kHz}$$

$$f_r = \frac{1}{2\pi\sqrt{LC}} = \frac{0.159}{\sqrt{0.166 \times 10^{-3} \times 500 \times 10^{-12}}}$$
$$= \frac{0.159}{\sqrt{830 \times 10^{-16}}} = \frac{0.159}{28.9 \times 10^{-8}} = 0.55 \times 10^6$$
$$= 550 \text{ kHz}$$

The tuning dial of the radio in the example would be connected to the capacitor. Then the dial would be calibrated for resonant frequencies across the broadcast band. A similar effect can be achieved by tuning the inductor rather than the capacitor. In practical radio circuits, the band would be slightly broader than the example shows to compensate for tolerances in the components.

11.5 Filters

Filters are used in electronic circuits to separate ac and dc voltages and to separate frequencies. AC signals, for instance, can be taken from dc levels by means of capacitor or transformer coupling. The capacitor blocks the dc while passing the ac. The primary of a transformer may have ac and dc, but only the ac can be induced in the secondary.

[11.5] Frequency Considerations

Capacitive coupling is shown in Figure 11–12. The internal resistance of the generator is assumed to be small in the example so that the battery voltage of 2 v is dropped across R_1. A sine wave is superimposed on the 2-v dc level and causes the dc level to fluctuate or follow the sine wave. The capacitor charges to the dc level, but the charge follows the ac component which is dropped across R_2. The dc voltage is blocked by the capacitor. Therefore, the reference level upon which the ac is superimposed changes from a $+2$-v level to a 0-v level. The amplitude of the ac still varies ± 1 v, but now it varies around the 0-v level.

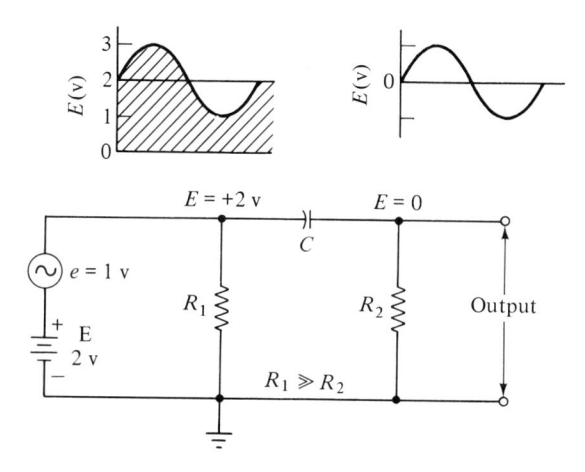

FIGURE 11–12 *Basic ac Coupling Circuit*

The transformer in Figure 11–13 has 10 v dc across its primary winding. The ac component from the generator causes the 10-v level to shift according to the frequency and amplitude of the signal. Since the windings are isolated, except for a common ground, the dc is not allowed to affect the level of the secondary winding. The lower terminal of the secondary is at a ground or zero reference level. So only the ac from the primary is induced in the secondary. Notice that the phase is shifted 180°.

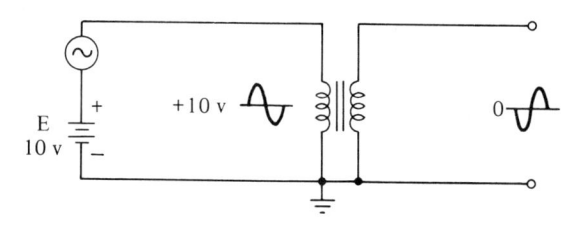

FIGURE 11–13 *Separation of ac from dc*

However, the output connections can be reversed for zero phase shift of the signal. And the signal may be increased or decreased through the transformer, depending upon the turns ratio.

Consider an example in which the dc is to be preserved while removing the ac component. The circuit in Figure 11-14 illustrates a 1-v ac sine wave superimposed on a 5-v dc level. The dc of 5 v is present on either side of R since the output is shown as an open circuit. The capacitor simply charges to the 5-v dc value, but acts as a short circuit to the ac signal. Thus, the output is $+5$ v dc without the ac component.

Because capacitive reactance is frequency dependent, the value of capacitance affects the coupling or bypass characteristics of a capacitor. At a given frequency, a high value of capacitance has a low value of X_C. Therefore, the frequency is more easily passed or bypassed. In Figure 11-12, the values of C and R_2 (the filter section of the circuit) determine the low frequency cutoff for the output of the circuit. That is, the C and R_2 ratio determines the lowest frequency that can be passed by C. Low frequency cutoff, f_L, is defined as the point at which the output is 70.7 percent of the input. Since the phase angle is 90°, the 70.7 percent point occurs when $X_C = R_2$. The cutoff point is found by the formula

$$f_l = \frac{1}{2\pi RC} \qquad (11\text{-}14)$$

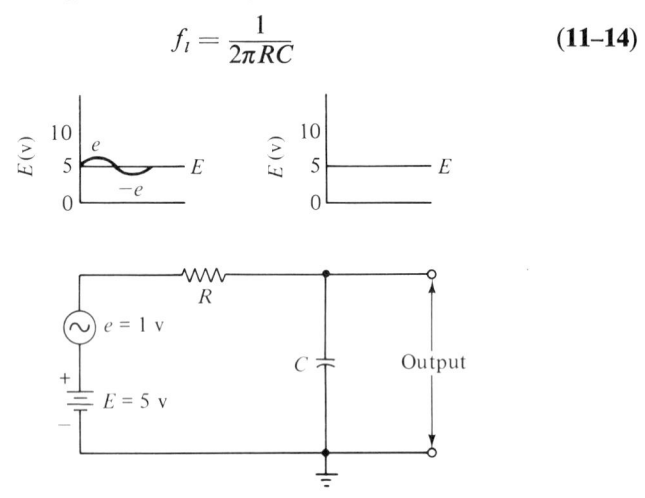

FIGURE 11-14 *Separation of dc from ac*

Example 11-10 Determine the low frequency cutoff for the filter portion of Figure 11-12 if $C = 0.05 \ \mu f$, and $R_2 = 2$ k ohms.

Solution
$$f_L = \frac{1}{2\pi RC} = \frac{0.159}{2 \times 10^3 \times 0.05 \times 10^{-6}}$$
$$= \frac{0.159}{0.1 \times 10^{-3}} = 1.59 \times 10^3 = 1.59 \text{ kHz}$$

The solution to Example 11-10 indicates that all frequencies from 1.59 kHz and higher will be 70.7 percent or more of the input to the filter when measured at the output. Also, all frequencies below 1.59 kHz will be less than 70.7 percent of the input to the capacitor when measured at the output. Note that for this level of discussion, the value of R_1 is much greater than the value of R_2.

Formula (11-14) can also be applied to the circuit in Figure 11-14. The inverse of what occurred in Example 11-10 occurs here. If the same values hold ($R = 2$ k ohms, and $C = 0.05$ μf) all frequencies below 1.59 kHz will be 70.7 percent or more of the input (e) when measured at the output. The higher frequencies, of course, will be shunted to ground.

The filter in Figure 11-12 is called a **high-pass filter** because it passes the higher frequencies. Figure 11-14 is a **low-pass filter** because it passes the lower frequencies.

Examples of high-pass filters are shown in Figure 11-15. Circuit (a) is the *RC* type already discussed. In (b) a choke is used in place of a resistor. The capacitor has a low X_C for high frequencies while the coil has a high X_L. Chokes, though more expensive, are more effective than resistors because

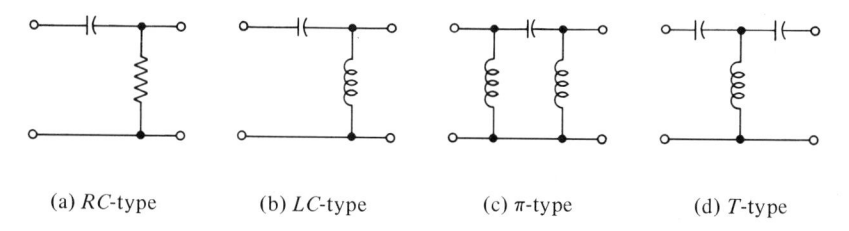

(a) *RC*-type (b) *LC*-type (c) π-type (d) *T*-type

FIGURE 11-15 *High-Pass Filters*

the reactance changes with frequency. For example, at lower frequencies X_C is higher, and X_L is lower so that the divider between *C* and *L* reduces the signal. As the frequency increases, X_C becomes smaller, and X_L becomes larger so that the reactance of *C* is only a very small part of the total reactance of *C* and *L*. Thus, nearly all of the signal from the input reaches the output.

Figure 11-15(c) illustrates a π-type filter which has an inductor placed on each side of a capacitor. The filter operates in a way similar to the way the one in (b) does. The T type in (d) is also similar to (b). A point to remember is that the *LC* type filters can resonate when $X_L = X_C$. This point will be discussed later.

Low-pass filters are shown in Figure 11-16. The one in (a) is a simple *RC* filter (see Figure 11-14). In (b) the coil offers a high reactance to high frequencies while the capacitor offers a low reactance. In terms of a divider, nearly all of the amplitude of the high frequencies is dropped

242   Frequency Considerations   [11.5]

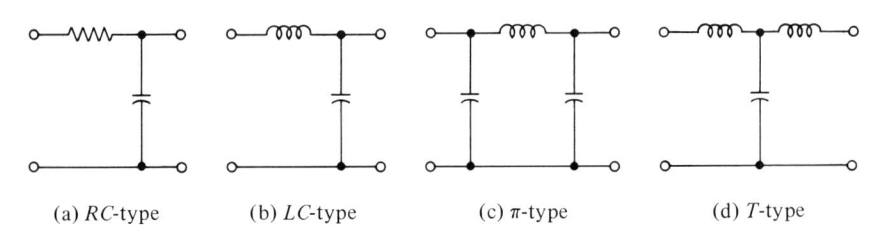

(a) *RC*-type   (b) *LC*-type   (c) π-type   (d) *T*-type

FIGURE 11–16 Low-Pass Filters

across the coil, so little if any high frequency signals reach the output. The filters in (c) and (d) are generally more effective because they offer additional attenuation of high frequencies due to the additional capacitor or coil.

An example of high-pass filter design is shown in Figure 11–17. Since the output in (a) is taken across X_L, the output will be

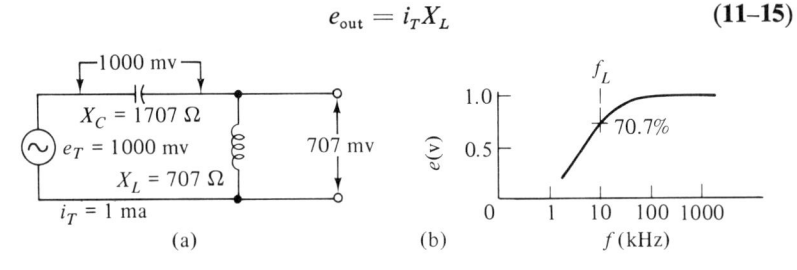

$$e_{\text{out}} = i_T X_L \qquad (11\text{–}15)$$

FIGURE 11–17 High-Pass Filter $f_L = 10$ kHz

However, the 70.7 percent point occurs when $e_{\text{out}} = 0.707 \times e_T$. Because the phase of the reactances is 180°, the net reactance is the difference, which is then used to determine i_T. The general expression for determining e_{out} for a high-pass filter is

$$e_{\text{out}} = \frac{e_T X_L}{X_C - X_L} \qquad (11\text{–}16)$$

e is in volts and X is in ohms. Notice that $e_T/(X_C - X_L)$ is for the total current i_T, which is then used to satisfy formula (11–15). The voltages e_T and e_{out} are known (following the 0.707 rule), so one of the reactances must be pre-selected. For example, the design of Figure 11–17 began with $X_L = 707$ ohms. Rearranging formula (11-16) gives

$$X_C = \frac{e_T X_L}{e_{\text{out}}} + X_L \qquad (11\text{–}17)$$

X_C is determined:

$$X_C = \frac{e_T X_L}{e_{\text{out}}} + X_L = \frac{1 \times 707}{0.707} + 707 = 1000 + 707$$
$$= 1707 \ \Omega$$

The total current and output voltage are

$$i_T = \frac{e_T}{X_C - X_L} = \frac{1}{1707 - 707} = 1 \text{ ma}$$
$$e_{\text{out}} = i_T X_L = 1 \text{ ma} \times 707 \, \Omega = 707 \text{ mv}$$

In the example, a cutoff frequency of 10 kHz was selected [see Figure 11–17(b)]. Values for L and C are

$$L = \frac{X_L}{2\pi f} = \frac{707}{2\pi \times 10 \times 10^3} \approx 12 \text{ mh}$$
$$C = \frac{1}{2\pi f X_C} = \frac{1}{2\pi \times 10 \times 10^3 \times 1707} \approx 0.009 \, \mu\text{f}$$

A low-pass filter is shown in Figure 11–18. The source voltage is 1000 mv, and i_T is 1 ma at $f_H = 1000$ kHz. The same procedure used with high-pass filters is used here. Now, however, the X_L value is the larger reactance as the output is taken from across X_C. Formula (11–16) becomes

$$e_{\text{out}} = \frac{e_T X_C}{X_L - X_C} \quad \quad \textbf{(11–18)}$$

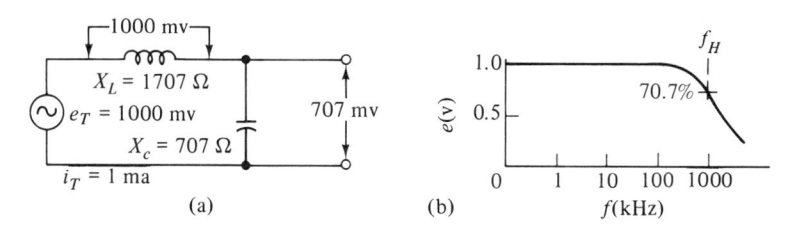

FIGURE 11–18 *Low-Pass Filter* $f_H = 100$ *kHz*

To determine X_L when X_C is known, formula (11–18) is rearranged so that

$$X_L = \frac{e_T X_C}{e_{\text{out}}} + X_C \quad \quad \textbf{(11–19)}$$

The calculations for X_L, L, and C for $f_H = 1000$ kHz are as follows.

$$X_L = \frac{e_T X_C}{e_{\text{out}}} + X_C = \frac{1 \times 707}{0.707} + 707 = 1707 \, \Omega$$
$$L = \frac{X_L}{2\pi f} = \frac{1707}{2\pi \times 1 \times 10^6} \approx 270 \, \mu\text{h}$$
$$C = \frac{1}{2\pi f X_C} = \frac{1}{2\pi \times 1 \times 10^6 \times 707} \approx 500 \text{ pf}$$

Notice that calculations for high- and low-pass filters are very much the same. The two figures, 11–17 and 11–18, were purposely designed with the same voltages, currents, and reactances to illustrate the similarity.

Band-pass is the term given to a band of frequencies that are passed by a filtering device. The filter rejects those frequencies outside of the band. For example, suppose the high and low filters of Figure 11-17 and 11-18 are combined as in Figure 11-19. The high-pass filter passes frequencies above 10 kHz, and the low-pass filter passes frequencies below 1 MHz.

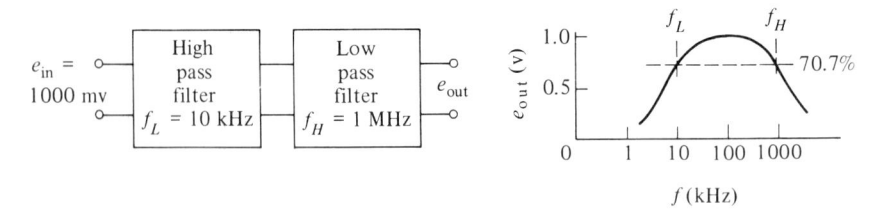

FIGURE 11-19 *Band-Pass Filtering*

The graph is a composite of the two earlier ones [Figures 11-17(b) and 11-18(b)] showing a band-pass of 10 to 1000 kHz.

Resonant circuits can also be used for band-pass filters. Figure 11-9 was a response curve for $f_r = 13$ MHz. The band-pass would be the same as the bandwidth, since both are measured at 70.7 percent of maximum amplitude.

A series resonant circuit has minimum impedance at f_r. Therefore, if a series resonant circuit is placed in series with the input signal, the line current will be maximum, allowing maximum current to be developed across the load. An illustration is given in Figure 11-20(a). In (b) the series

(a) Band-pass filter

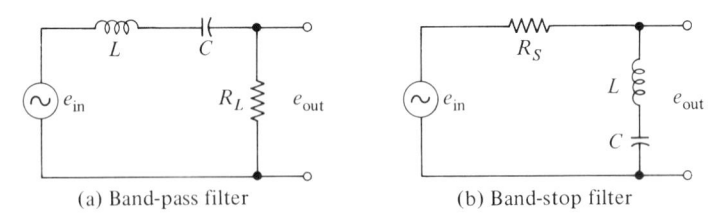

(b) Band-stop filter

FIGURE 11-20 *Series Resonant Filters*

resonant components are used to bypass a selected band of frequencies to the common line on the bottom, which shorts the output for the selected band. At resonance, the current will maximize while the impedance of L and C will minimize. Thus, the signal voltage will be developed across R_S with nearly zero signal output (e_{out}) at f_r. This type of filter is called a **band-stop filter**. The pass-bandwidth and stop-bandwidth are determined by the Q of the the circuit as described in Section 11.4.

A parallel resonant circuit has maximum impedance and minimum current at resonance. If used in parallel with the load, as in Figure 11–21(a), it becomes a band-pass filter. R_S is selected so that at resonance, the impedance of the tuned circuit is at least ten times as great as the resistance of R_S. Then nearly all of the input is transferred to the output. Outside of resonance and beyond the bandwidth, the impedance decreases to a minimum value much less than that of R_S. Since the current increases outside of resonance, the signal is dropped across R_S so that the output is nearly zero.

(a) Band-pass filter

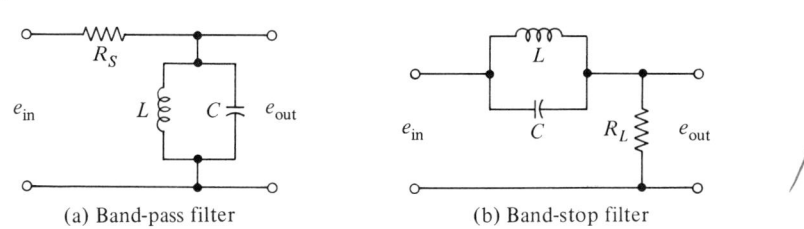

(b) Band-stop filter

FIGURE 11–21 *Parallel Resonant Filters*

In Figure 11–21(b), a parallel resonant circuit is placed in series with the signal. The ratio of R_L and the impedance of the tuned circuit are selected so that at resonance the impedance is many times more than that of R_L. The filter is then a band-stop circuit, which rejects the band of frequencies around f_r (BW). Outside the resonant band, the impedance is many times less than that of R_L, and the current is maximum. Therefore, the higher current develops a signal across R_L for e_{out} that is essentially equal to e_{in}.

11.6 Transient Circuits

The term **transient**, as it applies to electronics, has several definitions.

1. A sudden change from one voltage level to another, or a sudden change from one current level to another
2. A current surge resulting from a voltage surge
3. A non-cyclic or irregular change in signal—noise or interference pulses

Transient signals are used in many electronic devices. For example, pulsating dc (Chapter 7) is used in radar systems, digital computers, and various control systems. The response of a transient signal is called its **transient response**. It is a measure of how the input signal is reproduced at the output of a circuit. Often, however, it is desirable to change the response for special applications.

Transient response is illustrated in Figure 11–22. The inputs (a) are

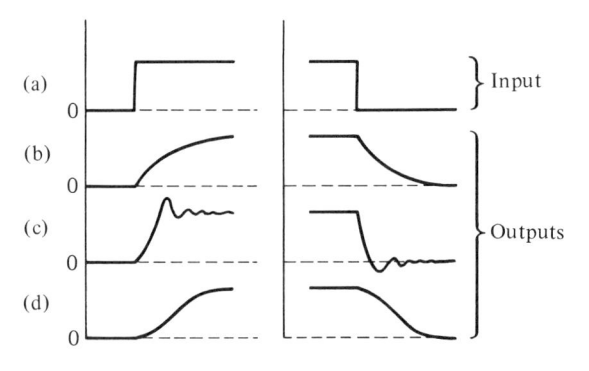

FIGURE 11-22 *Transient Response*

step functions, abruptly changing from one level to another. The remainder of the left hand column shows leading edge response at an output while the right shows trailing edge reponse at an output. The examples represent typical signals that might be seen at the output of various circuits, depending upon the type and design of the circuits. An *RC* network, for example, would respond exponentially as in (b). A tuned circuit might overshoot and ring as in (c). In (d) the signal starts slowly and tapers off slowly. This is typical of many semiconductor switching devices. The speed of a mechanical device also responds in this manner; beginning slowly as it accelerates, then slowing down to a stop.

Several special circuits are described for Figures 11-23 through 11-25. Figure 11-23 is a simple *RC* circuit in which the input is a square wave. The

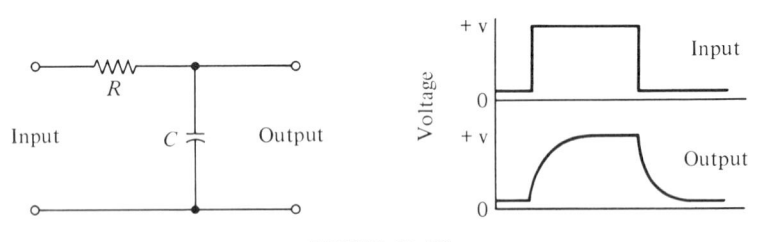

FIGURE 11-23

output voltage changes exponentially as a function of the capacitor charge and discharge ($t = RC$). In Figure 11-24 the input is a square wave of current while the output current responds exponentially due to the *LR* time constant.

The circuit in Figure 11-25 is called a **differentiating network**. The output responds to the sudden change at the leading edge. However, once the signal is at its maximum positive level, the capacitor discharges exponentially through the resistor at a rate determined by the *RC* time constant

[11.6]   Frequency Considerations   247

FIGURE 11-24

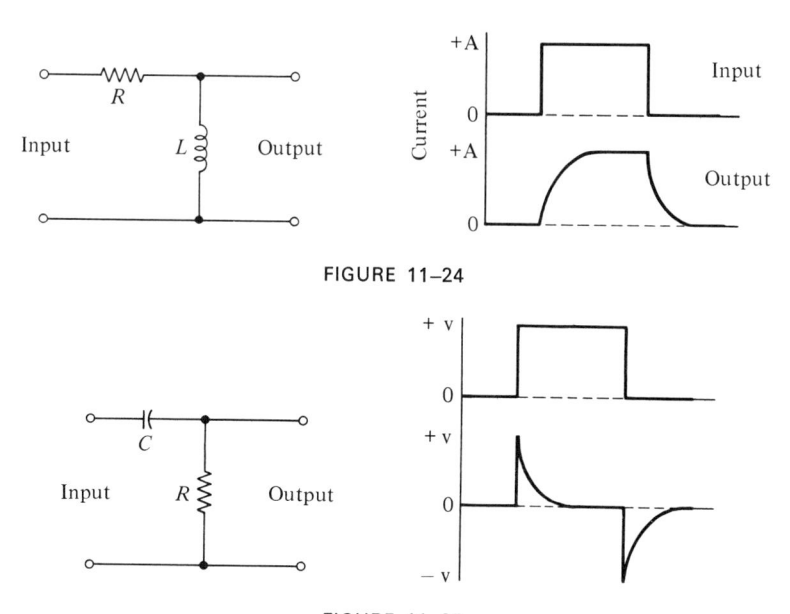

FIGURE 11-25

of the circuit. When the input changes negatively toward zero at the trailing edge, the output swings in a negative direction. The capacitor discharges to zero through R. The output then changes with every transition of the input: positive when the input switches toward positive, and negative when the input switches toward negative (toward zero in the diagram).

Suppose the entire frequency spectrum were to be investigated. One would find signals all through it, from the lowest frequencies to the highest. For example, the common household or automobile AM radio has signals (radio stations) from 550 to 1600 kHz. FM radio covers a band from 88 to 108 MHz. There are television bands, police radio bands, military bands, amateur radio bands, and many others. In addition, noise or static exists across the band, and it often comes in at levels that exceed the intended signal. In any case, selective tuning is required in order to separate signals and attenuate unwanted noise.

A band of 0.1 to 1000 kHz is illustrated in Figure 11-26. No special attention is being paid to specific radio signals because the particular spectrum represents a sample situation in which the levels vary according to the signal strength at the input. The transitions could represent radio signals or general noise conditions. Notice the transient noise spikes at various places across the band. If a radio signal is to be selected at 50 kHz, and the noise level is to be kept to a minimum, special tuning is required.

248 Frequency Considerations [11.6]

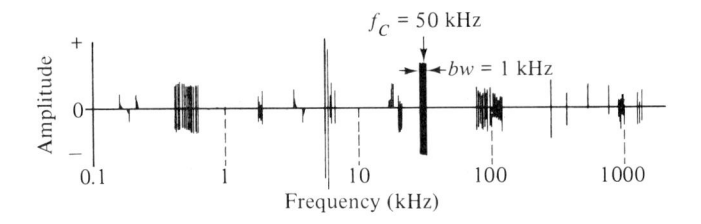

FIGURE 11-26 *Transient Signals and Noise for Example Spectrum*

Otherwise, various signals and transient noise across the band could be received and amplified, which would interfere with the reception of the desired signal.

Since a narrow band is desired with the exclusion of frequencies above and below, a band-pass filter is required. The response curve in Figure 11–27(a) has a bandwidth of 1 kHz, centering around the desired signal of 50 kHz. Notice that the curve fits the selected band at 50 kHz (Figure 11–26) and excludes all other points on the spectrum.

Circuitry for the filter is shown in Figure 11–27(b). The *LC* combination forms a series resonant circuit. Because the bandwidth and center frequency (f_c or f_r) are known, the Q of the circuit is calculated from formula (11–13).

$$Q = \frac{f_r}{\text{BW}} = \frac{50 \times 10^3}{1 \times 10^3} = 50$$

A coil is selected with a ratio of 50 between the internal resistance and inductive reactance. Since $R_C = 12.56$ ohms in the example, X_L must be

$$X_L = 50 R_C = 50 \times 12.56 = 628$$

The inductance, then, is found from a rearranged version of formula (11–1).

(a) Band-pass curve

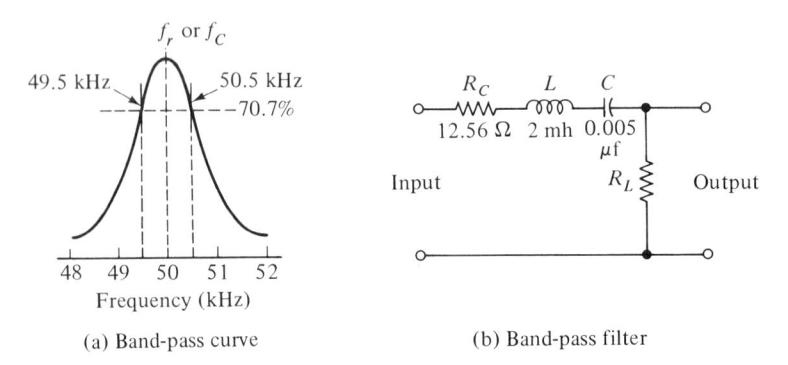

(b) Band-pass filter

FIGURE 11-27 *Band-Pass Filter Design*

$$L = \frac{X_L}{2\pi f} = \frac{628}{6.28 \times 50 \times 10^3} = \frac{100 \times 10^{-3}}{50}$$
$$= 2 \times 10^{-3} = 2 \text{ mh}$$

The capacitance can be found from formula (11–5), but X_C must equal X_L at resonance, so formula (11–2) can be applied instead.

$$C = \frac{1}{2\pi f X_C} = \frac{0.159}{50 \times 10^3 \times 628} = \frac{0.159}{31.4 \times 10^6}$$
$$= 0.0051 \times 10^{-6} \approx 0.005 \, \mu\text{f}$$

11.7 Non-Linear Circuits

A linear circuit is one which produces or reproduces a linear signal. Figure 11–28(a) is an example of a linear signal in which inductive reactance changes in direct proportion to frequency. Because capacitive reactance is inversely proportional to changes in frequency, a non-linear curve is produced as shown in Figure 11–28(b).

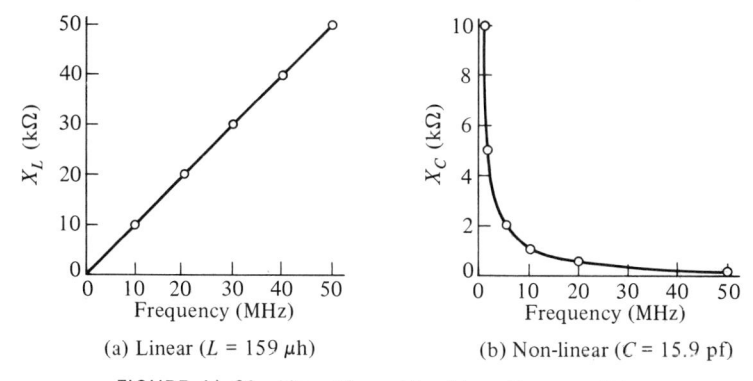

(a) Linear (L = 159 μh)   (b) Non-linear (C = 15.9 pf)

FIGURE 11–28 *Linear Versus Non-Linear Response Curves*

Any curve that is not a straight line is considered non-linear. Other examples of non-linear circuits were shown in Figures 11–22 to 11–25. The circuits in Figures 11–23 through 11–25 produce non-linear waveforms due to time constants in the circuits. Most filters are non-linear because the amplitude of signals at various frequencies changes on a curve, as shown in Figures 11–17 to 11–19. Thus, the basic elements of non-linear circuits are materials in which the impedance (or resistance) varies as a function of current, voltage, or frequency. The relationship is expressed by the following formula:

$$I = kE^n \qquad (11\text{--}15)$$

I is in amperes, E is in volts across the resistance, K is a constant equal to the initial conductance when E is nearly zero, and n is an exponent which defines the non-linearity of the resistance. If $n = 1$, the circuit is linear, thus n must be greater or less than one for a non-linear circuit.

Thyrites are examples of non-linear resistances where the value of n is greater than unity. Typical values of conductance (k) are from 0.0005 to 0.005. The resistance of a thyrite decreases as either the current or voltage increases. Table 11–3 summarizes characteristics for an imaginary thyrite. For simplification the exponential n was given a value of 2.

The table lists k as 0.002 and values of E from 1 to 10 v. E^2 is simply squaring the values in the E column.

TABLE 11–3

k	E (v)	E^2 (v²)	I (ma)	R (Ω)
0.002	1	1	2	500
0.002	2	4	8	250
0.002	3	9	18	167
0.002	4	16	32	125
0.002	6	36	72	83
0.002	8	64	128	62
0.002	10	100	200	50

$I = kE^n$ ($n = 2$ for example), and $R = E/I$

The current is calculated from formula (11–15).

$$I = kE^n = 0.002 \times E^2$$

The resistance is calculated from Ohm's law.

$$R = \frac{E}{I}$$

As an example for $E = 6$ v in Table 11–3,

$$I = kE^n = 0.002 \times (6 \text{ v})^2 = 0.002 \times 36 = 72 \text{ ma}$$

$$R = \frac{E}{I} = \frac{6 \text{ v}}{72 \text{ ma}} \approx 83 \, \Omega$$

The data in Table 11–3 are used to plot the curves in Figure 11–29. In (a) current is plotted as a function of voltage. Notice the exponential rise in the slope, demonstrating the non-linear characteristics of thyrites. The resistance is plotted in (b) as a function of voltage. Note the non-linear characteristic of the exponential decline in resistance as the voltage increases. If current instead of voltage were plotted on the horizontal axis of (b), the resistance curve would still be exponential and similar to the one in (b).

[11.7] Frequency Considerations 251

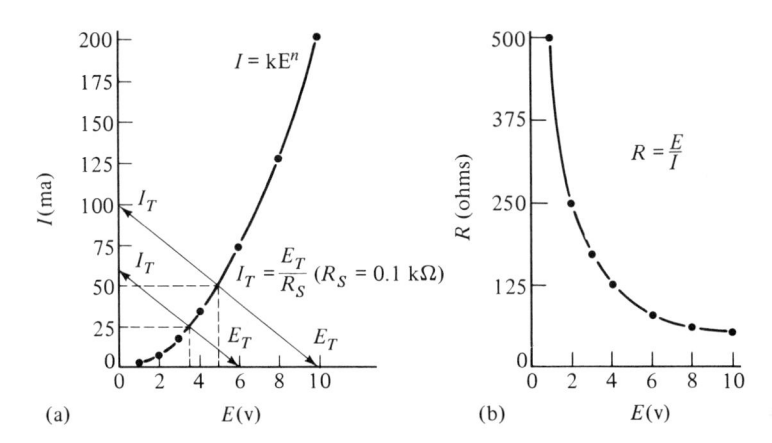

FIGURE 11-29 *Thyrite Characteristics*

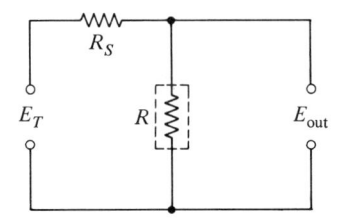

(a) $I = kE^n$, $I_T = \dfrac{E_T}{R_S}$ ($R_S = 0.1$ kΩ)

(b) $R = \dfrac{E}{I}$

FIGURE 11-30 *Non-Linear Resistance Circuit*

The circuit in Figure 11–30 can be used to illustrate an application of the thyrite. R_S is a series resistor normally larger than the R (resistance) of the thyrite. As E_T increases, E_{out} attempts to rise. The resistance of the thyrite consequently decreases. The current increases so that the IR drop across R_S increases and *tends* to prevent E_{out} from increasing significantly. If E_T decreases, E_{out} decreases, which causes R to increase so that less voltage is dropped across R_S. Thus, the output *tends* to increase. The circuit acts somewhat as a voltage regulator, but not as efficiently because the characteristics of a thyrite are not such that it will wholly compensate for voltage variations. Also, a thyrite does not react quickly enough to handle transient pulses.

Load lines are drawn in Figure 11–29(a) to demonstrate the changes in E_{out} that occur as a result of input changes to a thyrite. A load line is a graphic method of showing the absolute operating range of a circuit. For example, when $R_S = 0.1$ k ohms, and $E_T = 10$ v, the absolute maximum voltage that can occur at E_{out} is 10 v (if $R = \infty$). Therefore, a point is noted at $E = 10$ v on the graph. The other extreme is the absolute maximum current, which is the case if $R =$ zero.

$$I_T = \frac{E_T}{R_S} = \frac{10 \text{ v}}{0.1 \text{ k}\Omega} = 100 \text{ ma}$$

The second point is noted at $I = 100$ ma. A straight line is drawn between the two points, $E = 10$ v and $I = 100$ ma. E_{out} must operate within these limits. Since the characteristic curve intercepts the load line at $E = 5$ v (a dotted line), E_{out} is 5 v. The current in the circuit is 50 ma as indicated on the vertical current scale for I. If the input decreased from 10 v to 6 v, a new load line would be drawn between 6 v and 60 ma.

$$I_T = \frac{E_T}{R_S} = \frac{6 \text{ v}}{0.1 \text{ k}\Omega} = 60 \text{ ma}$$

The new load line intercepts the curve at $E = 3.5$ v and $I = 25$ ma. Notice the output voltage varied 1.5 v for a 4 v decrease of E_T.

Figure 11–31 demonstrates improved control of E_{out} with an R_S that is significantly larger than the resistance of the thyrite. The horizontal scale is expanded, and R_S is 1 k ohm. The curve is transposed from Figure 11–29(a). For an E_T of 100 v,

$$I_T = \frac{E_T}{R_S} = \frac{100 \text{ v}}{1 \text{ k}\Omega} = 100 \text{ ma}$$

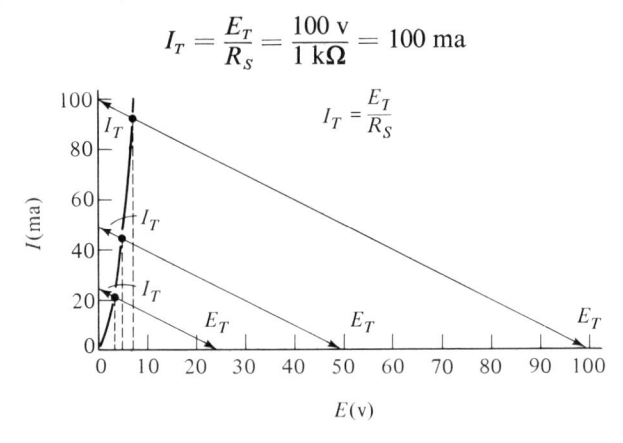

FIGURE 11–31 Load Lines for Thyrite Circuit ($R_S = 1 \text{ K}\Omega$)

The intercept point occurs at approximately 7 v. If E_T changes by half (50 v), E_{out} decreases to about 5 v, which is a 2-v change for a 50-v change at the input. If E_T drops to 25 v, E_{out} drops to about 3 v. Increasing R_S to a value much greater than that of the thyrite allows a circuit to maintain closer control of the output for large changes at the input.

Another type of non-linear circuit is shown in Figure 11–32(a). A variable resistor, such as a volume control on a radio or TV, is shunted with another resistance. The shunt, R_S, could be the input load resistance of an amplifier or any type of circuit for that matter. The effect on overall resistance as a function of R_S is summarized in Table 11–4 which lists the series/parallel resistance for R_{in} (the load for a preceding circuit), and the parallel resistance for R_{out} (the load for following circuits).

Frequency Considerations 253

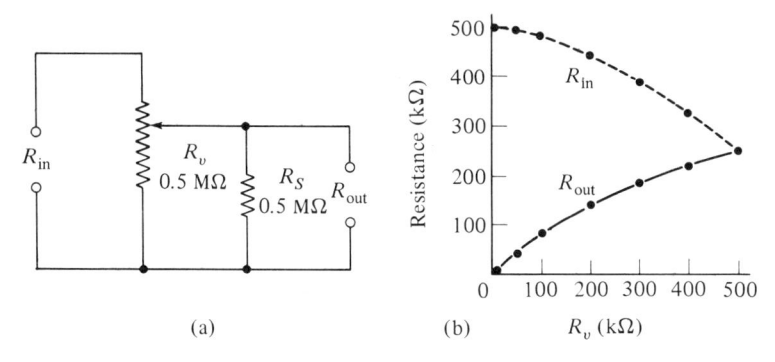

FIGURE 11-32 Non-Linear Characteristics from Variable Resistance Circuit

TABLE 11-4

Resistance in k ohms

R_v	R_S	R_{in}	R_{out}
10	500	499.8	9.8
50	500	495.5	45.5
100	500	483.0	83.0
200	500	443.0	143.0
300	500	387.0	187.0
400	500	322.0	222.0
500	500	250.0	250.0

When the slider of R_v is near the bottom, 10 k ohms, the value of R_{in} becomes the series resistance of 500 k − 10 k = 490 k, plus the parallel resistance of $R_v = 10$ k ohms and $R_S = 500$ k ohms. R_{out} is the parallel resistance of $R_v = 10$ k, and $R_S = 500$ k ohms. The procedure is carried on for increasing amounts of R_v, until $R_v = 500$ k ohms, so that now, both R_{in} and R_{out} are the parallel resistances of R_v and R_S.

The characteristics of R_{in} and R_{out} are plotted in Figure 11-32(b). In both cases the curves are non-linear, a fact which demonstrates the effect of fixed resistance across a variable resistance.

QUESTIONS

11-1 What is the phase relationship between X_L and X_C in terms of leading and lagging?

11-2 When frequency decreases, do X_L and X_C increase and/or decrease?

254   Frequency Considerations

11–3 Define resonant frequency.

11–4 In a tuned circuit, how does voltage react when the frequency approaches series resonance?

11–5 What is the resonant frequency of a 10-mh coil and a 16-pf capacitor?

11–6 If C is 0.015 μf, what value of L is required for a resonant frequency of 1.2 MHz?

11–7 If L is 36 mh, what value of C is required for a resonant frequency of 275 kHz?

11–8 In Figure 11–4, if $i_T = 7$ ma at resonance, what are the voltage drops across the components? How much is e?

11–9 In Figure 11–3, what value must C have to produce maximum current at 1 MHz? What are the voltage across L and C, and the total current at resonance?

11–10 In a parallel resonant circuit, what is the relationship between impedance and current?

11–11 when a parallel LC circuit resonates, what accounts for the total current flow in the circuit?

11–12 In Figure 11–6, what is the total current in the circuit at resonance if $C = 318$ pf?

11–13 What is the Q of a series resonant circuit if the internal resistance of the coil is 30 ohms, and X_L is 4.8 k ohms?

11–14 If the Q of a parallel resonant circuit is 200, what is the internal resistance of a 300-μh coil at 4.2 MHz?

11–15 If the input voltage to a series resonant LC circuit is 50 mv, what is the voltage across the capacitor if $X_L = 1$ k ohm, and the internal resistance of the coil is 10 ohms?

11–16 The i_T of a parallel resonant circuit is 7 ma. What is the capacitive current if $Q = 300$?

11–17 If the total impedance of a parallel resonant circuit is 210 k ohms, what is the value of X_L if the Q of the coil is 175?

11–18 What is the bandwidth of Question 11–17 if $f_r = 350$ MHz?

11–19 In Figure 11–6, what is the internal ac resistance of L if the BW is 1.25 kHz?

11–20 Explain at least three key points regarding capacitive and transformer coupling.

Frequency Considerations   255

11-21 What is the low-frequency cut-off point of a 0.047-μf coupling capacitor into a 20-k ohm load?

11-22 What value of coupling capacitor is required to achieve a low-frequency cut-off of 1 kHz into a 53-k ohm load?

11-23 In Figure 11-14, if frequencies above 20 kHz are to be shunted to ground, and X_C is 1 k ohm at 20 kHz, what are the values of R and C?

11-24 Explain the difference between high- and low-pass filters in terms of frequency and circuitry.

11-25 Refer to Figure 11-17. Design a high-pass filter for $f_H = 30$ kHz, and $C = 0.001$ μf. What are the values of X_L, X_C, and L?

11-26 Design a low-pass filter using Figure 11-18 as a reference. Use $e_T = 400$ mv, $X_C = 1200$ ohms, and $f_H = 0.5$ MHz. What are the values of C, L, and X_L?

11-27 What is the difference between band-pass and band-stop filters?

11-28 What happens at resonance when a series LC circuit is placed across the load (in parallel with the output)?

11-29 What is the effect of placing a parallel LC circuit across the output?

11-30 What is the meaning of the term transient response?

11-31 Name two techniques for reducing high frequency transients that occur above a specific broadcasting band.

11-32 Name several considerations in the design of a band-pass filter for the purpose of selecting a specific band of transient signals.

11-33 Refer to Figure 11-27. What are the values for R_c, L, and C for selecting a transient within a bandwidth of 10 kHz and center frequency of 7 MHz? Assume X_L is 1100 ohms.

11-34 How much current flows through a thyrite that has a conductance of 0.0008, an exponential characteristic of 2, and a dc voltage of 30 v? What is its dc resistance for these conditions?

11-35 In Figure 11-30, if E_T is 8 v, and R_S is 40 ohms, what are the values of E_{out}, the current through the thyrite, and the resistance of the thyrite? Use the thyrite curve in Figure 11-29 for reference.

11-36 Using thyrite characteristics of $k = 0.002$, and $n = 2$, how much voltage is dropped across the thyrite in Figure 11-30 if $E_T = 75$ v, and $R_S = 1$ k ohm?

11–37 If the circuit in Figure 11–32 were used to control the audio signal amplitude in a radio, the sound would be controlled in a non-linear manner (as shown in the non-linear curves). Name a method of controlling the volume linearly.

11–38 What would be the effect of reversing R_{in} and R_{out} in Figure 11–32?

12 Circuit Analysis

This chapter describes some of the various methods used to analyze and solve circuit problems. The subjects are approached for the most part with algebraic and trigonometric functions. Simultaneous equations are necessary for solving loop equations, but their use here is restricted to the basic method of eliminating unknown quantities.

12.1 Loop Equations

In earlier chapters, networks included various combinations of series and parallel components such as resistors, inductors, and capacitors. The networks were solved by applying special rules and formulas for reducing series and parallel components to less complex circuits. For example, combinations of resistances were reduced to a single resistance, then the circuit was solved for current, voltage, resistance, and power. This section discusses methods of setting up loop equations for networks and then solving them mathematically.

The theory of setting up loop equations is based on the principles of Kirchhoff's laws. In terms of voltage, the algebraic sum of the voltages around a closed loop is equal to zero. This means that the sum of the voltage drops is equal to the sum of the applied voltages (around a closed loop). In terms of current, the algebraic sum of the currents in and out of any point in a circuit is equal to zero, the sum of the currents into a point is equal to the sum of the currents out of that point.

Figure 12–1 illustrates voltage loops. In (a) the direction of current is from the negative terminal of the battery, around the loop, and back into the positive terminal. Starting at the battery,

$$30 \text{ v} - 15 \text{ v} - 5 \text{ v} - 10 \text{ v} = 0$$

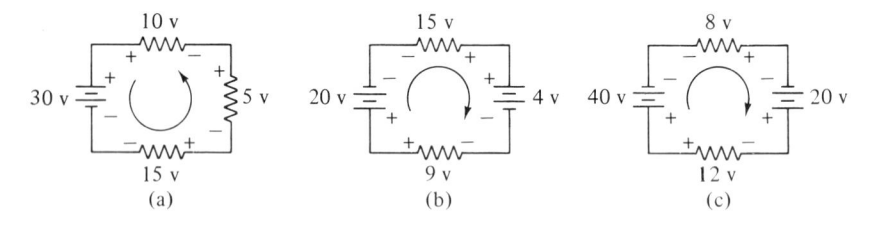

FIGURE 12–1 *Kirchhoff's Voltage Loops*

In terms of applied voltage:

$$30 \text{ v} = 15 \text{ v} + 5 \text{ v} + 10 \text{ v} = 30 \text{ v}$$

In (b) current flows from the negative terminals of the batteries which are additive.

$$20 \text{ v} - 15 \text{ v} + 4 \text{ v} - 9 \text{ v} = 0$$
or
$$20 \text{ v} + 4 \text{ v} = 15 \text{ v} + 9 \text{ v} = 24 \text{ v}$$

The loop in (c) shows the two batteries opposing each other. The direction of current was arbitrarily chosen to favor the higher of the two applied voltages, so that

$$40 \text{ v} - 8 \text{ v} - 20 \text{ v} - 12 \text{ v} = 0$$
or
$$40 \text{ v} - 20 \text{ v} = 8 \text{ v} + 12 \text{ v} = 20 \text{ v}$$

Assigning proper signs to a loop becomes important when algebra is used. The simplest way is to adhere to electron flow rules, and when more than one source voltage is used, favor the combinations that apply the most voltage in a given direction.

Figure 12–2 concerns currents in and out of a node. In (a) the currents flow in the same direction. But the algebraic sum of those entering and leaving must be

$$2 \text{ A} + 4 \text{ A} - 6 \text{ A} = 0$$
or
$$2 \text{ A} + 4 \text{ A} = 6 \text{ A}$$

And in (b)

$$10 \text{ A} - 3 \text{ A} - 5 \text{ A} - 2 \text{ A} = 0$$
or
$$10 \text{ A} = 3 \text{ A} + 5 \text{ A} + 2 \text{ A} = 10 \text{ A}$$

[12.1]   Circuit Analysis   259

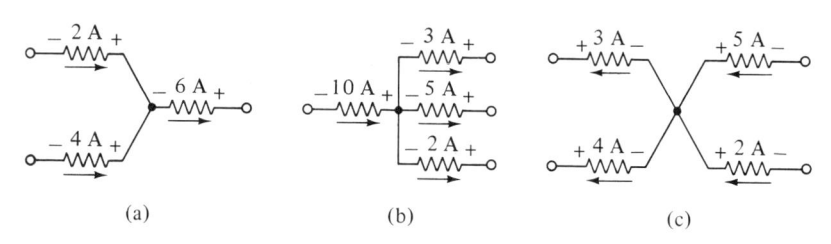

FIGURE 12-2 *Kirchhoff's Current Nodes*

The diagram in (c) has two branches on each side of the node, and the currents are the reverse of those in the two previous examples.

$$3A + 4A - 5A - 2A = 0$$

or

$$3A + 4A = 5A + 2A = 7A$$

In setting up loop equations, all loops must be in the same direction—clockwise or counterclockwise. The choice of direction is not important. An easy way to keep track of polarities is to start on one side of a source voltage, write down the value, and continue around the loop applying polarities as they come.

In Figure 12-3(a), the resistors are easily polarized, and the ends closest to the battery terminals are marked accordingly. Since the batteries oppose, the currents through R_3 are apparently in the same direction

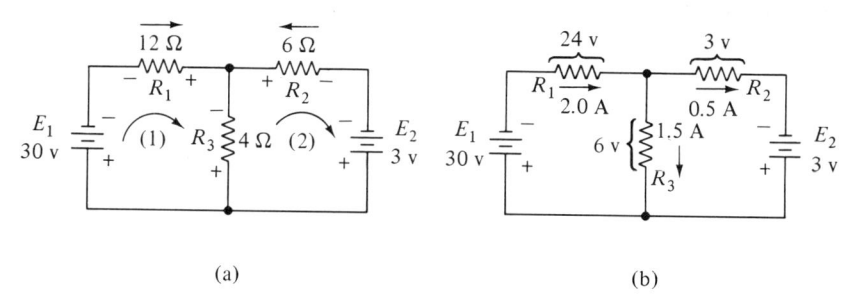

FIGURE 12-3 *Opposing Batteries for Loop Equations*

though not really true in the figure. It does not matter which way the currents actually flow in order to set up an equation. In setting up an equation, polarities are assigned as they come while working around a loop. For example, a resistor has a polarity of (+) at the end that is first encountered; its sign in the equation is (+). If a battery is encountered, the polarity of the first terminal is assigned. Beginning with E_1, the voltage in Figure 12-3(a) is

$$(1) \quad E_1 - I_1 R_1 - I_3 R_3 = 0$$

For E_2 the voltage loop is

(2) $\quad -E_2 + I_3R_3 + I_2R_2 = 0$

The IR drops are converted in the following manner.

$$I_3 = I_1 + I_2$$
$$I_1R_1 = 12I_1$$
$$I_3R_3 = (I_1 + I_2)4 = 4I_1 + 4I_2$$
$$I_2R_2 = 6I_2$$

Voltage loops (1) and (2) are rearranged with the source voltages E_1 and E_2 on the right side of the equations, and then the converted IR drops are substituted on the left side.

(1) $\quad -I_1R_1 - I_3R_3 = -E_1$
$\quad\quad -12I_1 - 4I_1 - 4I_2 = -30$
$\quad\quad\quad -16I_1 - 4I_2 = -30$

(2) $\quad I_3R_3 + I_2R_2 = E_2$
$\quad\quad 4I_1 + 4I_2 + 6I_2 = 3$
$\quad\quad\quad 4I_1 + 10I_2 = 3$

The loops, (1) and (2), are next arranged in the form of simultaneous equations, so that each has two unknowns. These unknowns are solved by multiplying loop (2) by the quantity 4, and then summing the two.

(2) $\quad 4I_1 + 10I_2 = 3$
$\quad\quad 16I_1 + 40I_2 = 12$

(1) $\quad -16I_1 - 4I_2 = -30$
(2) $\quad \underline{16I_1 + 40I_2 = 12}$
$\quad\quad\quad\quad 36I_2 = -18$
$$I_2 = \frac{-18}{36} = -0.5 \text{ A}$$

I_1 is determined by substituting the value of I_2 back into one of the earlier equations. For example, the value of I_2 would be substituted into loop (2) prior to multiplying it by 4.

(2) $\quad 4I_1 + 10(-0.5) = 3$
$\quad\quad\quad 4I_1 = 3 + 5$
$$I_1 = \frac{8}{4} = 2.0 \text{ A}$$

Notice that I_2 is -0.5 A. The negative sign indicates that the direction of current is the opposite of the original assumption in Figure 12–3(a). The circuit in (a) is redrawn in (b). It shows current flowing in the opposite

[12.1] Circuit Analysis 261

direction through R_2, as it should be. The current through R_3 was originally assumed to be $I_1 + I_2$. Algebraically, the assumption is still valid.

$$I_3 = I_1 + I_2 = 2.0 + (-0.5) = 1.5 \text{ A}$$

In the redrawn circuit of Figure 12-3(b), current flows back through the 3-v source, a fact that was not so obvious before working out a solution. The IR drops are derived from Ohm's law, now that the currents have been determined.

Figure 12-4 is an example that has the batteries in an aiding direction. The two loops are shown clockwise, with an arbitrarily assigned polarity for R_3. The voltage loops are as follows:

(1) $\quad E_1 - I_1 R_1 - I_3 R_3 = 0$
$\quad\quad\quad -I_1 R_1 - I_3 R_3 = -E_1$

(2) $\quad E_2 + I_3 R_3 - I_2 R_2 = 0$
$\quad\quad\quad I_3 R_3 - I_2 R_2 = -E_2$

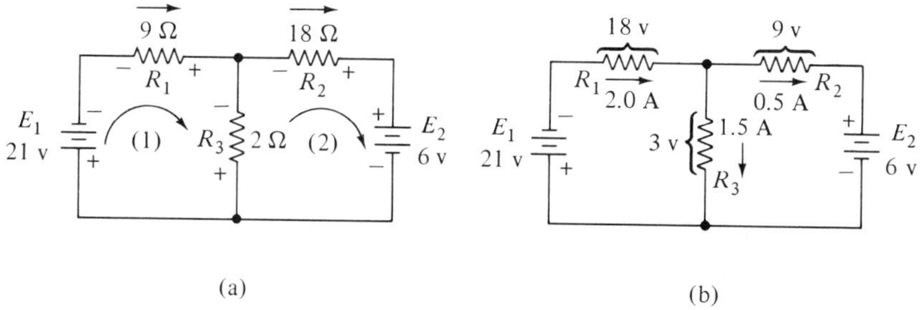

FIGURE 12-4 *Aiding Batteries for Loop Equations*

And the IR drops are converted,

$$I_3 = I_1 - I_2$$
$$I_1 R_1 = 9I_1$$
$$I_3 R_3 = (I_1 - I_2)2 = 2I_1 - 2I_2$$
$$I_2 R_2 = 18 I_2$$

Notice that I_3 is the difference between I_1 and I_2 because the currents through R_3 are opposite. The IR drops are now substituted back into the loops for E_1 and E_2.

(1) $\quad -I_1 R_1 - I_3 R_3 = -E_1$
$\quad\quad -9I_1 - 2I_1 + 2I_2 = -21$
$\quad\quad -11 I_1 + 2I_2 = -21$

(2) $\quad I_3 R_3 - I_2 R_2 = -E_2$
$\quad\quad 2I_1 - 2I_2 - 18 I_2 = -6$
$\quad\quad 2I_1 - 20 I_2 = -6$

The simultaneous equations are solved by multiplying loop (1) by 10. This makes it easier to eliminate the I_2 unknown.

$$\begin{array}{rl} (1) & -110I_1 + 20I_2 = -210 \\ (2) & 2I_1 - 20I_2 = -6 \\ \hline & -108I_1 \qquad = -216 \end{array}$$

$$I_1 = \frac{-216}{-108} = 2 \text{ A}$$

I_2 is determined by substituting $I_1 = 2$ A into either of the loop equations.

$$\begin{array}{rl} (1) & -11I_1 + 2I_2 = -21 \\ & -11(2) + 2I_2 = -21 \\ & -22 + 2I_2 = -21 \\ & 2I_2 = -21 + 22 \\ & I_2 = \frac{1}{2} = 0.5 \text{ A} \end{array}$$

Since I_3 was assumed to be $I_1 - I_2$, it is calculated:

$$I_3 = I_1 - I_2 = 2.0 \text{ A} - 0.5 \text{ A} = 1.5 \text{ A}$$

Both examples, Figures 12-3 and 12-4, resulted in currents that were in the same directions and of the same amounts. The source voltages and resistances were purposely selected to yield that result in order to illustrate the effects of careful selection. The examples contained only the basic elements for loop equations, which can be expanded to include additional components and/or source voltages.

12.2 Thévenin's Theorem

Thévenin's theorem is a valuable tool for reducing complex resistance networks to a single resistance and a single voltage source. It is especially useful for analyzing varying load conditions because complex equations, then, are not necessary for each variation. Figure 12-5 illustrates a box that may contain many voltage sources and resistances. Its equivalent is

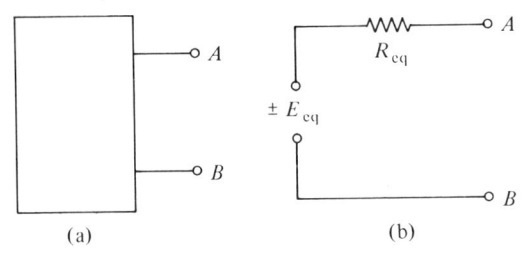

FIGURE 12-5 *Thévenin's Equivalent Circuit*

[12.2] Circuit Analysis 263

also shown. Notice that E_{eq} can be plus or minus, depending upon polarities within the "box."

Thévenin's theorem can be applied with the load connected or disconnected. The usual practice is to disconnect the load, reduce the circuit, then analyze the effect of loading. To implement the theorem, the open circuit voltage is determined by removing the load. Then the equivalent resistance is determined by shorting all voltage sources.

Figure 12–6(a) is a basic circuit with one voltage source, a two-resistor voltage divider, and a load that is connected across the lower half of the

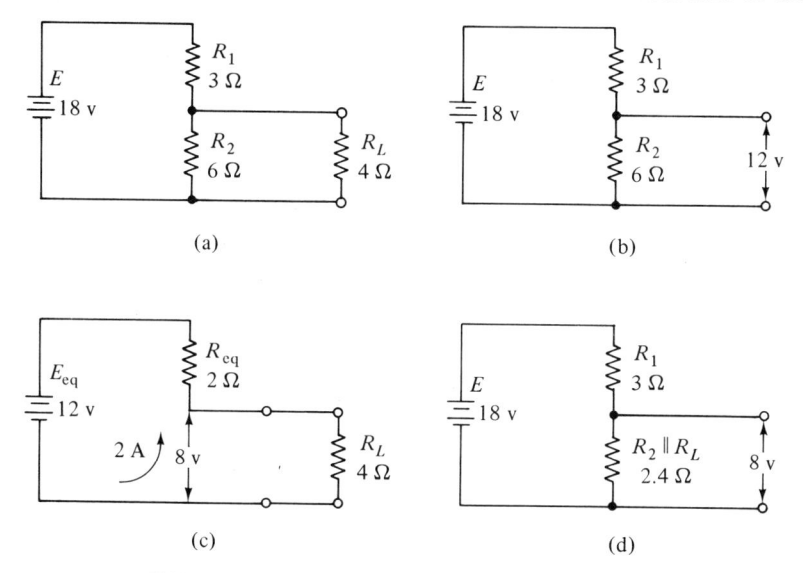

FIGURE 12–6 *Basic Application of Thévenin's Theorem*

divider. In (b) the load is disconnected, and the open-circuit voltage is determined.

$$E_{eq} = E \times \frac{R_2}{R_1 + R_2} = 18 \times \frac{6}{3 + 6} = 12 \text{ v}$$

When E is shorted, the output has a parallel combination of R_1 and R_2

$$R_{eq} = \frac{R_1 R_2}{R_1 + R_2} = \frac{3 \times 6}{3 + 6} = 2 \, \Omega$$

So Figure 12–6(c) illustrates a source of $E_{eq} = 12$ v, and $R_{eq} = 2$ ohms. The load is then connected across the open terminals, and the current through R_L and the voltage across R_L can be calculated.

$$I = \frac{E_{eq}}{R_{eq} + R_L} = \frac{12}{2 + 4} = 2 \text{ A}$$

$$E_{RL} = IR_L = 2 \times 4 = 8 \text{ v}$$

or
$$E_{RL} = E_{eq} \times \frac{R_L}{R_{eq} + R_L} = 12 \times \frac{4}{2+4} = 8 \text{ v}$$

The method can be verified by equating the circuit in (a) to the one in (d) in which the parallel combination of R_2 and R_L is 2.4 ohms. The voltage across the combination is

$$E_{RL} = E \times \frac{R_2 \| R_L}{R_2 \| R_L + R_1} = 18 \times \frac{2.4}{5.4} = 8 \text{ v}$$

The current through R_L in Figure 12-6(a) is

$$I = \frac{E_{RL}}{R_L} = \frac{8}{4} = 2 \text{ A}$$

which agrees with the current in (c).

Figure 12-7(a) is an example of two source voltages, each having a series resistor, that are connected across a load resistor (R_L). The circuit is redrawn in (b) with the negative terminal (E_1) at the top, and the positive terminal

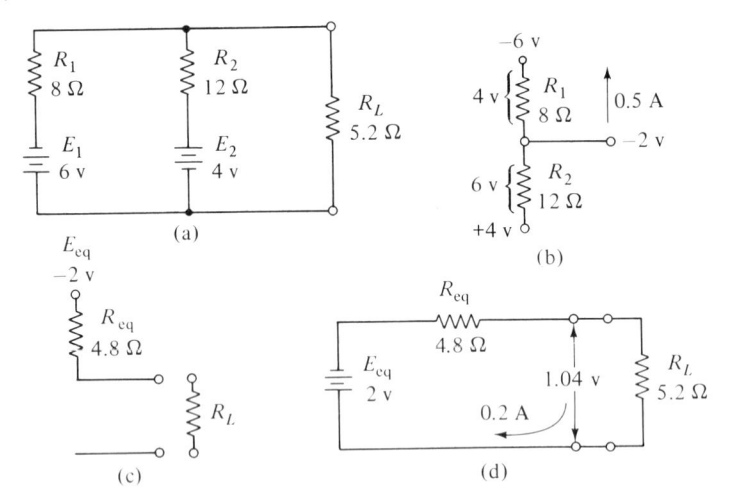

FIGURE 12-7 Thévenin's Application for Two Voltage Sources

(E_2) at the bottom. Since the bottom line of (a) is common to the sources and load, the drawing in (b) can assume the same common without drawing it. Removing the load, the open-circuit voltage E_{eq} of (b) is calculated through the following sequence.

$$I = \frac{E_2 - E_1}{R_1 + R_2} = \frac{4+6}{8+12} = 0.5 \text{ A}$$
$$E_{eq} = E_2 - IR_2 = 4 - (0.5 \times 12) = -2 \text{ v}$$

[12.2] Circuit Analysis 265

The equivalent resistance in (c) is found by shorting the two voltage sources in (b).

$$R_{eq} = \frac{R_1 R_2}{R_1 + R_2} = \frac{8 \times 12}{8 + 12} = 4.8 \, \Omega$$

The circuit can be redrawn as in (d) with the load in series with a single supply and resistor. The total circuit current is

$$I = \frac{E_{eq}}{R_{eq} + R_L} = \frac{2}{4.8 + 5.2} = 0.2 \, \text{A}$$

The IR drop for R_L is

$$E_{RL} = IR_2 = 0.2 \times 5.2 = 1.04 \, \text{v}$$

The principles of Thévenin's theorem can be applied to Figures 12-3 and 12-4 for ease of analysis. Figure 12-3 is redrawn in Figure 12-8(a)

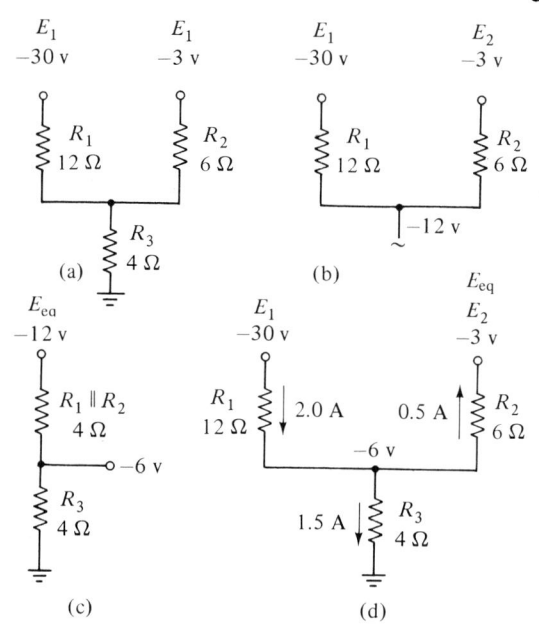

FIGURE 12-8 *Thévenin's Analysis of Figure 12-3*

using ground as common. In (b) the open-circuit voltage is determined as follows.

$$I = \frac{E_2 - E_1}{R_1 + R_2} = \frac{-3 + 30}{12 + 6} = -1.5 \, \text{A}$$
$$E_{R2} = IR_2 = -1.5 \times 6 = -9 \, \text{v}$$
$$E_{eq} = E_2 + E_{R2} = -3 - 9 = -12 \, \text{v}$$

R_1 and R_2 are paralleled in Figure 12–8(c). The series voltage divider equally divides E_{eq} so that the center voltage is -6 v. This value is shown at the junction (top of R_3) in circuit (d). Now that each junction has been assigned a voltage, the currents can be calculated directly. Each current flows toward the least negative terminal, which agrees with Figure 12-3(b).

The circuit in Figure 12-4 is analyzed in Figure 12-9. The open-circuit equivalent is shown in (a). E_{eq} is found as follows.

$$I = \frac{E_2 - E_1}{R_1 + R_2} = \frac{6 + 21}{9 + 18} = 1 \text{ A}$$

$$E_{eq} = E_2 - IR_2 = 6 - (1 \times 18) = -12 \text{ v}$$

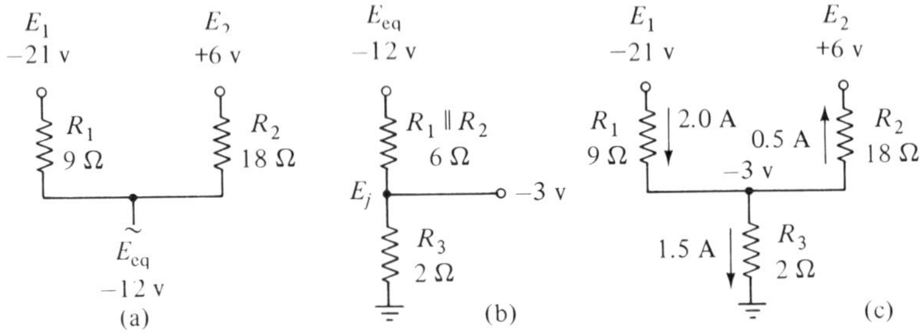

FIGURE 12-9 *Thévenin's Analysis of Figure 12–4*

The equivalent resistance shown in (b) is

$$R_{eq} = \frac{R_1 R_2}{R_1 + R_2} = \frac{9 \times 18}{9 + 18} = 6 \text{ }\Omega$$

And the junction voltage at the top of R_3 is

$$E_j = E_{eq} \times \frac{R_3}{R_1 \| R_2 + R_3} = -12 \times \frac{2}{6 + 2} = -3 \text{ v}$$

Diagram (c) shows the complete circuit with the junction voltages and currents. The direction and magnitude of the currents are the same as those in Figure 12-4.

12.3 Norton's Theorem

Norton's theorem is a method of reducing complex resistance networks to a single resistance and a single current source, so that regardless of the number of sources, the sources can be equated to one current source. The basic circuit is illustrated in Figure 12–10. Notice that the basic two-terminal box in (a) is the same as the Thévenin's two-terminal box in Figure 12–5(a). The equivalent Norton's circuit is shown in Figure 12–10(b), which consists of a single current source that is shunted by a single resistor. The arrow in the current source I indicates the direction of electron current flow, which can be in either direction as required.

Circuit Analysis

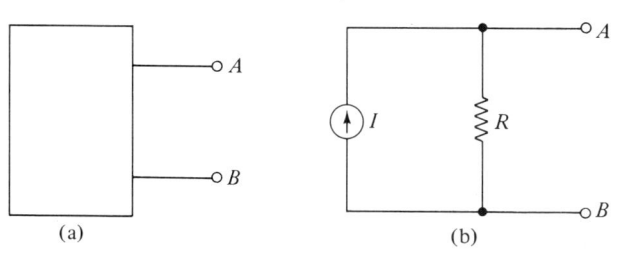

FIGURE 12-10 *Norton's Equivalent Circuit*

The theorem is illustrated in Figure 12-11. The basic circuit is in (a). The load is removed, and the output is shorted as shown in (b). The equivalent current is

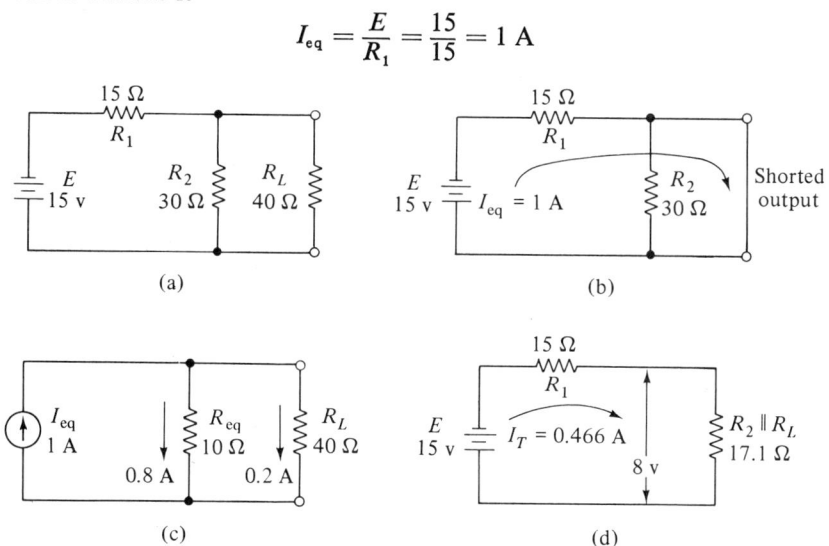

$$I_{eq} = \frac{E}{R_1} = \frac{15}{15} = 1 \text{ A}$$

FIGURE 12-11 *Basic Application of Norton's Theorem*

The equivalent resistance is calculated in the same way as it is in Thévenin's theorem. The source is shorted, and the combination of resistors (excluding R_L) is reduced to a single resistor, which is the resistance when looking back into the output.

$$R_{eq} = \frac{R_1 R_2}{R_1 + R_2} = \frac{15 \times 30}{15 + 30} = 10 \text{ } \Omega$$

The equivalent Norton circuit is shown in (c) with a current source of 1 A and a shunt resistance of 10 ohms. R_L is connected across the shunt resistor. The current through the load is calculated:

$$I_{RL} = I_{eq} \frac{R_{eq}}{R_{eq} + R_L} = 1 \times \frac{10}{10 + 40} = 0.2 \text{ A}$$

The *IR* drop across R_L is
$$E_{RL} = I_{RL}R_L = 0.2 \times 40 = 8 \text{ v}$$
As a cross-check, the current through R_{eq} is
$$I = I_{eq}\frac{R_L}{R_L + R_{eq}} = 1 \times \frac{40}{40 + 10} = 0.8 \text{ A}$$

The circuit in Figure 12–11(d) is an equivalent of the one in (a). Its purpose is to show the validity of Norton's theorem. The parallel resistance of R_2 and R_L is 17.1 ohms, and the total current in the circuit is
$$I_T = \frac{E}{R_1 + R_2 \| R_L} = \frac{15}{32.1} = 0.466 \text{ A}$$
The voltage drop across R_L is
$$E_{RL} = I_T \times R_2 \| R_L = 0.466 \times 17.1 = 8 \text{ v}$$
Returning to circuit (a), the current through R_L is
$$I_{RL} = \frac{E_{RL}}{R_L} = \frac{8}{40} = 0.2 \text{ A}$$

This agrees with Norton's solution in circuit (c).

A second example of Norton's theorem involves two voltage sources. The circuit of Figure 12–7, previously solved with Thévenin's theorem, is repeated in Figure 12–12(a). The load is removed, and its terminals are

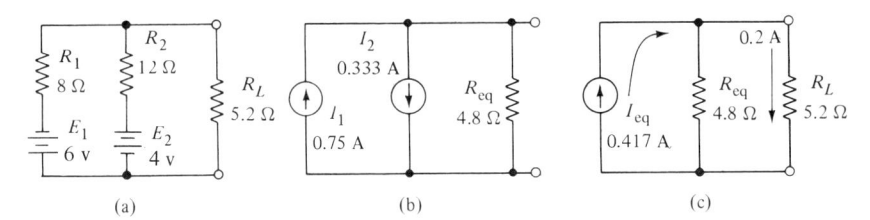

FIGURE 12–12 *Norton's Application for Two Voltage Sources*

shorted. Each of the two currents in (b) flows in a direction that leaves the negative terminal of its source battery. The calculations are
$$I_1 = \frac{E_1}{R_1} = \frac{6}{8} = 0.75 \text{ A}$$
$$I_2 = \frac{E_2}{R_2} = \frac{4}{12} = 0.333 \text{ A}$$

The equivalent resistance, as seen at the open-circuited output terminals with both batteries shorted, is calculated in the same way as it was in Thévenin's theorem.
$$R_{eq} = \frac{R_1 R_2}{R_1 + R_2} = 4.8 \text{ }\Omega$$

Now the two source currents can be reduced to one. Since the source currents oppose each other, the equivalent current is simply the difference and it flows in a direction favoring the higher of the two (I_1).

$$I_{eq} = I_1 - I_2 = 0.75 - 0.333 = 0.417 \text{ A}$$

The final equivalent circuit is shown in Figure 12-12(c). The load current is calculated from the ratio of the resistances.

$$I_{RL} = I_{eq} \frac{R_{eq}}{R_{eq} + R_L} = 0.417 \times \frac{4.8}{4.8 + 5.2} = 0.2 \text{ A}$$

Note that the load current of 0.2 A agrees with the solution to Figure 12-7.

Figure 12-13 shows a comparison of the Thévenin and Norton equivalent circuits. Conversions from one to the other are completed in the following manner.

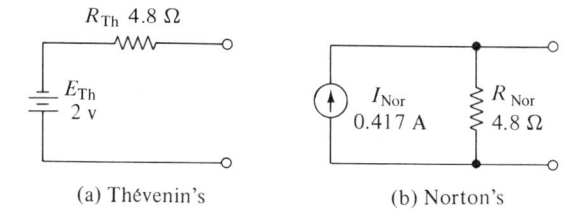

$$R_{TH} = R_{NOR}$$
$$E_{TH} = I_{NOR} R_{NOR}$$
$$I_{NOR} = \frac{E_{TH}}{R_{TH}}$$

(a) Thévenin's   (b) Norton's

FIGURE 12-13 *Comparison of Thévenin's and Norton's Equivalent Circuits*

The voltage and current of Figure 12-13 are converted as follows.

$$E_{TH} = I_{NOR} R_{NOR} = 0.417 \times 4.8 = 2 \text{ v}$$
$$I_{NOR} = \frac{E_{TH}}{R_{TH}} = \frac{2}{4.8} = 0.417 \text{ A}$$

12.4 Superposition Theorem

The basic theorem states that the current or voltage in any part of a circuit is the algebraic sum of the individual currents or voltages. For a network that contains more than a single current or voltage source, the current anywhere in the network is the algebraic sum of the currents generated by the individual sources. In other words, the currents for each source are determined while shorting all other sources. Then the currents are summed algebraically. This analogy is valid only if the circuit is completely linear, and current is free to flow in any direction. To be linear, the values of the components must not change with varying current or voltage. And for current to flow in either direction, there cannot be any rectification in the circuit.

An example of the basic superposition theorem is described for Figure 12–14, which is a revision of Figure 12–7. The problem is to determine the IR drop and current for R_L. In Figure 12–14(a), E_2 (for Figure 12–17) is shorted, and the currents and voltage are calculated.

$$I_{T1} = \frac{E_1}{R_1 + R_2 \| R_L} = \frac{-6}{8 + 3.62} = -0.516 \text{ A}$$

$$I_{RL} = I_{T1} \frac{R_2}{R_2 + R_L} = -0.516 \times \frac{12}{12 + 5.2} = -0.36 \text{ A}$$

$$E_{RL} = I_{RL} R_L = -0.36 \text{ A} \times 5.2 = -1.87 \text{ v}$$

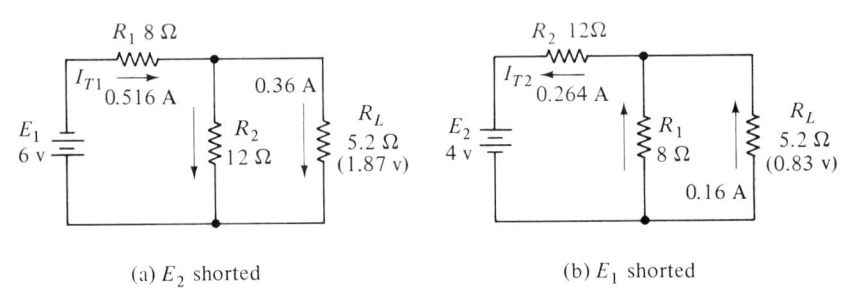

(a) E_2 shorted

(b) E_1 shorted

FIGURE 12–14 *Revised Drawing of Figure 12–7 for Superposition Analysis*

In Figure 12–14(b), E_1 (for Figure 12–17) is shorted, and the currents and voltage are calculated.

$$I_{T2} = \frac{E_2}{R_2 + R_1 \| R_L} = \frac{4}{12 + 3.16} = 0.264 \text{ A}$$

$$I_{RL} = I_{T2} \frac{R_1}{R_1 + R_L} = 0.264 \times \frac{8}{8 + 5.2} = 0.16 \text{ A}$$

$$E_{RL} = I_{RL} R_L = 0.16 \times 5.2 = 0.83 \text{ v}$$

The net current (I_N) for R_L is the algebraic sum of I_{RL} for (a) and I_{RL} for (b).

$$I_N = -0.36 \text{ A} + 0.16 \text{ A} = -0.2 \text{ A}$$

This answer is the same as the one found in the current analysis for Figure 12–12(c). The net current goes down through R_L in both cases.

In Figure 12–14, the net voltage (E_N) for R_L is the algebraic sum of E_{RL} for (a) and E_{RL} for (b).

$$E_N = -1.87 \text{ v} + 0.83 \text{ v} = -1.04 \text{ v}$$

This answer agrees with the voltage analysis for Figure 12–7(d). The negative side of the voltage is at the top of R_L in both Figures 12–14 and

[12.4]   Circuit Analysis

12-7. In Figure 12-14 the higher voltage is in circuit (a), which means that the top of R_L is negative when resistance values are approximately the same.

The superposition theorem can also be applied to ac/dc circuits (Chapter 7). When an ac signal is superimposed on a dc level, the total amplitude at any specific time is the algebraic sum of the two. In Figure 12-15 a sine wave is superimposed on a dc level of 3 v. At a time of $t = 3$ sec, the 3-v level begins to shift positively, so that at $t = 4$ sec, the level is +5 v. At $t = 6$ sec, the level has shifted negatively and is now +1 v. The level resumes its dc value of 3 v at $t = 7$ sec.

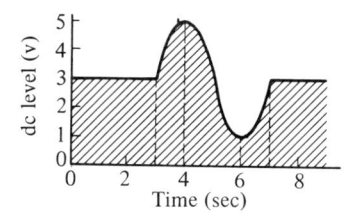

FIGURE 12-15 *Superposition of ac on dc*

Figure 12-16 illustrates the superposition of two ac signals. The signals in (a) (e_1 and e_2), are out-of-phase by 180°. The two signals have a canceling effect because the net signal (e_n) is less than the higher of the two original signals. For example, at the 90° point

$$e_n(90°) = e_1 + e_2 = 30 \text{ mv} + (-10 \text{ mv})$$
$$= 20 \text{ mv (peak)}$$

20 mv is the algebraic sum of the two signals. At the 270° point,

$$e_n(270°) = e_1 + e_2 = -30 \text{ mv} + 10 \text{ mv}$$
$$= -20 \text{ mv (peak)}$$

In Figure 12-16(b) the signals are in-phase. The net signals at 90° and 270° are

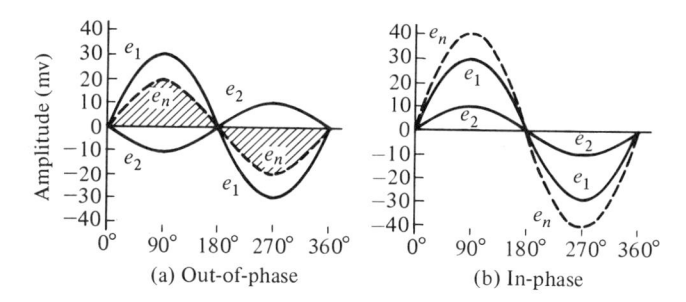

(a) Out-of-phase   (b) In-phase

FIGURE 12-16 *Superposition of ac Signals*

$$e_n(90°) = e_1 + e_2 = 30 \text{ mv} + 10 \text{ mv} = 40 \text{ mv (peak)}$$
$$e_n(270°) = e_1 + e_2 = -30 \text{ v} + (-10 \text{ mv}) = -40 \text{ mv (peak)}$$

The above examples were simplified for two signals of the same frequency. However, the superposition theorem can be applied to signals of any amplitude, any frequency, any phase, and at any time in their cycle using the formulas developed in Chapter 7.

12.5 *j* Factors

The *j* factor is a form of complex numbers which provides a mathematical method of solving complex ac circuits that contain resistance and reactance. In particular, the method is essential when both resistance and reactance are present in branching circuits. Chapter 10 described vector techniques for the more common circuits and introduced several formulas for solving specific arrangements of resistance and reactance. The application of both *j* factors and polar form equivalents (described in the next section) permits the solving of any arrangement of resistance and reactance.

In respect to resistance and reactance, resistance is represented individually by a number with a zero phase angle since voltage and current are in-phase. Reactances are represented individually by a number and a phase angle since inductive and capacitive reactance are 180° out-of-phase with each other. In Figure 12-17 the resistive component is on the horizontal axis, and it can be a positive or negative number. A positive number indicates zero phase shift, and a negative number indicates a 180° phase shift.

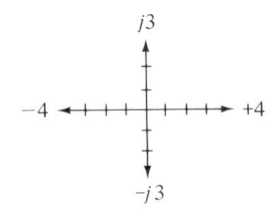

FIGURE 12-17 *R and j Numbers Versus Phase*

The vertical axis in Figure 12-17 represents reactive components, designated *j* factors. For instance, a positive reactance of 3 is *j*3, which also indicates a +90° phase shift from a positive resistive quantity which is +4 in the figure. A negative reactance of 3 is −*j*3, which indicates a −90° phase shift from a positive resistive quantity.

Further explanation of the j factor is found in Figure 12–18. At 0° the exponent for the j factor is 0, or j^0. Any quantity to the zero power is equal to one, so at 0° the factor is $+1$. At 90° the exponent is one, so $j^1 = +j$. $j^2 = -1$ at 180°, $j^3 = -j$ at 270°. The factors are also related: $j \times j = j^2$, and $j \times j^2 = j^3$.

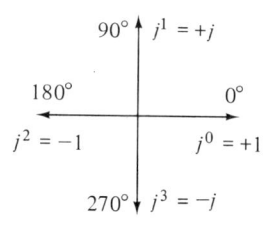

FIGURE 12–18 *Phase Relationship of j Numbers*

The application of j numbers to impedance vectors is shown in Figure 12–19. Resistance is depicted as a real number on the horizontal vectors. In (a) an X_L of three units is represented as $j3$, and it is vertical and positive to indicate a $+90°$ phase angle. Reactance is viewed in terms of the voltage across the inductor or capacitor, so that with an inductor the voltage leads (positive) the current by 90°, whereas with a capacitor, the voltage lags (negative) the current by 90°.

X_L E leads I by 90°
X_C E lags I by 90°

The vectors in Figure 12–19(a) are combined as $4 + j3$. The formula for impedance in *general form* is

$$Z = R \pm jX \tag{12-1}$$

j is impedance, R is resistance, and X is reactance (all of these are in ohms). In (b) the j factor is negative, or $-j3$. Applying the factor to formula (12-1) yields $Z = 4 - j3$. This form is known as the **rectangular form**.

The vectors in Figure 12–19 illustrate the series circuits in Figure 12–20. The total impedance in both cases fits the general impedance formula. For (a), $Z_T = 4 + j3$, and for (b), $Z_T = 4 - j3$.

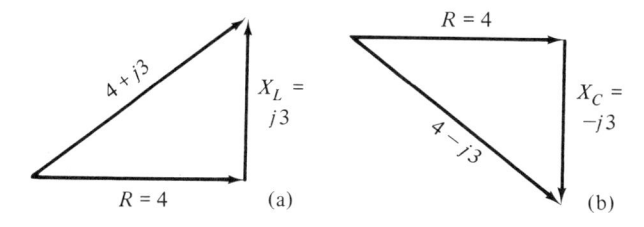

FIGURE 12–19 *Rectangular Form for Impedance Vectors*

274　　　　　　　　　　　Circuit Analysis　　　　　　　　　　　[12.5]

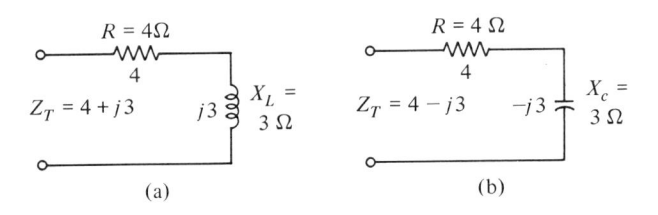

FIGURE 12–20 *Rectangular Form for Series Circuits*

The branch currents in parallel circuits are shown in Figure 12–21. The inductive current in (a) is a negative factor, and the capacitive current for (b) is a positive factor. The vector diagrams for Figure 12–21 are drawn in Figure 12–22. The vertical vectors follow the rule of negative for i_L and positive for i_C. The hypotenuse of each represents total current i_T.

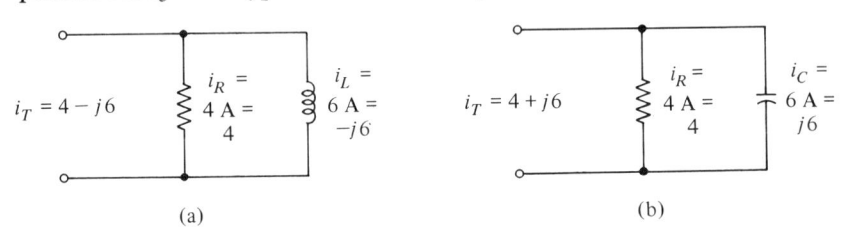

FIGURE 12–21 *Rectangular Form for Parallel Circuits*

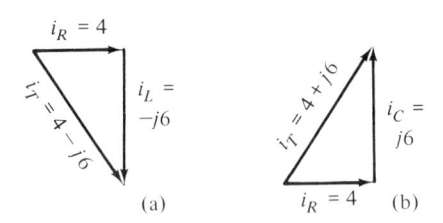

FIGURE 12–22 *Rectangular Form for Current Vectors*

Examples of the rectangular form for series components are:

$$R = 10 \quad X_L = 15 \quad Z_T = 10 + j15$$
$$R = 3\,k \quad X_L = 2\,k \quad Z_T = 3\,k + j2\,k$$
$$R = 45 \quad X_C = 80 \quad Z_T = 45 - j80$$
$$R = 0 \quad X_C = 12 \quad Z_T = 0 - j12$$
$$R = 2 \quad X_C = 0 \quad Z_T = 2 - j0$$

Examples for parallel currents are:

$$i_R = 50 \quad i_L = 40 \quad i_T = 50 - j40$$
$$i_R = 0.2 \quad i_L = 0.15 \quad i_T = 0.2 - j0.15$$
$$i_R = 0 \quad i_L = 4 \quad i_T = 0 - j4$$
$$i_R = 8 \quad i_C = 15 \quad i_T = 8 + j15$$
$$i_R = 22 \quad i_C = 0 \quad i_T = 22 + j0$$

Circuit Analysis

The rectangular form keeps the zero quantities so the circuit can be better described and so the general form for calculations involving multiple impedances or currents can be maintained. General impedance and current formulas are

$$Z_T = Z_1 + Z_2 + \cdots + Z_n \quad \text{(series)} \tag{12-2}$$

$$Z_T = \frac{1}{\frac{1}{Z_1} + \frac{1}{Z_2} + \cdots + \frac{1}{Z_n}} \quad \text{(parallel)} \tag{12-3}$$

or

$$\frac{1}{Z_T} = \frac{1}{Z_1} + \frac{1}{Z_2} + \cdots + \frac{1}{Z_n}$$

$$Z_T = \frac{Z_1 Z_2}{Z_1 + Z_2} \quad \text{(two in parallel)} \tag{12-4}$$

$$Z_T = \frac{Z_1}{2} = \frac{Z_2}{2} \quad \text{(two equal in parallel)} \tag{12-5}$$

$$i_T = i_1 + i_2 + \cdots + i_n \quad \text{(parallel)}$$

$$i_T = i_1 = i_2 = \cdots = i_n \quad \text{(series)}$$

The total magnitude of a complex number follows the same method as that used for vectors. Impedance and current are calculated:

$$Z_T = \sqrt{R^2 \times X^2} \tag{12-6}$$

$$i_T = \sqrt{i_R^2 + i_X^2} \tag{12-7}$$

And the phase angles are calculated:

$$\tan \theta = \frac{j}{R} = \frac{X}{R} \quad \text{(impedance)} \tag{12-8}$$

$$\tan \theta = \frac{j}{i_R} = \frac{i_X}{i_R} \quad \text{(current)} \tag{12-9}$$

Examples of the magnitude and phase formulas follow.

Example 12-1 For Figures 12-19 and 12-20:

(a) $$Z_T = 4 + j3 = \sqrt{4^2 + 3^2} = 5 \, \Omega$$

$$\tan \theta = \frac{j3}{4} = 0.75$$

$$\theta = 36.8°$$

(b) $$Z_T = 4 - j3 = \sqrt{4^2 - 3^2} = 5 \, \Omega$$

$$\tan \theta = \frac{-j3}{4} = -0.75$$

$$\theta = -36.8°$$

Example 12-2 For Figures 12-21 and 12-22:

(a) $i_T = 4 - j6 = \sqrt{4^2 - 6^2} = 7.2$ A

$\tan \theta = \dfrac{-j6}{4} = -1.5$

$\theta = -56.3°$

(b) $i_T = 4 + j6 = \sqrt{4^2 + 6^2} = 7.2$ A

$\tan \theta = \dfrac{j6}{4} = 1.5$

$\theta = 56.3°$

Other examples are:

(a) $Z_T = 5\text{ k} + j3\text{ k} = \sqrt{5\text{ k}^2 + 3\text{ k}^2} = 5.83\text{ k}\Omega$

$\tan \theta = \dfrac{j3\text{ k}}{5\text{ k}} = 0.6$

$\theta = 31°$

(b) $Z_T = 30 - j40 = \sqrt{30^2 - 40^2} = 50\ \Omega$

$\tan \theta = \dfrac{-j40}{30} = -1.33$

$\theta = -53.2°$

(c) $i_T = 2 + j1.5 = \sqrt{2^2 + 1.5^2} = 2.5$ A

$\tan \theta = \dfrac{j1.5}{2} = 0.75$

$\theta = 36.8°$

(d) $i_T = 8 - j14 = \sqrt{8^2 - 14^2} = 16.1$ A

$\tan \theta = \dfrac{-j14}{8} = -1.75$

$\theta = -60.3°$

Combining Numbers with j Factors. Special rules apply to complex numbers when it comes to arithmetic operations. The various terms cannot be handled directly due to the phase shift; the R and j terms are 90° out-of-phase.

When adding or subtracting, the R and j terms must be handled separately.

$(3 + j10) + (8 + j3) = 11 + j13$
$(4 - j3) + (4 + j7) = 8 + j4$
$(9 + j2) - (3 + j6) = 6 - j4$
$(5 - j4) - (3 - j18) = 2 + j14$

When multipling or dividing an R term by an R term, the numbers are handled in the same way as in normal arithmetic operations since j terms are not involved.

$$4 \times 5 = 20 \qquad 30 \div 6 = 5$$

To multiply or divide a j term by an R term it is necessary to multiply or divide the numbers directly.

$$2 \times j6 = j12 \qquad j20 \div 2 = j10$$
$$5 \times -j4 = -j20 \qquad -j40 \div 8 = -j5$$
$$-3 \times j10 = -j30 \qquad j15 \div -5 = -j3$$
$$-6 \times -j9 = j54 \qquad -j32 \div -4 = j8$$

When a j term is either multiplied or divided by a j term, the answer is in the form of an R term. For example, the multiplication of $j \times j$ shifts the term 90° to the R axis (see Figure 12-18).

$$j5 \times j4 = j^2 20 = (-1)20 = -20$$
$$-j10 \times j3 = -j^2 30 = -(-1)30 = 30$$
$$-j6 \times -j2 = j^2 12 = (-1)12 = -12$$

In division, the j terms cancel.

$$j4 \div j2 = 2 \qquad j30 \div -j10 = -3$$

When multipling or dividing a complex number by a complex number, the rules are the same as for algebra.

$$(4 + j3)(7 + j9) = 28 + j36 + j21 + j^2 27$$
$$= 28 + j57 - 27 = 1 + j57$$
$$(2 + j6)(3 - j2) = 6 - j4 + j18 - j^2 12$$
$$= 6 + j14 + 12 = 18 + j14$$
$$\frac{40 + j30}{4 + j2} = \frac{40 + j30}{4 + j2} \times \frac{4 - j2}{4 - j2}$$
$$= \frac{160 - j80 + j120 - j^2 60}{16 - j8 + j8 - j^2 4}$$
$$= \frac{160 + j40 + 60}{16 + 4} = \frac{220 + j40}{20}$$
$$= 11 + j2$$

The denominator is converted to an R term by multiplying the numerator and denominator by the conjugate of the denominator. In the above case, the conjugate of $4 + j2$ is $4 - j2$. In the example below, the conjugate is $4 + j3$.

$$\frac{20+j25}{4-j3} = \frac{20+j25}{4-j3} \times \frac{4+j3}{4+j3}$$
$$= \frac{80+j30+j100+j^275}{16+j12-j12-j^29}$$
$$= \frac{80+j130-75}{16+9} = \frac{5+j130}{25}$$
$$= 0.2+j5.2$$

12.6 Complex Numbers in Polar Form

The polar form is actually a statement of the vector addition and phase angle of a j factor. For example, the rectangular form for Figure 12-19 is $4+j3$. In polar form it is $5\ \underline{/\ 36.8°}$. The 5 represents impedance, and $\underline{/\ }$ represents the phase angle. For Figures 12-19 and 12-20 (reference Example 12-1):

(a) $4+j3 = 5\ \underline{/\ 36.8°}$
(b) $4-j3 = 5\ \underline{/\ -36.8°}$

For Figures 12-21 and 12-22 (reference Example 12-2):

(a) $4-j6 = 7.2\ \underline{/\ -56.3°}$
(b) $4+j6 = 7.2\ \underline{/\ 56.3°}$

Other examples are:

$$5\ \text{k}+j3\ \text{k} = 5.83\ \text{k}\ \underline{/\ 31°}$$
$$2+j1.5 = 2.5\ \underline{/\ 36.8°}$$
$$30-j40 = 50\ \underline{/\ -53.2°}$$
$$8-j14 = 16.1\ \underline{/\ -60.3°}$$

The polar form, then, shows the magnitude of the vector sum and the phase angle of the vectors. If one term of a vector is zero, the polar form carries the other term.

$$0+j4 = 4\ \underline{/\ 90°}$$
$$4+j0 = 4\ \underline{/\ 0°}$$

When an R term is zero, the j term is displaced 90°, and when the j term is zero, the phase is the same as for the R term, or zero.

Complex numbers in polar form must be converted to rectangular form (j factors) for adding or subtracting. But for multiplying or dividing, the operation is considerably easier in polar form.

Multiplying. The magnitudes are multiplied directly, and the angles are added algebraically.

$$10 \,\underline{/35°} \times 4 \,\underline{/20°} = 40 \,\underline{/55°}$$
$$-40 \,\underline{/20°} \times 3 \,\underline{/55°} = -120 \,\underline{/75°}$$
$$8 \,\underline{/-45°} \times 9 \,\underline{/80°} = 72 \,\underline{/35°}$$
$$12 \,\underline{/35°} \times 2 \,\underline{/-75°} = 24 \,\underline{/-40°}$$
$$3 \,\underline{/-10°} \times 6 \,\underline{/-42°} = 18 \,\underline{/-52°}$$

When multiplying by an *R* (or plain) number, only the magnitudes are multiplied.

$$5 \times 8 \,\underline{/32°} = 40 \,\underline{/32°}$$
$$-3 \times 10 \,\underline{/-40°} = -30 \,\underline{/-40°}$$

Dividing Complex Numbers By Complex Numbers. The magnitudes are divided directly, and the angles are subtracted algebraically.

$$40 \,\underline{/70°} \div 5 \,\underline{/40°} = 8 \,\underline{/30°}$$
$$24 \,\underline{/20°} \div 8 \,\underline{/50°} = 3 \,\underline{/-30°}$$
$$35 \,\underline{/-25°} \div 7 \,\underline{/35°} = 5 \,\underline{/-60°}$$
$$32 \,\underline{/-30°} \div 8 \,\underline{/-25°} = 4 \,\underline{/-5°}$$

Dividing Complex Numbers By R Numbers. To do this, convert the *R* numbers to complex form, then divide the magnitudes and subtract the angles.

$$50 \,\underline{/32°} \div 5 = 50 \,\underline{/32°} \div 5 \,\underline{/0°} = 10 \,\underline{/32°}$$
$$20 \,\underline{/-45°} \div 4 = 20 \,\underline{/-45°} \div 4 \,\underline{/0°} = 5 \,\underline{/-45°}$$

Dividing R Numbers By Complex Numbers. The same procedure used to divide complex numbers by *R* numbers is used.

$$30 \div 5 \,\underline{/20°} = 30 \,\underline{/0°} \div 5 \,\underline{/20°} = 6 \,\underline{/-20°}$$
$$15 \div 3 \,\underline{/-65°} = 15 \,\underline{/0°} \div 3 \,\underline{/-65°} = 5 \,\underline{/65°}$$

In the last section, complex numbers were explained in terms of *j* factors. Both addition and subtraction were accomplished by handling the terms separately. Multiplication and division involved considerable manipulation of algebraic terms. Therefore, it would be simplest to add and subtract in rectangular form and to multiply and divide in polar form.

Converting from Polar Form to Rectangular Form. This involves trigonometric relations of the sine/cosine functions. The conversion for impedance is illustrated in Figure 12–23. The functions are

$$R \text{ term} = Z \cos \theta \qquad (12\text{–}10)$$
$$j \text{ term} = Z \sin \theta \qquad (12\text{–}11)$$

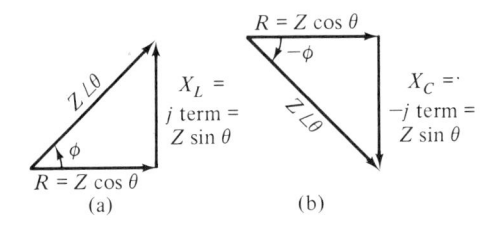

FIGURE 12-23 Polar Form for Impedance Vectors

For example, consider the polar form of impedance for R and X_L, $72\ \underline{/\ 56.3°}$.

R term $= Z \cos \theta = 72 \cos 56.3° = 72 \times 0.555 = 4$
j term $= Z \sin \theta = 72 \sin 56.3° = 72 \times 0.833 = 6$
$72\ \underline{/\ 56.3°} = 4 + j6$

Another example is for R and X_C, $20\ \underline{/-30°}$.

R term $= Z \cos \theta = 20 \cos -30° = 20 \times 0.866 = 17.3$
$-j$ term $= Z \sin \theta = 20 \sin -30° = 20 \times 0.5 = 10$

The rectangular form is

$$20\ \underline{/-30°} = 17.3 - j10$$

The functions for converting current from polar to rectangular form are similar to formulas (12–10) and (12–11).

$$R \text{ term} = i \cos \theta \qquad (12\text{–}12)$$
$$j \text{ term} = i \sin \theta \qquad (12\text{–}13)$$

The vectors in Figure 12–24 illustrate the conversion method. Notice that the examples for impedance can also be used for current. The only difference is that X_L is $+j = +90°$ where i_L is $-j = -90°$, and that X_C is $-j = -90°$ where i_C is $+j = +90°$.

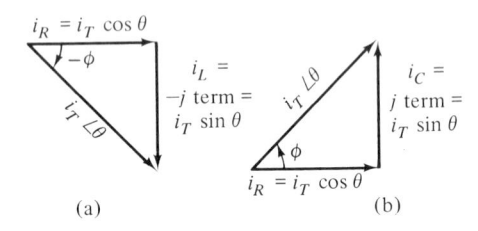

FIGURE 12-24 Polar Form for Current Vectors

12.7 Applications of Complex Numbers

To use complex numbers in solving ac circuits, equations must be set up to fit the elements in the circuit. Sometimes it becomes necessary to re-

Circuit Analysis

draw the circuit to fit the appropriate equation. Equations are initially fitted to the rectangular form. After the equations are assembled, they are solved with the methods described in the last two sections.

Example 12–3 Using the series circuit in Figure 12–25 (also see Figure 10–20), determine the total impedance of the circuit.

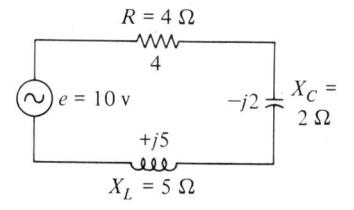

FIGURE 12–25 *Series Circuit for Example 12–3*

Solution The resistance can be included with either the capacitor or the inductor since it is a series circuit. The total impedance can be solved in rectangular form and solved with formulas (12–2), (12–6) and (12–8). For R and X_C, $Z = 4 - j2$. For $R = 0$ and X_L, $Z = 0 + j5$.

$$Z_T = Z_1 + Z_2 = 4 - j2 + 0 + j5 \quad (12\text{–}2)$$
$$= 4 + j3$$

$$Z_T = \sqrt{R^2 \pm X^2} = \sqrt{4^2 \pm 3^2} = 5\,\Omega \quad (12\text{–}6)$$

$$\tan \theta = \frac{X}{R} = \frac{3}{4} = 0.75 \quad (12\text{–}8)$$

$$\theta = 36.8°$$

Example 12–4 Refer to the parallel circuit in Figure 12–26(a) (also see Figure 10–21). Using complex numbers, determine the total impedance, the total current, and the equivalent series circuit.

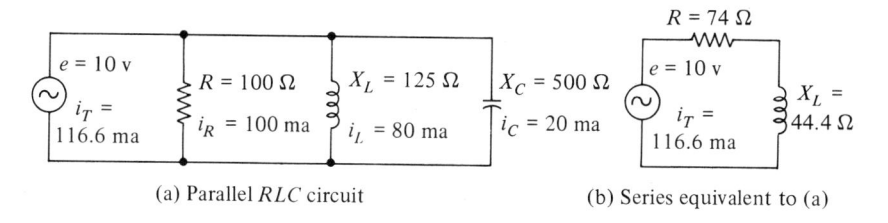

(a) Parallel RLC circuit

(b) Series equivalent to (a)

FIGURE 12–26 *Parallel Circuit for Example 12–4*

Solution The circuit is set up in rectangular form and fitted to formula (12–3).

$$R = 100 + j0$$
$$X_L = 0 + j125$$
$$X_C = 0 - j500$$

$$\frac{1}{Z_T} = \frac{1}{100 + j0} + \frac{1}{0 + j125} + \frac{1}{0 - j500} \quad (12\text{–}3)$$

For division, (12–3) is converted to polar form.

$$\frac{1}{Z_T} = \frac{1\,\underline{/0°}}{100\,\underline{/0°}} + \frac{1\,\underline{/0°}}{125\,\underline{/90°}} + \frac{1\,\underline{/0°}}{500\,\underline{/-90°}}$$

$$\frac{1}{Z_T} = 0.010\,\underline{/0°} + 0.008\,\underline{/-90°} + 0.002\,\underline{/+90°}$$

For adding, the numbers must be reverted to polar form.

$$\frac{1}{Z_T} = 0.010 + j0 + 0 - j0.008 + 0 + j0.002$$
$$= 0.010 - j0.006$$

$$Z_T = \frac{1}{0.010 - j0.006}$$

Once more the rectangular form is converted to polar for division.

$$Z_T = \frac{1\,\underline{/0°}}{0.01166\,\underline{/-31°}} = 86\,\underline{/31°}$$

To determine total current, the rectangular form is used.

$$i_T = 100 + j0 + 0 - j80 + 0 + j20$$
$$= 100 - j60$$

Converting to polar form provides the current and phase angle.

$$i_T = \sqrt{i_R^2 + i_X^2}$$
$$= \sqrt{(100\text{ ma})^2 + (60\text{ ma})^2} \quad (12\text{–}7)$$
$$= 116.6\text{ ma}$$
$$i_T = 0.1166\,\underline{/-31°} \text{ in polar form}$$

$$\tan\theta = \frac{i_X}{i_R} = \frac{60\text{ ma}}{100\text{ ma}} = 0.6 \quad (12\text{–}9)$$
$$\theta = \underline{/-31°}$$

The equivalent series circuit is determined by converting Z_T from polar to rectangular form in accordance with formulas (12–10) and (12–11).

[12.7]   Circuit Analysis   283

$$Z_T = 86 \:\underline{/31°}$$

R term $= Z_T \cos \theta = 86 \times 0.86 = 74$ (12-10)
j term $= Z_T \sin \theta = 86 \times 0.515 = 44.4$ (12-11)

$$Z_T = 74 + j44.4$$

The R term of 74 indicates a resistance of 74 ohms. The $+j$ term of 44.4 indicates an inductive reactance of 44.4 ohms. If the j term had been negative, it would have indicated capacitive reactance. The equivalent series circuit is shown in Figure 12-26(b).

Example 12-5   Refer to the series/parallel circuit in Figure 12-27 [also see Figure 10-23(a)]. Determine the total impedance using impedance methods, and verify the Z_T using an assumed e_T.

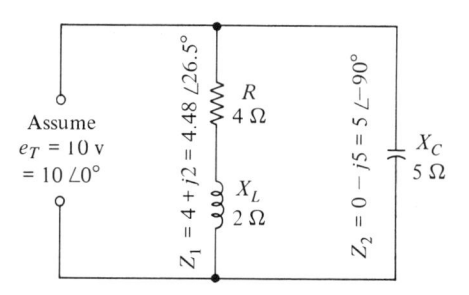

FIGURE 12-27   *Series/Parallel Circuit for Example 12-5*

Solution   The branches in rectangular form are:

$$Z_1 = 4 + j2 \qquad Z_2 = 0 - j5$$

With only two branches, formula (12-4) can be applied.

$$Z_T = \frac{Z_1 Z_2}{Z_1 + Z_2} = \frac{(4 + j2)(0 - j5)}{4 + j2 + 0 - j5}$$

For multiplication the numerator is converted to polar form while the denominator is left in rectangular form.

$$Z_T = \frac{(4.48 \:\underline{/26.5°})(5 \:\underline{/-90°})}{4 + j2 + 0 - j5} = \frac{22.4 \:\underline{/-63.5°}}{4 - j3}$$

And for division the denominator must be converted to polar form; the numerator is already in polar form.

$$Z_T = \frac{22.4 \:\underline{/-63.5°}}{5 \:\underline{/-36.8°}} = 4.48 \:\underline{/-26.7°}$$

The impedance is 4.48 ohms, and the phase angle is $-26.7°$.

Assume that $e_T = 10$ v. The branch currents are determined by setting e_T to polar form $10\ /0°$, and using the polar forms of Z_1 and Z_2.

$$i_1 = \frac{10\ /0°}{4.48\ /26.5°} = 2.23\ /-26.5°$$

$$i_2 = \frac{10\ /0°}{5\ /-90°} = 2\ /90°$$

Since i_T is the vector sum of i_1 and i_2, the currents must be converted to rectangular form.

For i_1 R term $= i_1 \cos \theta = 2.23 \times 0.896 = 2$ (12–12)
         j term $= i_1 \sin \theta = 2.23 \times 0.447 = 1$ (12–13)
         $i_1 = 2 - j1$

For i_2 R term $= i_2 \cos \theta = 2 \times 0 = 0$ (12–12)
         j term $= i_2 \sin \theta = 2 \times 1 = 2$ (12–13)
         $i_2 = 0 + j2$
         $i_T = i_1 + i_2 = 2 - j1 + 0 + j2$
         $ = 2 + j1$

In polar form $i_T = 2.23\ /26.5°$.

$$Z_T = \frac{e_T}{i_T} = \frac{10\ /0°}{2.23\ /26.5°} = 4.48\ /-26.5°$$

This is essentially the same as the Z_T calculated by the impedance method. The value of 10 v for e_T was chosen arbitrarily, but the resultant Z would be the same regardless of the assumed value for e_T.

Example 12–6 Refer to the series/parallel circuit in Figure 12–28 [also see Figure 10–23(b)], and determine the total impedance.

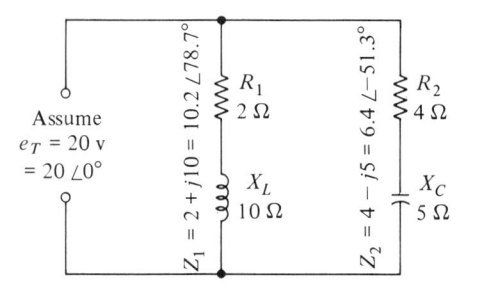

FIGURE 12–28 *Series/Parallel Circuit for Example 12–6*

[12.7] Circuit Analysis 285

Verify the impedance by assuming a value of e_T and solving by current methods.

Solution   The branches in rectangular form are:
$$Z_1 = 2 + j10 \qquad Z_2 = 4 - j5$$
Applying formula (12–4) gives
$$Z_T = \frac{Z_1 Z_2}{Z_1 + Z_2} = \frac{(2 + j10)(4 - j5)}{2 + j10 + 4 - j5}$$
$$= \frac{(2 + j10)(4 - j5)}{6 + j5}$$

For multiplication and division, the numerator and denominator are converted to polar form.
$$Z_T = \frac{10.2 \underline{/78.7°} \times 6.4 \underline{/-51.3°}}{7.8 \underline{/39.8°}}$$
$$= \frac{65.3 \underline{/27.4°}}{7.8 \underline{/39.8°}} = 8.36 \underline{/-12.4°}$$

To solve by currents, assume $e_T = 20$ v ($20 \underline{/0°}$).
$$i_1 = \frac{20 \underline{/0°}}{10.2 \underline{/78.7°}} = 1.96 \underline{/-78.7°}$$
$$i_2 = \frac{20 \underline{/0°}}{6.4 \underline{/-51.3°}} = 3.125 \underline{/51.3°}$$

Converting to rectangular form (formulas 12–12 and 12–13) yields

For i_1
  R term $= i_1 \cos \theta = 1.96 \times 0.196 = 0.384$
  j term $= i_1 \sin \theta = 1.96 \times 0.98 = 1.92$
  $i_1 = 0.384 - j1.92$

For i_2
  R term $= i_2 \cos \theta = 3.125 \times 0.625 = 1.96$
  j term $= i_2 \sin \theta = 3.125 \times 0.78 = 2.44$
  $i_2 = 1.96 + j2.44$

  $i_T = i_1 + i_2 = 0.384 - j1.92 + 1.96 + j2.44$
     $= 2.344 + j0.52$
  $i_T = 2.4 \underline{/12.4°}$ in polar form

$$Z_T = \frac{e_T}{i_T} = \frac{20 \underline{/0°}}{2.4 \underline{/12.4°}} = 8.36 \underline{/-12.4°}$$

This agrees with the Z_T calculated by the impedance method.

QUESTIONS

12-1 What is the voltage loop equation for Figure 12-1(b) if the battery terminals are reversed, the 20-v battery is 15 v, and the 9-v resistor is 4 v?

12-2 What is the voltage loop equation for Figure 12-1(c) if the 40-v battery is reversed, and 20 v is added to the drop across each resistor?

12-3 Assume that a wire is connected between the nodes of Figure 12-2(a) and (b). What is the current loop equation?

12-4 Refer to Figure 12-29. How much current flows through each resistor, and what is the voltage drop across each resistor?

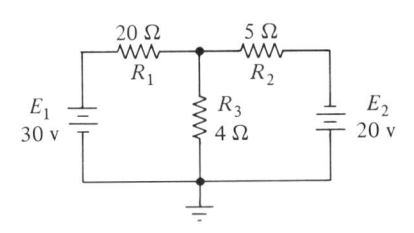

FIGURE 12-29

$E_1 = 30$ v, $R_1 = 20\,\Omega$, $R_2 = 5\,\Omega$, $R_3 = 4\,\Omega$, $E_2 = 20$ v

12-5 Repeat Question 12-4 for Figure 12-30.

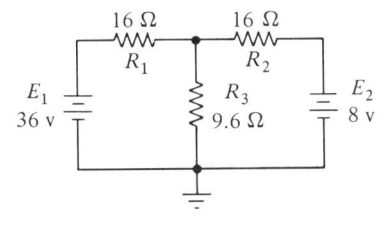

FIGURE 12-30

$E_1 = 36$ v, $R_1 = 16\,\Omega$, $R_2 = 16\,\Omega$, $R_3 = 9.6\,\Omega$, $E_2 = 8$ v

12-6 Refer to Figure 12-6. Assume the following values: $E = 5$ v, $R_1 = 12$ ohms, $R_2 = 8$ ohms, and $R_L = 1.6$ ohms. What are the Thévenin equivalent voltage (E_{eq}), the Thévenin equivalent resistance (R_{eq}), and the current through R_L?

12-7 Repeat Question 12-6 for $R_2 = 18$ ohms and $R_L = 2.8$ ohms.

12-8 Refer to Figure 12-7. Reverse the battery polarities and exchange the values of E_1 and E_2. What is the value of E_{eq}?

12-9 For Figure 12-31, what are the values of E_{eq}, I_{RL}, and E_{RL}?

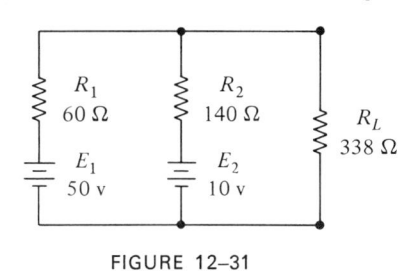

FIGURE 12-31

12-10 Solve Question 12-4 for currents and IR drops using Thévenin's theorem.

12-11 Solve Question 12-5 for currents and IR drops using Thévenin's theorem.

12-12 Refer to Figure 12-11. Solve the problem for I_{eq}, R_{eq}, I_{RL}, and E_{RL} with Norton's theorem. $E = 18$ v, $R_1 = 30$ ohms, $R_2 = 20$ ohms, and $R_L = 24$ ohms.

12-13 Solve the circuit in Figure 12-29 for I_{R3} and E_{R3} using Norton's theorem.

12-14 Repeat Question 12-13 for Figure 12-30.

12-15 Refer to Figure 12-31. Reverse the polarity of E_1 and determine I_{RL} and E_{RL} using Norton's theorem.

12-16 Convert the Thévenin's equivalent circuit of Figure 12-7 to a Norton's equivalent circuit.

12-17 Convert the Norton's equivalent circuit of Figure 12-11 to a Thévenin's equivalent circuit.

12-18 Solve the circuit in Figure 12-29 for I_{R3} and E_{R3} using the superposition theorem.

12-19 Reverse the polarity of E_2 in Figure 12-31. Using the superposition theorem, determine the values of I_{RL} and E_{RL}.

288   Circuit Analysis

12–20 Repeat Question 12–19 for Figure 12–31 with the circuit as shown.

12–21 Show the rectangular form for the following.
- a. $R = 450$ $\quad X_C = 48$
- b. $R = 0$ $\quad X_L = 75$
- c. $R = 42$ $\quad X_L = 12$
- d. $R = 120$ $\quad X_C = 0$

12–22 Show the rectangular form for the following.
- a. $i_R = 0.2$ $\quad i_L = 0.95$
- b. $i_R = 1.7$ $\quad i_C = 2$
- c. $i_R = 0$ $\quad i_C = 72$
- d. $i_R = 4.2$ $\quad i_L = 14$

12–23 Solve the following for impedance and phase angle.
- a. $Z_T = 50 + j40$
- b. $Z_T = 7 - j12$

12–24 Solve the following for current and phase angle.
- a. $i_T = 12 + j9$
- b. $i_T = 22 - j14$

12–25 What are the sum and difference of $(40 + j25)$ and $(17 - j22)$?

12–26 Solve the following problems.
- a. $7 \times j1.5$
- b. $j80 \div 4$
- c. $j8 \times -j12$
- d. $j32 \div j64$

12–27 Arrange the following in polar form.
- a. $Z_T = 125$ $\quad \theta = 25°$
- b. $Z_T = 1050$ $\quad \theta = 80°$
- c. $i_T = 15$ $\quad \theta = -42°$
- d. $i_T = 0.12$ $\quad \theta = 30°$

12–28 Solve the following in polar form.
- a. $6\,\underline{/20°} \times 11\,\underline{/41°}$
- b. $17\,\underline{/72°} \times 3\,\underline{/-40°}$
- c. $120\,\underline{/10°} \div 30\,\underline{/-35°}$
- d. $16\,\underline{/45°} \div 50\,\underline{/13°}$

12–29 Solve the following in polar form
- a. $20 \times 6\,\underline{/-51°}$
- b. $72\,\underline{/37°} \div 12$
- c. $150 \div 30\,\underline{/62°}$

12–30 Convert the following impedances to rectangular form.
- a. $6\,\underline{/35°}$
- b. $32\,\underline{/-26.5°}$

12–31 Refer to Figure 12–25. Assume $R = 20$ ohms, $X_C = 50$ ohms, and $X_L = 35$ ohms. Determine Z_T, i_T, and θ using complex numbers.

12-32 Refer to Figure 12-26. Assume $R = 250$ ohms, $X_L = 500$ ohms, and $X_C = 125$ ohms. Determine Z_T and θ with complex numbers using parallel impedance methods.

12-33 Repeat Question 12-32, but solve for i_T and θ using parallel current methods.

12-34 Calculate the impedance, phase angle, and total current for Figure 12-27 for $e_T = 10$ v, $R = 20$ ohms, $X_L = 45$ ohms, and $X_C = 25$ ohms. Use complex numbers.

12-35 Repeat Question 12-34 using complex numbers and parallel current methods.

12-36 Calculate the impedance, phase angle, and total current for Figure 12-28 for $e_T = 20$ v, $R_1 = 120$ ohms, $R_2 = 60$ ohms, $X_L = 150$ ohms, and $X_C = 70$ ohms. Use complex numbers.

12-37 Repeat Question 12-36 using complex numbers and parallel current methods.

12-38 What are the impedance, phase angle, and total current for Figure 12-32?

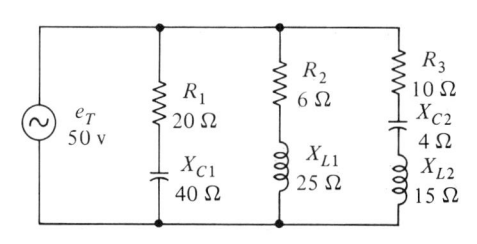

FIGURE 12-32

13 Vacuum Tubes

The first twelve chapters of this text have been devoted to the basic elements of electronic circuits. They included resistance, capacitance, inductance, and the variations in parameters as related to dc and ac levels. A knowledge of these elements is necessary in the study of any electronic device. One such device is called the **vacuum tube.**

The various types of vacuum tubes, sometimes referred to as electron tubes, are all somewhat similar to each other in their basic operation. When properly biased, electrons flow in a given direction within the tube. This characteristic lends itself to rectification which occurs when current passes only in a single direction. Current flow in the vacuum tube can be controlled by an element known as a **control grid**. Small signals at the control grid cause large current changes at the output of the tube that result in an amplified signal. Thus, a very small signal can be amplified many times its original amplitude. The larger signals, then, can be put to useful work.

In earlier chapters, many circuits were driven with an ac voltage signal represented as e, e_g, or e_{in}. The schematic symbol is a circle with a single sine wave in its center. These ac signals could have come from a vacuum tube, or they could have been used to drive a vacuum tube. In any case, the characteristics of the basic electronic circuits are necessary for the operation of vacuum tubes.

In general, vacuum tubes are cylindrical in shape, and vary from one to five inches in height and three-quarters to two inches in diameter. Photographs of several types are shown in Figure 13-1. Most vacuum tubes today are housed in glass envelopes. The operating elements are within the envelope, which is evacuated of air; hence, the term vacuum tube. Some are housed in metal envelopes, and some special purpose tubes are purposely designed to operate with gas inside the envelope.

FIGURE 13-1 *Typical Vacuum Tubes*

The internal electrodes of vacuum tubes are connected to pins in their bases. Modern vacuum tubes have seven, eight, or nine pins. The 7- and 9-pin tubes are approximately 1.5 inches high. The 8-pin tubes are generally 4 or 5 inches high. Some of the special purpose vacuum tubes may have as few as four pins or as many as twelve.

13.1 Diodes

The basic principle of vacuum tube operation is electron current flow. A diode, for example, has two metal elements inside its envelope. One element is called the **filament** or **cathode**. When this element is heated, the electrons become extremely mobile due to heat energy. If the heat is high enough, the electrons accelerate to such an extent that they leave the surface of the element. The act of leaving the surface is known as **thermionic emission**. The electrons that have left the surface form a space charge around the filament or cathode. The charge polarity for electrons is negative.

If a second element, called a **plate**, is in close proximity to the emitting

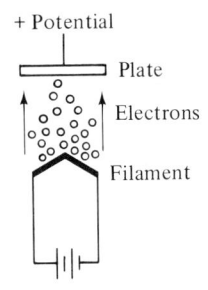

FIGURE 13-2 *Electron Flow from Heated Cathode*

electrons, and if the plate has a positive charge, the electrons will be attracted to it. There is electron flow from filament (or cathode) to plate. Figure 13-2 illustrates the current flow. The filament is heated from a battery in the same way a flashlight bulb is heated. The filament could just as well be energized by an ac voltage. A positive potential is applied to the plate, and the electrons are shown leaving the filament and traveling to the plate. Electron flow can be equated to current flow as discussed in earlier chapters.

Filaments may be connected in several ways. Two methods are shown in Figure 13-3. In (a) the particular filament is rated at 6.3 v for proper thermionic emission. A battery of 6.3 v or an ac supply of 6.3 v rms is required. In (b) a 12.6-v center-tapped filament is shown for comparison. It could be connected to 12.6 v ac or paralleled to a 6.3-v supply (as it is in the figure). The construction of each is shown in Figure 13-4. Both

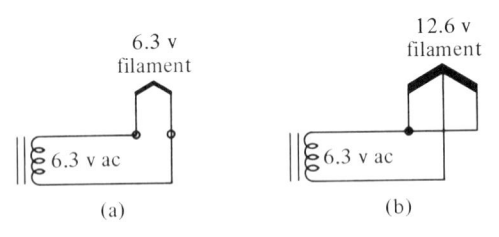

FIGURE 13-3 *Filament Connections*

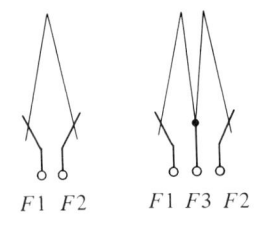

FIGURE 13-4 *Filament Construction*

filaments are suspended on wire leads. Sometimes a filament is constructed in coil fashion, particularly when it is used as a heater for a separate cathode element.

A separate cathode is generally a metal sleeve that fits over the filament. Thermionic emission is the result of heating the cathode sleeve to the specified temperature of the material. The drawing in Figure 13-5 illustrates the filament and cathode relationship. The cathode is usually coated with the oxides of barium or strontium since their efficiency is extremely high at reasonably low temperatures. Typical temperatures for these materials are in the area of 1000°K (727°C). Tungsten is another emitting material, but tungsten requires temperatures in the range of 2500 to 2600°K (about 2200 to 2300°C).

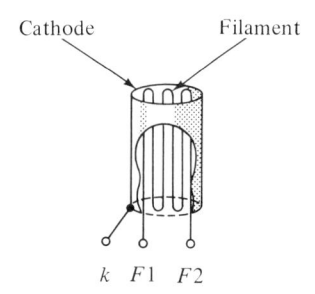

FIGURE 13-5 *Cathode-Filament Construction*

Directly heated cathodes, where emission is from the filament, have several advantages over indirectly heated cathodes. The warm-up time is less, and the efficiency is greater since the heat loss is less. Indirectly heated cathodes lose some heat due to flow around the electrodes. A distinct disadvantage of directly heated cathodes arises when ac voltage is used to power the filaments. The ac fluctuations result in emission variations along the length of the filament. The ac produces a small variation in current flow which can be troublesome in small signal circuits by imposing a "hum" component on the signal.

The construction of a directly heated cathode (filament) and a cylindrical plate electrode is shown in Figure 13-6. In this example, a positive potential is applied to the plate, and when the filament reaches its emission temperature, electrons pass through the vacuum to the plate. The indirectly heated version is shown in Figure 13-7. Notice that the filament is inside the cathode sleeve which in turn is inside the cylindrical plate electrode. In all of the illustrations, Figures 13-5 through 13-7, each electrode is connected with a lead that is brought out of the vacuum tube envelope. In this way,

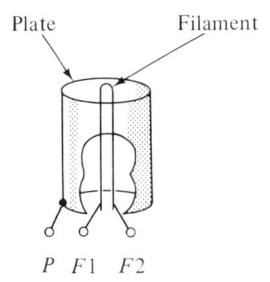

FIGURE 13-6 *Plate-Filament Construction*

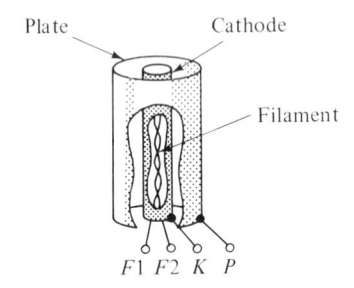

FIGURE 13-7 *Diode Construction with Separate Filament*

the electrodes can be attached to appropriate points within an electronic circuit.

The proper operating temperature for any particular filament is reached with a voltage-current balance that meets the requirement of the filament material. For example, a longer filament (same cross-sectional area) would require a higher voltage. Manufacturers specifications list the voltage and current requirements for the various vacuum tubes. Typical filament voltages range from 1.4 v to 117 v. A 117-v filament, for example, can conveniently be placed across the output of a common household outlet. Some types are purposely designed to be connected in series so that the total voltage drop is compatible with household outlets. When connected in series, however, the current requirements must be the same because the current in a series circuit is equal in all series components.

As an example of filament currents, typical 12.6-v tubes require 150 ma. The resistance of the filament, when heated to its operating temperature, is

$$R_{\text{hot}} = \frac{V}{I} = \frac{12.6 \text{ v}}{150 \text{ ma}} = 81$$

Typical 6.3-v filaments operate at 300 ma. The voltage rating of a vacuum tube is either dc or ac (rms), recalling that rms is equivalent to its dc value.

Plate Characteristics. Up to this point, electron flow in a diode resulted from thermionic emission of a heated cathode. Electron flow reached the plate by virtue of its positive potential. If the diode is connected in a continuous circuit, electron flow is continuous from the plate, through the circuit, and back to the cathode. Such a circuit is shown in Figure 13-8. The cathode is indirectly heated with a filament—its connections (two x's) are assumed to be connected to a suitable supply. The battery provides the positive potential for the plate. The negative side of the battery and cathode provides the return path for current flow (I). The two meters are for measuring plate current (ma) and plate voltage (v). The dc supply is variable, as indicated by the arrow through the battery.

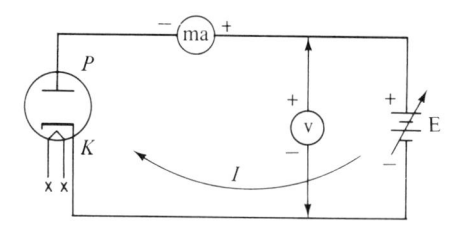

FIGURE 13-8 *Diode Bias Circuit*

A typical plate characteristic curve for the circuit (Figure 13-8) is shown in Figure 13-9(a). The plate current is plotted as a function of changing plate voltage. With zero plate voltage, the current through the diode is zero. The heated cathode releases electrons, but the electrons form a space charge which does not conduct without a positive potential on the plate. The current rises slowly at first, then accelerates until the tube reaches its saturation region. Saturation occurs when the limit is reached for thermionic emission of the cathode. That is, additional plate voltage is unable to

(a) $V - I$ characteristics (Diode)

(b) $V - R$ characteristics (Diode)

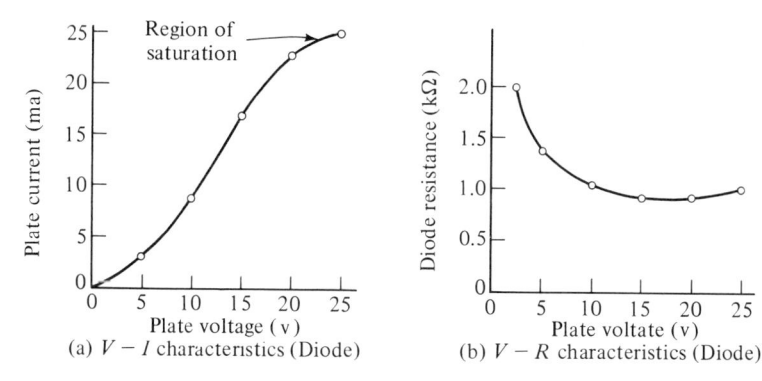

FIGURE 13-9 *Diode Characteristic Curves*

produce more current unless the cathode temperature is increased to provide additional emission.

The curve in Figure 13–9(b) is a plot of resistance as a function of changing plate voltage. The values are tabulated in Table 13-1. Plate voltage and current are taken directly from the graph in Figure 13–9(a). Diode resistance is simply plate voltage divided by plate current. The resistance curve is non-linear, and the vacuum tube diode is a non-linear device.

TABLE 13–1

Plate Voltage (v)	Plate Current (ma)	Diode Resistance (kΩ)
0	0	—
2.5	1.25	2.0
5	3.0	1.67
10	8.75	1.14
15	17	8.8
20	23	8.8
25	25	1.0

The V-I characteristic curve of Figure 13–9(a) illustrates a saturation region in which the emission of the cathode was limited by the operating temperature of the filament. Changes in operating temperature are illustrated in Figure 13–10. The center curve is typical for oxide coated cathodes at 727°C or 1000°K. Lower temperatures result in lower plate currents, and higher temperatures result in higher plate currents, assuming a constant plate voltage.

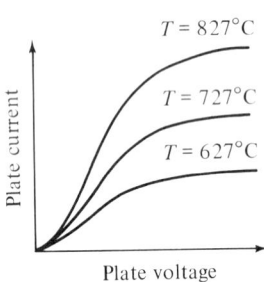

FIGURE 13–10 *Characteristics of Plate Current Versus Cathode Temperature*

Rectification. Because diodes emit current only when the plate potential is positive with respect to the cathode, current can only flow in one direction, from cathode to plate. Therefore, the diode makes a natural rectifying device for alternating voltages. For example, if the cathode is connected at a common ground with the input signal, positive pulses on the plate will allow the tube to conduct current during the time the pulse is present. A negative pulse on the plate will not allow the tube to conduct. If the plate is connected at a common ground with the input signal, only negative pulses (or voltages) will allow the tube to conduct.

An example of these connections is shown in Figure 13-11. In (a) the cathode is common to the input (e_{in}) and the output. The input is a sine

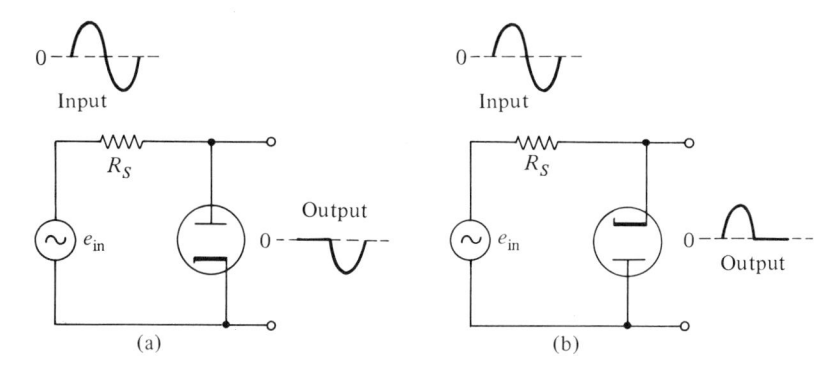

FIGURE 13-11 *Diode Clipping Circuit*

wave, and R_S is a current limiting resistor which prevents e_{in} from being shorted when the diode conducts current. During the positive cycle of the sine wave, the diode has a positive potential on its plate. Therefore, the positive cycle is essentially shorted, and the output is approximately zero for the first half cycle. During the negative cycle, the diode is reverse-biased and cannot conduct current. Since the diode now appears as an open circuit, the output reproduces the negative cycle. In (b) the plate is common to the input and output. This circuit operates in a manner which is opposite to the one in (a). Negative portions of a cycle cause the diode to conduct so that the output is zero. During positive half cycles, the tube is back-biased which causes it to appear as an open circuit. The output, then, reproduces the positive portions of a sine wave. Circuits of this type are used to select pulses for special applications like timing pulses.

In Figure 13-12 the diodes are connected in series with the input and output. In (a) the cathode is still dc connected to the bottom line through R_L which is common to the input. A positive signal at the plate allows the diode to conduct so that the output develops the same positive cycle across R_L. A negative cycle prevents the diode from conducting so that the output remains at zero. The circuit in (b) is the same as in (a), except the diode is reversed.

One way to visualize the operation of the circuits is to assume that the

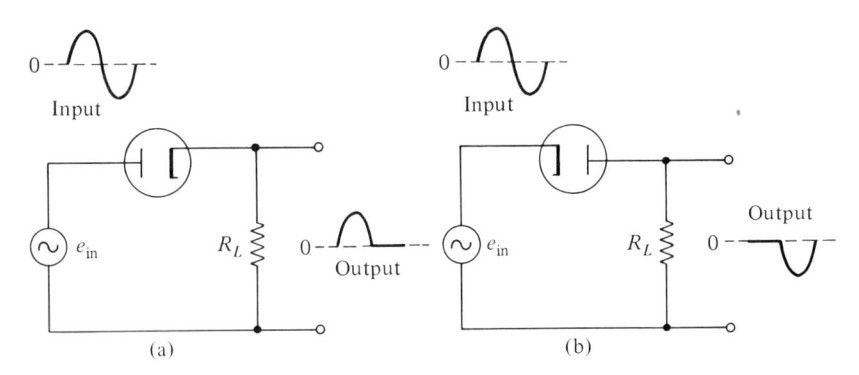

FIGURE 13-12 *Diode Rectifier Circuit*

diode is shorted when conducting and open when not conducting. Realistically, the diode will exhibit some resistance when conducting. But the effect of the diode resistance is negligible if R_S or R_L is large in comparison (greater than ten times the diode resistance).

An additional application of diodes is in dc power supplies. The input sine wave from a power transformer is rectified so that only the positive pulses (or negative for a $-E$ supply) are used. Heavy capacitors and inductive filtering are used to smooth the ripples and produce a dc level. The details of power supply operation are covered in Chapter 16.

13.2 Triodes

A triode has the same basic structure as a diode, and the theory of operation is also much the same. However, the triode has an additional electrode which is placed between the cathode and plate. This electrode is called a **control grid**. It consists of a thin screen of wires, normally suspended on two posts, that encircle the cathode (shown in Figure 13-13). The grid is intentionally placed close to the cathode. The distance from grid to plate is several times greater than the distance from cathode to grid, causing

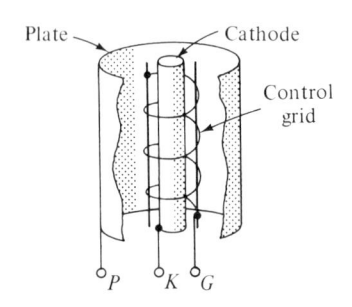

FIGURE 13-13 *Triode Construction*

the grid to exert more control over the cathode than the plate does.

All electrons that are emitted by the cathode must pass through the openings in the control grid. Suppose the control grid were placed at a fairly high negative potential. Because the electrons are negative, the grid would repel them. Few, if any, could pass the grid so that the space charge around the cathode would be re-strained. In fact, many of the electrons would be forced to re-enter the cathode. Since the grid is very close to the cathode, small values of negative voltage (-1 v to -8 v are typical) have sufficient control to overcome large values of positive plate voltage (typically 100 v to 300 v).

The voltage potential applied to a control grid is called **bias voltage**. The bias voltage is always measured in respect to the cathode because their potential difference is what controls current flow. Each of the various types of triodes has a specific negative potential for the grid for any given plate voltage. This will prevent current flow. For example, the bias voltage needed to cut off a tube with a 300-v plate voltage might be specified at -8 v. This would be the **bias-cutoff** point or the **grid-cutoff** voltage. Bias voltages that are less than -8 v would allow some amount of plate current to flow.

In general, maximum plate current flows when the bias is close to zero. Positive amounts of bias can cause more cathode current to flow, but the grid would attract some of the electrons causing current to flow in the grid circuit. Normally, grid current is not desired since the grid is not designed to carry current, except in special cases.

Vacuum tube symbols are summarized in Table 13-2. The table also lists

TABLE 13-2 *Vacuum Tube Symbols*

Symbol	Plate
E_{bb}	DC supply voltage
E_b	DC plate-to-cathode voltage
I_b	Average dc current
e_b	Instantaneous fluctuating dc voltage
i_b	Instantaneous total current
e_p	Instantaneous ac signal voltage
i_p	Instantaneous ac signal current
	Control Grid
E_{cc}	DC bias voltage
E_c	DC grid-to-cathode voltage
I_c	Average dc current
e_c	Instantaneous fluctuating dc voltage
i_c	Instantaneous total current
e_g	Instantaneous ac signal voltage
i_g	Instantaneous ac signal current

symbols for the plate and grid. Notice that dc symbols for the plate have a b subscript, and dc symbols for the control grid have a c subscript. The supply voltage for each is shown with a double subscript. Instantaneous values are indicated with lower-case letters. An instantaneous dc value is the total dc voltage or current that exists on the electrode. An instantaneous ac value is the signal component part of the total voltage or current. The ac plate values carry a p subscript, and the ac grid values carry a g subscript.

Figure 13-14 shows the circuit connections for a triode. The control grid is illustrated by a dashed line between the cathode and plate. Positive

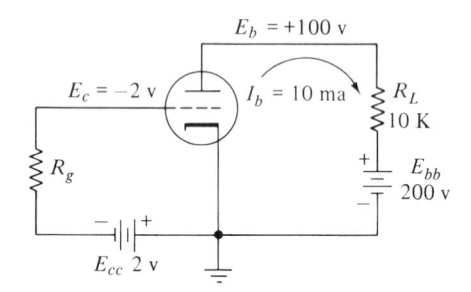

FIGURE 13-14 *Grid Bias for Triode Vacuum Tube*

potential is applied to the plate by the battery E_{bb}. Negative bias is applied to the control grid by the battery E_{cc}. The battery bias of -2 v is also the bias on the grid since the grid electrode is not allowed to draw current. In this example, it is assumed that the tube draws about half the available current from E_{bb}/R_L with a bias of -2 v applied to the grid. Assuming the plate current is 10 ma with $E_c = -2$ v, the plate voltage is calculated

$$E_b = E_{bb} - I_b R_L = 200 - (10 \times 10^{-3} \times 10 \times 10^3)$$
$$= 200 - 100 = 100 \text{ v}$$

If the tube were in full conduction and near zero resistance, the highest possible plate current would be

$$I_b = \frac{E_{bb}}{R_L} = \frac{200 \text{ v}}{10 \text{ }\Omega} = 20 \text{ ma}$$

Phase inversion of the input signal is one of the characteristics of triode circuits. The signal at the plate is 180° out-of-phase with the input. An analogy is: as the grid becomes positive, more plate current flows, so that the IR drop across R_L increases. This causes the plate voltage to decrease. A positive voltage pulse at the grid results in a negative voltage pulse at the plate. As the grid becomes negative, less plate current flows, so that the IR drop across R_L decreases. This causes the plate voltage to increase. A negative voltage pulse at the grid results in a positive voltage pulse at the plate.

Amplification of ac signals is one of the key features of a vacuum tube circuit. For that reason, the circuit in Figure 13-14 is considered to be a vacuum tube *amplifier*. A small voltage change on the control grid causes a plate current change which results in a fairly large voltage change. One of the ratings for a vacuum tube is its *amplification factor*. The amplification factor is an indication of the tube's ability to amplify signals. Typical amplification factors for triodes are 10 to 40. This means the ac plate voltage (e_p) can be nearly as much as 10 to 40 times greater than the ac grid voltage (e_g). If the internal resistance of the tube were zero, the amplification factor would be the same as plate-to-grid voltage ratio (10 to 40 for triodes). The voltage ratio is actually called voltage gain (A_v).

Table 13-3 summarizes changing values for Figure 13-14 and illustrates

TABLE 13-3

E_c (v)	I_b (ma)	$I_b R_L$ (v)	E_b (v) $E_{bb} - I_b R_L$	ΔE_c (v)	ΔE_b (v)	A_v $\Delta E_b / \Delta E_c$
−1	13	130	70			
−2	10	100	100	2	60	30
−3	7	70	130			

$E_{bb} = 200$ v $R_L = 10$ kΩ

the amplification or voltage gain (A_v) of a triode amplifier. The quiescent values of the circuit are shown on the second line for $E_c = -2$ v. The rest of the table demonstrates the effects produced by varying the control grid voltage. On the basis that a 1-v change in E_c causes a 3-ma change in I_b, the rest of the values are simple calculations. For example, when $I_b = 13$ ma, the voltage drop across R_L is

$$E_{RL} = I_b R_L = 13 \text{ ma} \times 10 \text{ k}\Omega = 130 \text{ v}$$

With the higher current, more voltage is dropped across R_L so that the plate voltage E_b decreases 30 v to a value of 70 v. When $E_c = -3$ v, E_b increases to 130 v. The voltage gain of the amplifier is the quotient of E_b and E_c since the plate voltage changed a total of 60 v for a 2-v change at the grid. Voltage gain is

$$A_v = \frac{\Delta E_b}{\Delta E_c} \tag{13-1}$$

Under the conditions in Table 13-3,

$$A_v = \frac{\Delta E_b}{\Delta E_c} = \frac{60 \text{ v}}{2 \text{ v}} = 30$$

In Figure 13-14 the resistor (R_g) in the grid circuit is required if an ac signal is to be superimposed on the control grid. As the circuit stands, the grid does not draw current, so the grid is at the same potential as the bias supply E_{cc}. If the bias supply were connected directly to the grid, an ac signal would be shorted across the battery. Therefore, the resistance allows signals to be placed at the grid without significantly loading previous stages or circuits.

Cathode bias is a method of obtaining grid bias without using a second power supply. The premise is that if the grid is held at zero potential, and the cathode is biased at some positive value, then the control grid is at a negative potential in respect to the reference potential of the cathode. That means that the electrons emitted from the positive cathode must pass through a grid that is less positive than the cathode. The control grid effect on emitted electrons is exactly the same as though the bias were applied directly to the grid.

TABLE 13-4 *Vacuum Tube Symbols*

Symbol	Cathode
E_{ff}	DC supply voltage
E_f	DC cathode-to-common (ground) voltage
I_f	Average dc current

To illustrate the method of cathode bias the basic bias elements are shown in Figure 13-15. The symbols E_{ff} and E_f are explained in Table 13-4. In the figure, E_{ff} and E_f have the same value, +2 v. Cathode current

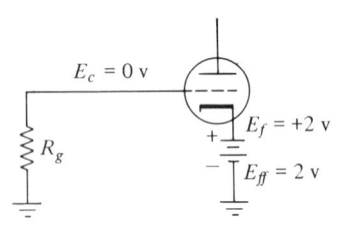

FIGURE 13-15 *Cathode Bias for Triode Circuit*

(I_f) has the same value as I_b since the control grid does not draw current. When the cathode is referenced at +2 v, and the grid is referenced at 0 v, their difference is 2 v. The grid, then, is more negative than the cathode by 2 v.

Using the same analogy, the circuit in Figure 13-16 can be constructed from the one in Figure 13-14. It has already been assumed that the plate current is 10 ma with a bias of −2 v, and that $I_f = I_b = 10$ ma. If the

[13.2] Vacuum Tubes 303

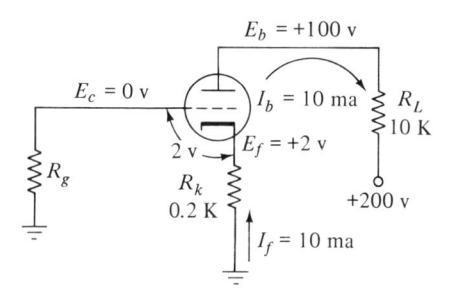

FIGURE 13–16 *Self Cathode Bias for Triode Circuit*

current through the cathode resistor R_K is used to develop a positive voltage on the cathode, then the bias requirement of -2 v is satisfied. By Ohm's law,

$$R_K = \frac{E_f}{I_f} = \frac{2 \text{ v}}{10 \text{ ma}} = 0.2 \text{ k}\Omega \tag{13-2}$$

If the bias voltage were specified at some other value, formula (13–2) could be used in the same way to derive a new value for the cathode resistor.

An important consideration in the use of cathode bias is the effect of instantaneous ac changes on the control grid. As previously noted, the grid controls plate current by the potential difference between it and the cathode. When the grid receives an ac signal, the positive portion shifts the grid to a more positive position. The first reaction is an increase in plate current. When the plate current starts to rise, the cathode current also starts to rise. As the cathode current rises, the cathode voltage increases in a positive direction. With both the grid and cathode rising in the positive direction, the potential difference between them is less than if the cathode were held at a constant dc level. Therefore, the plate current will not be able to change as much.

A negative transition on the grid reduces the plate current, which reduces the cathode current. This causes the cathode voltage to decrease. Again, the potential difference from grid-to-cathode is less than if the cathode were held constant.

The normal method of holding the cathode at ac ground is to bypass R_K with a capacitor. The capacitor should be large enough so that its reactance at the lowest signal frequency is less than a tenth of the resistance of R_K. In Figure 13–17, C_K is $\leq 0.1 \times R_K$, and since $R_K = 200$ ohms, an X_C of 20 ohms would satisfy the circuit. For simplicity of calculation, a slightly smaller value of 15.9 ohms is chosen. A minimum frequency of 1 kHz is arbitrarily chosen for the lower end of the band.

$$C = \frac{0.159}{fX_C} = \frac{0.159}{1 \times 10^3 \times 15.9} = 10 \text{ }\mu\text{f}$$

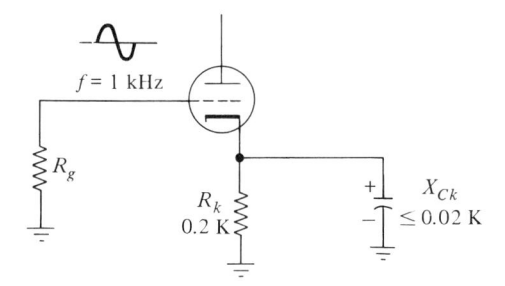

FIGURE 13-17 *Cathode Bypass Circuit*

A 10-μf capacitor across the 200-ohm cathode resistor is sufficient enough to bypass 1-kHz excursions and hold the cathode at approximately ac zero. A voltage rating for the capacitor must be higher than the maximum voltage expected on the cathode with reference to ground. Typical dc ratings are 25 v and 50 v.

13.3 Characteristic Curves

Characteristic curves are a family of curves which describe the operational characteristics of a vacuum tube. Each type of vacuum tube has its own set of curves since the characteristics tend to be different from one type to the next. In Section 13.1 typical plate characteristics were plotted for a diode vacuum tube. The curves demonstrated the non-linearity of vacuum tubes. Because of this non-linear factor and variations between tube types, curves for a vacuum tube become necessary if the tube's characteristics are to be accurately described.

A family of triode curves is shown in Figure 13-18. Notice that various values of grid bias (e_c) are plotted against plate voltage (E_b) and plate current (I_b). Plate voltage is defined as the voltage from plate-to-cathode,

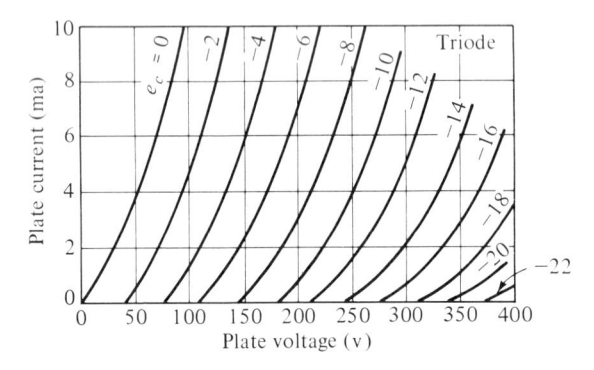

FIGURE 13-18 *Plate Characteristics—Triode*

which is normally less than the supply voltage (E_{bb}). If the tube were at cutoff and had zero plate current, it would resemble an open circuit. The plate voltage would be the same amount as the supply voltage.

As an example of what the curves mean, assume that the control grid is biased at -4 v with a battery to remain constant. Plate current can be determined for any plate voltage that comes within the range of the -4-v curve. Working within the confines of the graph, the upper end of the -4-v curve indicates a maximum E_b of just over 175 v, found by a vertical line from top to bottom. Observing the left-hand scale, I_b is 10 ma. The lower end of the e_c line (-4 v) indicates $E_b = 75$ v for zero plate current. The range for $e_c = -4$ v is $E_b = 75$ v to 175 v, and $I_b = 0$ to 10 ma.

Example 13–1 What is the plate current in Figure 13–18 if $e_c = -12$ v, and $E_b = 300$ v?

Solution The E_b and e_c lines intercept at a point that is horizontally out from the 5-ma point, so $I_b = 5$ ma.

Example 13–2 What is the cutoff bias in Figure 13–18 for $E_b = 275$ v?

Solution An I_b of 0 and an E_b of 275 v intercept the bias curve of $e_c = -16$ v. Therefore, e_c (cutoff) $= -16$ v.

Other factors can be extracted from characteristic curves, such as the amplification factor (μ), the mutual transconductance (g_m), and the plate resistance (r_p).

The **amplification factor** is calculated by the formula

$$\mu = \frac{\Delta e_b}{\Delta e_c} \text{ (with } i_b \text{ constant)} \tag{13-3}$$

Figure 13–19 illustrates an application of formula (13–3). For a given set of curves, draw a horizontal line from an arbitrary value of plate current

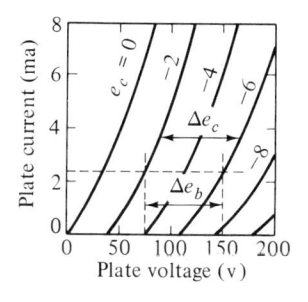

FIGURE 13-19 $\mu = \dfrac{\Delta e_b}{\Delta e_c}$

(approximately 2.4 ma in the figure). Draw two vertical lines from points where the i_b line intersects two different e_c lines. The figure shows intersecting points for $e_c = -2$ v and -6 v. The vertical lines cross the base line at $E_b = 75$ v and 150 v. The change in e_c is the difference between -2 v and -6 v, which is $\Delta e_c = 4$ v. The change in E_b is $\Delta e_b = 150 - 75 = 75$ v. The amplification factor of the tube is calculated from formula (13–3).

$$\mu = \frac{\Delta e_b}{\Delta e_c} = \frac{75 \text{ v}}{4 \text{ v}} = 18.75$$

The amplification factor is dependent upon the internal construction of the tube. It represents the maximum amount of gain that the tube is capable of achieving. The voltage gain (A_v) of a tube is always less than the amplification factor.

The **mutual tranconductance** is calculated by the formula

$$g_m = \frac{\Delta i_b}{\Delta e_c} \text{ (with } e_b \text{ constant)} \tag{13-4}$$

Because the i/e ratio is the reciprocal of e/i for ohms, units for transconductance are in mhos. It is conventional to refer to them in terms of 10^{-6} mhos, or μ mhos. The symbol for mho is an inverted ohm symbol, ℧.

Figure 13–20 illustrates the use of formula (13–4). An arbitrary vertical line is drawn which intercepts two different e_c curves. The vertical line in

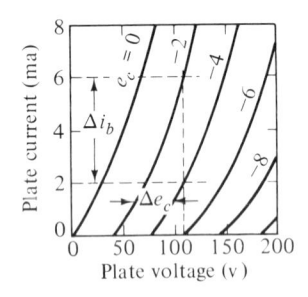

FIGURE 13–20 $\quad g_m = \dfrac{\Delta i_b}{\Delta e_c}$

the figure is at $E_b = 110$ v. Two horizontal lines are drawn through the points of interception. In the figure, the points are $e_c = -2$ v for $i_b = 6$ ma and $e_c = -4$ v for $i_b = 2$ ma. Taking the differences, $\Delta e_c = 2$ v and $\Delta i_b = 4$ ma. Therefore, the transconductance is

$$g_m = \frac{\Delta i_b}{\Delta e_c} = \frac{4 \times 10^{-3}}{2} = 2 \times 10^{-3} = 2000 \; \mu \text{ mhos}$$

[13.3]   Vacuum Tubes   307

Plate to cathode resistance is calculated:

$$r_p = \frac{\Delta e_b}{\Delta i_b} \text{ (with } e_c \text{ constant)} \tag{13-5}$$

The unit for r_p is ohms since e/i is a direct measure of resistance. Figure 13-21 illustrates the method of calculating the plate to cathode resistance. Since e_c is to be held at a constant value, two points are chosen along one

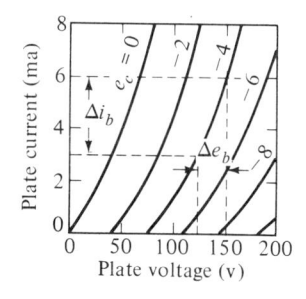

FIGURE 13-21 $r_p = \frac{\Delta e_b}{\Delta i_b}$

of the e_c curves. In the figure, the two points coincide with $E_b = 125$ v and 153 v. Horizontal lines from the two points coincide with $I_b = 3$ ma and 6 ma. The differences are

$$\Delta e_b = 153 - 125 = 28 \text{ v}$$
$$\Delta i_b = 6 - 3 = 3 \text{ ma}$$

The resistance is calculated from formula (13-5).

$$r_p = \frac{\Delta e_b}{\Delta i_b} = \frac{28}{3 \times 10^{-3}} = 9.33 \times 10^3 = 9.38 \text{ k}\Omega$$

The three factors, μ, g_m, and r_p, are related by the expression

$$\mu = g_m r_p \tag{13-6}$$

One important drawback in the relationship is trying to achieve accuracy. First, the factors must be taken from the same area of the curves to reduce errors due to the non-linear characteristic of the curves. Second, it is quite difficult to transcribe figures by graphical means which will stand up to close cross-checking. For example, the data taken from Figures 13-19 through 13-21 were from arbitrary portions of each graph, and the numbers were generally rounded. In any event, formula (13-6) can be applied to see how much error was involved.

$$\mu = g_m r_p = 2000 \times 10^{-6} \times 9.33 \times 10^3 = 18.7$$

The graphic μ was 18.75, which is surprisingly close. However, practical applications can easily produce errors as high as 10 percent.

Load line analysis is shown in Figure 13–22 for the circuit of Figure 13–23. A load line is intended to graphically cover the range of all possible operating points of a given circuit. The circuit has a load resistance of 30 k ohms. The maximum possible plate current that can flow if the tube were a short-circuit is $E_b = 0$. Therefore, the first point is

$$I_b = \frac{E_{bb}}{R_L} = \frac{300}{30 \times 10^3} = 10 \times 10^{-3} = 10 \text{ ma}$$

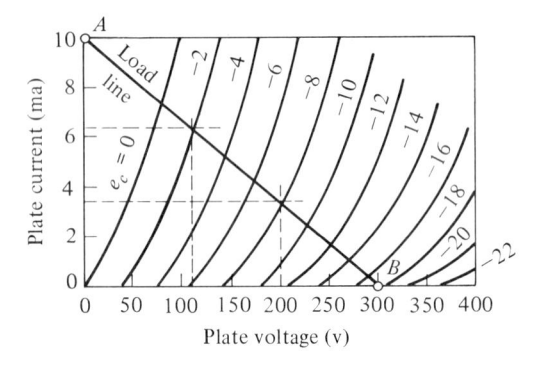

FIGURE 13–22 *Load Line Analysis for Triode Curves*

Point A is shown with $E_b = 0$, and $I_b = 10$ ma. The maximum plate voltage occurs only when the tube is an open-circuit and $E_b = E_{bb}$. Point B occurs where $I_b = 0$, and $E_b = 300$ v. A straight line can be drawn between points A and B, and the operating characteristics of the type of vacuum tube for those curves in that circuit will correspond to the load line. For example, if $e_c = -8$ v, the plate voltage is 200 v, straight down from the intersection of -8 v and the load line. Taking a horizontal line left to the plate current scale yields an I_b of about 3.4 ma.

The circuit in Figure 13–23 can be cross-checked. If $E_b = 200$ v, the IR drop across R_L is $E_{bb} - E_b = 300$ v $- 200$ v $= 100$ v.

$$I_b = \frac{E_{RL}}{R_L} = \frac{100}{30 \times 10^3} = 3.33 \times 10^{-3} = 3.33 \text{ ma.}$$

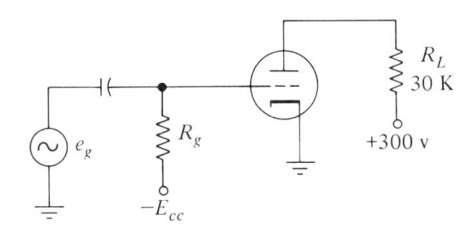

FIGURE 13–23 *Basic Triode Amplifier Circuit*

As a second example, the plate voltage is about 112 v for $e_c = -2$ v. The corresponding plate current is about 6.3 ma. Cross-checking the circuit gives

$$E_{RL} = E_{bb} - E_b = 300 \text{ v} - 112 \text{ v} = 188 \text{ v}$$

$$I_b = \frac{E_{RL}}{R_L} = \frac{188}{30 \times 10^3} = 6.3 \times 10^{-3} = 6.3 \text{ ma}$$

Referring back to the load line in Figure 13-22, the operating range of the grid is from 0 v to -17 v. The operating range of the plate voltage is from 80 v to 300 v. Within the operating range, plate current can vary from 0 to better than 7 ma. It takes an e_c of -18 v to completely cut off the circuit.

Once the amplification factor (μ) and plate to cathode resistance of a vacuum tube are known, the voltage gain of a circuit can be calculated.

$$A_v = \mu \frac{R_L}{R_L + r_p} \tag{13-7}$$

The values of μ and r_p were previously calculated from graphs that are identical to Figure 13-22. Thus, their values are the same: $\mu = 18.75$ and $r_p = 9.33$ k ohms. The A_v of Figure 13-23 is

$$A_v = \mu \frac{R_L}{R_L + r_p} = 18.75 \times \frac{30 \times 10^3}{30 \times 10^3 + 9.33 \times 10^3}$$

$$= 18.75 \times \frac{30 \times 10^3}{39.33 \times 10^3} = 18.75 \times 0.764 = 14.3$$

The voltage gain is smaller than the amplification factor. Formula (13-7) indicates that the ratio approaches unity only if $R_L \ggg r_p$. It becomes obvious that the voltage gain of an amplifier increases when R_L is increased. The increase is not nearly as much as one might expect. Table 13-5 summarizes examples of A_v for various R_L values. The numbers are based on formula (13-7) and previously calculated values of μ and r_p.

TABLE 13-5

μ	r_p	R_L	$\frac{R_L}{R_L + r_p}$	A_v
18.75	9.33 kΩ	10 kΩ	0.52	9.7
		20 kΩ	0.68	12.8
		30 kΩ	0.76	14.3
		60 kΩ	0.87	16.2
		100 kΩ	0.91	17.1

13.4 Tetrodes and Pentodes

The triode vacuum tube becomes ineffective in very high frequency applications. Capacitance exists between the control grid and plate since the electrodes are metal and separated by a dielectric (vacuum). At high frequencies, the reactance can be low enough to couple a significant amount of signal back to the grid which is out-of-phase. The effect of the coupling is to cancel part or all of the input signal at the grid so that the tube can cease to be an amplifier. Typical capacitance for triodes is in the range of 1 to 5 pf.

If another grid were placed between the control grid and plate, it would act as a shield or screen. Then the capacitance from control grid to plate could be reduced by a factor of 10^2 to 10^3, depending upon the actual construction. A tetrode is actually a triode with a second grid called a **screen grid**. A drawing of a tetrode is shown in Figure 13-24. The screen grid (grid 2 or $G2$) is closer to the plate than the control grid (grid 1 or $G1$). Since the control grid is closer to the cathode, it still exercises control over cathode emission.

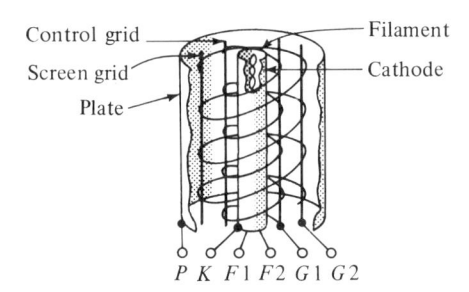

FIGURE 13-24 *Tetrode Construction*

Bias procedures for tetrodes are similar to those for triodes. The screen grid must be at a positive potential that is preferably less than the plate potential. Electrons are accelerated toward the screen at a high velocity. The screen allows most of them to pass on through to the plate, though some electrons are absorbed by the screen. As a result, some screen current is necessary, even though it does not do useful work. Screen grid current is typically 10 to 30 percent of the amount of current the plate requires. Furthermore, the screen and plate currents both come from the cathode, so that the cathode current is the sum of the two. If the control grid draws current as the result of positive bias, the cathode current is the sum of both grids and the plate.

$$I_f = I_c + I_s + I_b \tag{13-8}$$

Circuit connections for a tetrode are illustrated in Figure 13–25. The cathode, grid, and plate connections are similar to those of a triode. The bias voltage levels are illustrative and not necessarily the same for other tetrodes. Assuming that a -2-v grid bias produces 15 ma of I_b, the plate voltage is

$$E_b = E_{bb} - I_b R_L = 350 - (15 \times 10^{-3} \times 10 \times 10^3) = 200 \text{ v}$$

For the conditions stated, the screen grid current is 4 ma.

$$E_s = E_{bb} - I_s R_s = 350 - (4 \times 10^{-3} \times 62.5 \times 10^3) = 100 \text{ v}$$

The total cathode current is calculated from formula (13-8).

$$I_f = I_c + I_s + I_b = 0 + 4 \text{ ma} + 15 \text{ ma} = 19 \text{ ma}$$

Notice in Figure 13–25 that $I_c = 0$ since the control grid is biased negatively and does not draw current. Also, the bypass capacitor C_s from the screen grid to ground maintains the screen grid at an ac ground potential.

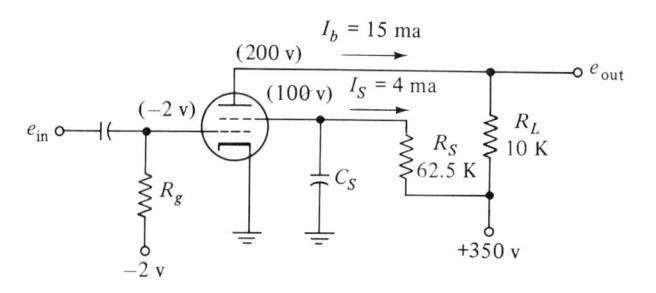

FIGURE 13–25 *Tetrode Bias Connections*

Variations of the input voltage can vary the screen grid current along with the plate current, though the screen grid current varies to a lesser degree. Allowing the screen to vary would result in lost signal energy from the plate. Bypassing the screen prevents it from varying. The value of the capacitor is usually chosen for a reactance that is approximately one-tenth the resistance of R_s at the lowest frequency expected. Thus, the reactance for higher frequencies will always be lower than one-tenth of R_s.

The addition of a screen grid has several other effects. The amplification factor generally increases, and the plate resistance significantly increases (10 to 100 times as high). Secondary emission has a very detrimental effect. A high positive potential on the plate receives highly accelerated electrons. So many, in fact, that the metal plate releases some which are then called **secondary electrons**. In a triode, the positive plate simply attracts them again and does not cause any particular problem. In a tetrode, the plate will re-attract the secondary electrons, as long as the plate potential is higher than the screen potential. The screen is usually held steadily at a potential that is lower than that of the plate. When a signal is amplified and produced at the plate, the plate potential can swing lower than the screen. When that happens, the screen can draw more current than the plate.

Figure 13-26 shows the effect of secondary emission with an illustrative V-I characteristic curve for a tetrode. The cathode current is the sum of I_s and I_b with zero I_c. When the plate voltage is lowered close to the screen

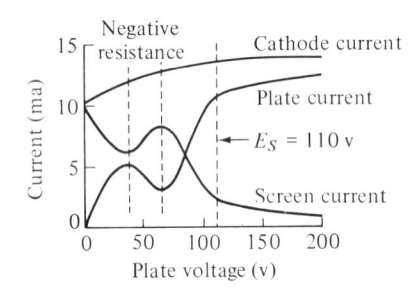

FIGURE 13-26 V-I Characteristics (Tetrode)

voltage (110 v), the plate current tends to drop off while the screen current tends to increase. When the plate voltage is zero, ($I_b = 0$), the screen current is equal to the cathode current because the screen is attracting the electrons from the cathode. As E_b is increased, plate current increases with a mirror-image decrease in I_s. However, at $E_b = 40$ v in the example, secondary emission begins and remains effective to about 70 v. During this period, the plate curve has reversed direction, and I_b has decreased while E_b has increased. This region is called the **negative resistance** part of the curve. Above the 70-v point, secondary emission begins to lose its effect as the plate current finally takes over when E_b passes E_s.

The pentode is constructed in the same way as the tetrode (refer to Figure 13-24) except one more grid is added. The added electrode is called a **suppressor grid** (grid 3 or G3). The suppressor is a screen-like material that fits between the plate and screen electrodes. It is normally connected to the cathode (most often internally) and acts as an electrostatic shield between the screen and plate. Electrons leaving the cathode are accelerated toward the positive potential on the screen. They pass on through the openings in the screen and are immediately collected by the higher potential on the plate. Their velocity is also sufficient to pass on through the openings in the suppressor grid. Secondary electrons leaving the plate are repelled by the suppressor grid and recollected by the plate. The negative resistance region of a tetrode no longer exists. Even if the plate potential drops below that of the screen, the electrons are primarily attracted to the plate because their high velocity causes most of them to pass the screen. Obviously, if the plate voltage drops far enough, the screen instead of the plate will attract electrons. However, secondary emission will not occur.

The electrode connections for a pentode are labeled in Figure 13-27. Notice that the suppressor is internally connected to the cathode, and the arrangement of electrodes is the same as it was for the tetrode in Figure 13-24.

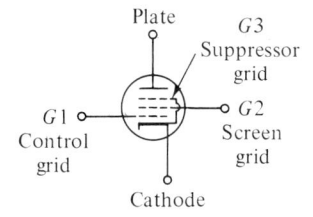

FIGURE 13-27 *Pentode Connections*

Plate characteristic curves for pentodes are considerably different than those for triodes (see Figure 13-28). The e_c curves are non-linear at low values of plate voltage—somewhere below 100 v. Above $E_b = 100$ v, the curves tend to flatten so that there is only a very small increase in I_b for large increases in E_b. The plate current is stabilized because of the constant dc potential on the screen.

The relative flatness of pentode characteristic curves indicates that the control grid exercises close control over a rather extensive range of plate voltage. For example, a pentode is sometimes known as a **sharp-cutoff device**. An e_c of about -5.0 v will cut off the tube (see Figure 13-28) over a large range of plate voltage.

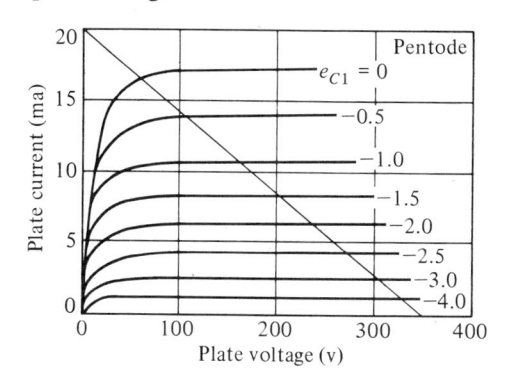

FIGURE 13-28 *Plate Characteristics—Pentode*

The flatness of pentode curves suggests other effects. Since a very small change in control grid voltage affects a large change in plate voltage, amplification factors are very high. Factors of 1000 to 4000 are not uncommon. Mutual conductance values are only a little higher as a general rule. They range from 2000 to 8000 μ mhos. Plate resistance is extremely high since the plate current changes little with large changes in plate voltage. Values range from 100 k ohms to over 1 M ohm.

Due to the small changes in the curves, beyond the non-linear portions, accurate graphical analysis is almost impossible with the curves most often available. However, to illustrate the parameters of a pentode, approximate calculations are made for Figure 13-28. The region surrounding $E_b = 100$ to 200 v and $e_c = -0.5$ to -1.0 v is selected since the curves are more easily evaluated there. Formulas (13-3) to (13-5) are applied in the following manner.

$$\mu = \frac{\Delta e_b}{\Delta e_c} = \frac{200 - 100}{-0.5 - (-0.55)} = \frac{100}{0.05} = 2000 \tag{13-3}$$

$$g_m = \frac{\Delta i_b}{\Delta e_c} = \frac{(13.5 - 10.5)(10^{-3})}{-0.5 - (-1.0)}$$
$$= \frac{3 \times 10^{-3}}{0.5} = 6000 \ \mu \text{ mhos} \tag{13-4}$$

$$r_p = \frac{\Delta e_b}{\Delta i_b} = \frac{200 - 100}{(13.5 - 13.2)(10^{-3})}$$
$$= \frac{100}{0.3 \times 10^{-3}} = 333 \text{ k}\Omega \tag{13-5}$$

Using formula (13-7) in which voltage gain is some ratio times the amplification factor gives

$$A_v = \mu \frac{R_L}{R + r_p}$$

The plate resistance can have a significant effect on the gain. For instance, if $r_p \gg R_L$, the voltage gain is only a small percentage of the amplification factor. On the other hand, if $r_p \ll R_L$, the voltage gain approaches the amplification factor. However, the voltage gain is always less than the amplification factor. In a pentode, plate resistances are quite high, so that plate supply resistors (R_L) must be carefully considered if the high gain characteristic of a pentode is to be achieved. Too high an R_L results in a very low plate current for a given plate voltage, making a high negative control grid voltage necessary. As the bias increases, the curve becomes even flatter, which in turn causes the plate resistance to increase dramatically. Because of these compensating effects, the voltage gain is not nearly as high for a pentode as the amplification factor might suggest.

[13.4]   Vacuum Tubes   315

Figure 13-29 illustrates a circuit with a fairly high voltage gain when R_L is not especially large. A load line for the circuit can be drawn in Figure 13-28. The ends of the line are $E_{bb} = 350$ v and $I_b = 20$ ma. If E_b is to be 200 v, I_b will be 8.5 ma (junction of the load line and a vertical line from the 200-v point). e_c will be about -1.4 v as indicated by the junction of the load line and $E_b = 200$ v. Under those conditions:

$$R_L = \frac{E_{bb} - E_b}{I_b} = \frac{350 - 200}{8.5 \times 10^{-3}} = 17.5 \text{ k}\Omega$$

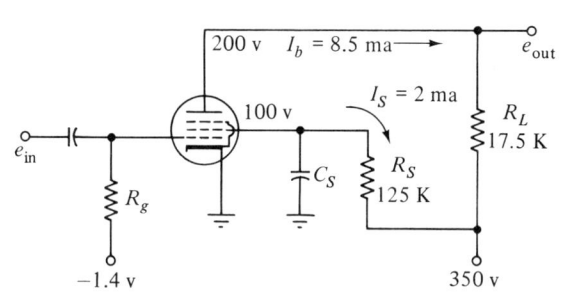

FIGURE 13-29   *Pentode Amplifier Circuit*

For this example, I_s is given as 2 ma for $E_s = 100$ v.

$$R_s = \frac{E_{bb} - E_s}{I_s} = \frac{350 - 100}{2 \times 10^{-3}} = 125 \text{ k}\Omega$$

The voltage gain is calculated from formula (13-7).

$$A_v = \mu \frac{R_L}{R_L + r_p} = 2 \times 10^3 \times \frac{17.5 \times 10^3}{(17.5 \times 10^3) + (333 \times 10^3)}$$
$$= 2 \times 10^3 \times 0.05 = 100$$

Cathode bias for a pentode is calculated in much the same way as for a triode. The cathode current is

$$I_f = I_c + I_s + I_b$$

If the control grid does not draw current, which is usually valid for reverse (negative) bias situations, then the cathode current is the sum of I_s and I_b. In Figure 13-29,

$$I_f = I_s + I_b = 2 \text{ ma} + 8.5 \text{ ma} = 10.5 \text{ ma}$$

And for a bias of -1.4 v, the required cathode resistor is

$$R_K = \frac{E_f}{I_f} = \frac{1.4}{10.5 \times 10^{-3}} = 0.133 \times 10^3 = 133 \text{ }\Omega$$

A bypass capacitor is required if the gain of 100 is to be maintained. To draw a load line, the plate current is calculated when the vacuum tube resistance is zero (short-circuit). That means that the cathode resistance is included with R_L for the total resistance. But the cathode resistance is normally so much smaller than R_L that it becomes insignificant. Even so, its effect must not be forgotten since there can be cases where it is high enough to affect the load line. If cathode bias is considered for Figure 13-29, the total resistance and current in the loop for the plate circuit are

$$R_T = R_k + R_L = 0.133 \text{ k} + 17.5 \text{ k} = 17.68 \text{ k}\Omega$$

$$I_b \text{ (max)} = \frac{E_{bb}}{R_T} = \frac{350}{17.68 \times 10^3} = 19.8 \text{ ma}$$

A **remote cutoff** pentode is designed to cut off at a greater bias voltage than the sharp cutoff versions that typically cut off at -4 v to -6 v. Typical cutoff voltages range from -20 v to -30 v for remote types. To achieve high ranges of -20 v to -30 v, the control grid is made with the screen wiring closely spaced at the ends, and widely spaced at the center. When the control grid is negative, the closely spaced screens tend to cut off some of the plate current. But the wide spacings in the center require a greater negative voltage to cut off the plate current. Tubes of this construction are sometimes called variable mu tubes.

Summarizing, a pentode (compared to a triode) has a very high amplification factor, slightly higher mutual conductance, and extremely high plate resistance. Higher voltage gains are possible because of the high amplification factor. Pentodes are more effective as high frequency amplifiers due to lower inter-electrode capacitances.

13.5 Specifications

Vacuum tubes are listed according to their type numbers in booklets called **Vacuum Tube Manuals**. These manuals cover such parameters as voltages, currents, power, inter-electrode capacitance, amplification factor, mutual conductance, and plate resistance. Diagrams are ordinarily included that describe the circuit connections, base-pin connections, and characteristic curves. In addition to maximum allowable voltages and currents, average or service values are included that describe parameters for one or two typical operating situations.

Rectifiers, diodes, triodes, tetrodes, and pentodes have all been discussed in terms of a single circuit for each vacuum tube. These tubes can be and are generally packaged in groups of two or three within an envelope. Thus, a vacuum tube may contain two diodes, two triodes, a pentode and a triode, a triode and two diodes, or other combinations.

Filament Voltage and Current. Voltage is usually given as an ac value or rms dc value. For example, the voltage for a 6AU6 tube is given as 6.3 v.

This tube is normally operated with a 6.3-v filament transformer, or it can be operated from a 6-v dc battery. Filament voltages can vary considerably without serious effect on the operating characteristics of the tube. More voltage increases the tube's emission while less voltage decreases its emission. However, increased voltage can shorten the life of the tube.

Filament current is specified in milliamperes for most of the everyday radio and TV tubes. Currents of 150 ma and 300 ma are typical. The current is the amount required by the tube for the rated filament voltage and is important when considering the size of a filament supply. For example, if 10 tubes each draw 150 ma and all of them are to be connected in parallel across the same transformer winding, the total current requirement is

$$I_T = 10 \times 150 \text{ ma} = 1.5 \text{ A}$$

Therefore, the transformer winding must be rated for 1.5 A or more to carry the load.

Plate Wattage. The power that can be dissipated by the plate itself is specified as a maximum value. The limit is a function of the internal construction of the plate electrode.

Maximum Plate and Screen Voltages. These are maximum values that the electrodes can withstand without internal arcing between electrodes.

Inter-Electrode Capacitance. Three parameters are usually of interest. Input capacitance describes the total parallel capacitance between the grid and all other elements except the plate. Output capacitance is the total parallel capacitance between the plate and all other elements except the grid. Grid to plate capacitance is also specified, though normally smaller than the input and output capacitances.

Average or Service Ratings. As discussed earlier, these are given for a set of typical operating conditions.

E_b—dc plate voltage
E_s—dc screen voltage
I_b—plate current
I_s—screen current
E_c—control grid bias
r_p—plate-to-cathode resistance
g_m—mutual conductance
μ or mu—amplification factor
R_L—load resistor for rated output conditions
P_{out}—output power

All of the average or service ratings are based on a specific area of the operating region of the vacuum tube. Since considerable variations can occur, a second set of conditions is sometimes listed for a different operating region.

Rectifiers and Diodes. Rectifiers and diodes are usually used to rectify an ac voltage. Therefore, the plate and cathode would be reverse biased for half a cycle. The **peak inverse voltage (PIV)** is specified as the maximum reverse bias the tube will handle without breaking down. The **voltage drop across the tube** (plate to cathode) for a given dc current is shown so that voltage drops can be accounted for in a circuit application. The **maximum current** is the dc rating of the plate and cathode electrodes.

Base connections are diagrammed which show the connections for the various electrodes. Samples are drawn in Figure 13-30. Eight-pin tubes are shown in (a) which is a dual rectifier, and in (b) which is a triode. Diagram (c) is a 7-pin dual diode, (d) is a 9-pin dual triode, and (e) is a 7-pin pentode.

(a) 8-pin rectifier (b) 8-pin triode (c) 7-pin dual diode

(d) 9-pin dual triode (e) 7-pin pentode

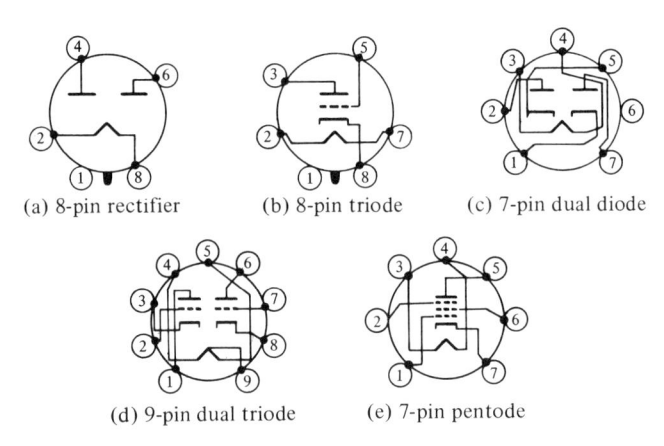

FIGURE 13-30 *Base Connections for Selected Vacuum Tubes*

Characteristic curves have been described for both triodes and pentodes. When these are shown for a particular tube type, they represent the average tube. Practical situations often find variations of a type that do not fit the curves as precisely as an average tube. Therefore, it is the responsibility of the designer to take into account possible variations so that circuits will continue to operate as tubes change with age or are replaced due to failures.

13.6 Special Amplifier Configurations

This section introduces two unique configurations: the grounded grid amplifier and the grounded plate amplifier.

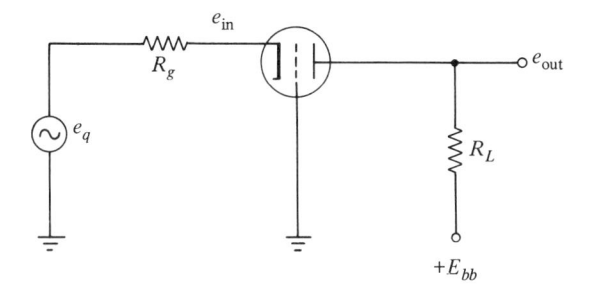

FIGURE 13-31 *Grounded Grid Amplifier*

The **grounded grid amplifier** is drawn in Figure 13-31. The grid is ac connected to the common point of both the input and the output. By placing the grid at an ac ground potential, a significant amount of shielding is provided between the plate and cathode. The shielding reduces the inter-electrode capacitance, which improves operation of the vacuum tube at higher frequencies. Recall that inter-electrode capacitance introduces a reactance that decreases the ability of a vacuum tube to amplify at higher frequencies.

As a signal amplifier, no phase inversion occurs. The output goes positive at the same time the input goes positive. Conversely, the output goes negative at the same time the input goes negative. For example, as e_g swings in a positive direction, the cathode becomes more positive in respect to the grid. This is the same as the grid becoming more negatively biased, which reduces the plate current. As the plate current is reduced, the IR drop across R_L decreases and allows the plate voltage to rise in a positive direction.

The voltage gain formula for the grounded grid amplifier is

$$A_v = \frac{(1 + \mu)R_L}{(1 + \mu)R_g + R_L + r_p} \tag{13-9}$$

The symbol μ is the amplification factor, R_g is the internal resistance of the voltage generator, R_L is the load resistance, and r_p is the plate resistance. As with Formula (13-7), the voltage gain will never be as high as the amplification factor (μ) of the vacuum tube. Formula (13-9) shows two factors in the denominator that prevent higher voltage gains: R_g and r_p. The voltage gain can approach the value of μ if R_L is very large in comparison to the sum of R_g and r_p.

Example 13-3 What is the voltage gain of a grounded grid amplifier if R_L is 56 k ohms, μ is 20, r_p is 8 k ohms, and R_g is 300 ohms?

Solution Using formula (13–9) gives

$$A_v = \frac{(1+\mu)R_L}{(1+\mu)R_g + R_L + r_p}$$
$$= \frac{(1+20)56 \times 10^3}{(1+20)300 + (56 \times 10^3) + (8 \times 10^3)}$$
$$= \frac{1176 \times 10^3}{70 \times 10^3} = 16.8$$

Grounded grid amplifiers are widely used as *RF* amplifiers due to their low-noise and low-interelectrode capacitance characteristics. However, a disadvantage is that the input requires a current driving capability since the cathode current is equal to the plate current.

The **grounded plate amplifier** is drawn in Figure 13–32. This configuration is more often called a **cathode follower**. Its basic characteristics are very high input resistance and low output resistance. Because of the high input resistance, the cathode follower does not appreciably load other circuitry. The low output resistance characteristic allows good impedance matching into low impedance circuits. The matching of impedances, of course, provides maximum power transfer.

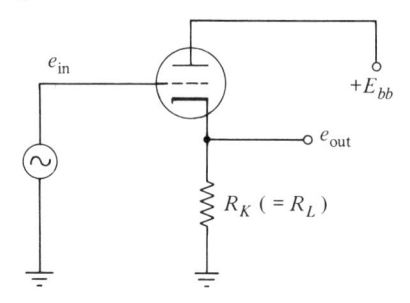

FIGURE 13–32 *Grounded Plate Amplifier (Cathode Follower)*

As with the grounded grid amplifier, the grounded plate amplifier (cathode follower) does not invert the input signal. As the input changes in a positive direction, the cathode current increases. An increasing cathode current causes the cathode voltage (e_{out}) to rise in a positive direction. The opposite occurs for negative signals.

Voltage gain for the grounded plate amplifier is calculated from the following formula:

$$A_v = \mu \frac{R_K}{(1+\mu)R_K + r_p} \quad (13\text{–}10)$$

Notice that the voltage gain will always be less than unity, even if r_p is small in comparison to R_K. The voltage gain can approach unity if R_K is very large in comparison to r_p.

Example 13–4 What is the voltage gain of a grounded plate amplifier if μ is 20, R_K is 15 k ohms, and r_p is 3.5 k ohms?

Solution From formula (13–10),

$$A_v = \mu \frac{R_K}{(1 + \mu)R_K + r_p}$$

$$= \frac{20 \times 15 \times 10^3}{(21 \times 15 \times 10^3) + (3.5 \times 10^3)}$$

$$= \frac{300 \times 10^3}{318.5 \times 10^3} = 0.94$$

13.7 Special Purpose Tubes

This section summarizes the key characteristics of some of the special purpose tubes. In general, these tubes follow the same fundamentals as the vacuum tubes already discussed.

Beam power tubes are intended for power applications such as the driving of audio output transformers. The tube is a pentode type, but the suppressor grid is replaced with internal plates that form the electron beam. The suppressor plates are directly connected to the cathode. This type of construction permits low screen current. The screen current of regular pentodes is 10 to 30 percent of the plate current requirement; beam power screens require less than 10 percent.

Another feature of these tubes is that the screen voltage can be about the same value as the plate voltage. This feature has a distinct advantage in low voltage, low cost amplifiers. First, it maintains the screen voltage at a level high enough for effective operation from low voltage power supplies, such as those in a 117-v ac/dc radio which uses power supply voltages that are typically 110 v to 125 v. As shown in Figure 13–33, the direct connection to the same supply as the plate eliminates the need for a screen dropping resistor and bypass capacitor. The screen voltage is probably several volts higher than the plate voltage since the primary of the transformer

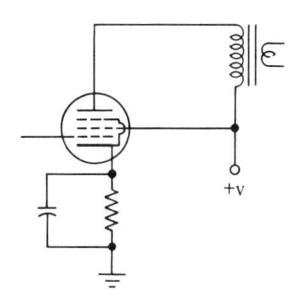

FIGURE 13–33 *Beam Power Pentode*

drops several volts. The difference is permissible in beam power tube applications and it is not permissible in regular pentode tubes.

Transmitter tubes are generally much larger than ordinary radio and television tubes because they handle large amounts of power. High power transmitter tubes are likely to produce thousands of watts. Since high power is a function of high voltage and current, the electrodes have to be separated enough to prevent inter-electrode arcing. And the electrodes have to be heavy enough to withstand high currents. In some of the larger transmitters, heat dissipation becomes a serious factor. Many high power tubes are designed with plates that are water-cooled. This is accomplished by pumping water through coils that surround the plate.

Because of the high potentials, high voltage tubes often have a special plate (or anode) connection on the top of the tube. A connection of this type separates the connections and prevents arcing to closely spaced connections in the socket.

Gas tubes operate with a controlled amount of low-pressure gas inside their envelope; vacuum tube elements are operated within a vacuum. Two types employed for general use are the cold-cathode and hot-cathode gas tubes.

The cold-cathode tube utilizes two electrodes. One is the anode (plate), and the other is the cathode, which is usually larger. A filament or heater is not used. Conduction occurs when the anode potential is high enough to allow the gas to ionize and conduct current.

The hot-cathode tube requires a heating element (filament). The cathode is heated to a temperature that will cause thermionic emission of electrons from the cathode. When the anode potential is high enough, the gas molecules will ionize by collision and allow full conduction of current. A distinct feature of gas tubes is their ability to regulate voltage. Once the gas is ionized, the ionizing potential tends to remain constant as long as a resistance is connected in series with the anode.

An example application is shown in Figure 13-34. A gas tube is selected

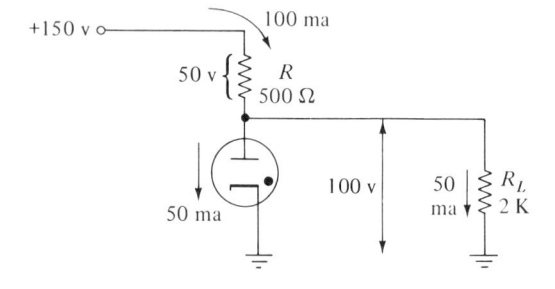

FIGURE 13-34 *Gas Tube*

to regulate its plate to cathode voltage at 100 v with 50 ma through the tube. If the average load is 50 ma, the series resistor must carry 100 ma. Since the supply voltage is 150 v, the series resistance is

$$R = \frac{E_{bb} - E_b}{I_t} = \frac{150 - 100}{100 \text{ ma}} = 500 \text{ }\Omega$$

If the voltage drop across the tube is constant at 100 v, then the current through R is constant. Any change in the load current must be compensated for by the gas tube so that the total current $(I_b + I_{RL})$ is constant at 100 ma. If I_{RL} increases, I_b decreases, and if I_{RL} decreases, I_b increases.

Phototubes are constructed of materials that change their characteristics when illuminated. Some materials change their resistance, and some emit photoelectrons. Those that emit photoelectrons use what is often called a **photocathode**. A positive potential at the plate causes these electrons to flow when the cathode electrode of the tube is illuminated by an external light source. The electrodes are vacuum sealed in a glass envelope in the same way as a vacuum tube. The glass, however, must be uncoated to allow light to fall on the cathode.

An example of a phototube's application is shown in Figure 13–35. Either type, resistance or current, will operate in this circuit configuration. If the resistance changes in the phototube, the output voltage will vary.

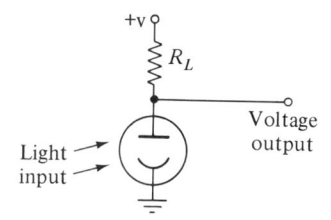

FIGURE 13–35 *Phototube*

If the tube conducts current, the voltage will also vary. For a photocurrent application, assume zero light input (dark). The tube does not conduct, and the output is the same as +v. When light shines on the tube, it conducts current which increases the *IR* drop across R_L. That, in turn, causes the output to decrease toward a ground level. The output can be used to drive relays or other circuits to indicate the presence or absence of light.

A **cathode ray tube** is usually referred to as a CRT. In a television, the CRT is called a picture tube. A CRT is a vacuum tube which utilizes beam forming techniques to place an electron spot on the end of the tube, called the **face**. For oscilloscope work, common face diameters are three or five inches, with some that are seven inches. Television picture tubes run as high as twenty-seven inches in diameter.

Figure 13-36 illustrates a typical CRT. The filament is enclosed in a cylindrical cathode sleeve that is closed on the end. The end is coated to permit electron emission. Like a vacuum tube, the control grid controls the electron flow from the cathode through an opening in its end. The second grid or **accelerating anode** is at a positive potential; 300 v is typical. Its purpose is to attract and accelerate the beam of electrons. The second grid tends to focus the beam. The first anode is at a more positive potential, around 1 kv, to further accelerate and shape the beam. The second anode is the **collection anode** or **plate**. The inner surface of the tube is coated to serve as a positive anode, and its potential may be anywhere from 10 kv to 30 kv.

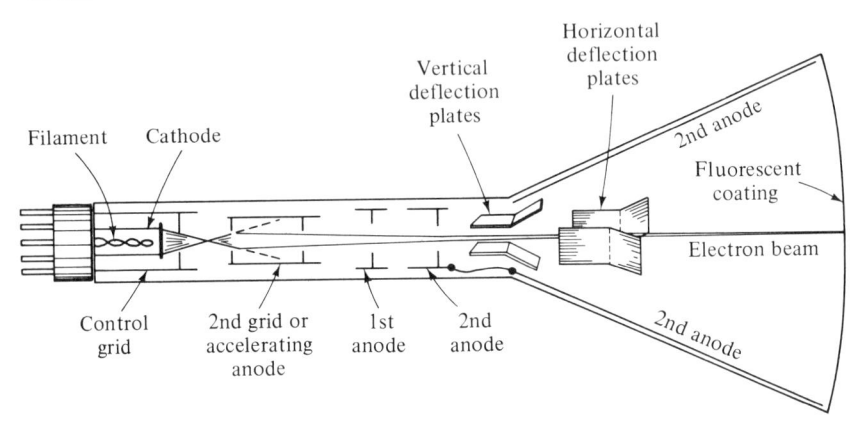

FIGURE 13-36 *Cathode Ray Tube*

The ultimate aim of a CRT is to produce a very small spot of light on the face of the tube. The various grids and anodes are designed to do that. The second grid and first anode, however, are often combined as one electrode. The inner face of the tube is coated with a fluorescent material that becomes illuminated when struck by electrons. The glass face allows the illuminated spot to be viewed externally. The vertical and horizontal deflection plates are used to deflect the beam via special signals.

A difference in potential across a pair of deflection plates will move the beam toward the more positive potential. This technique is called **electrostatic deflection**. In oscilloscopes a time-base generator moves the spot horizontally at a selected speed. The speed is adjustable in terms of frequency or time per unit distance. The vertical deflection plates are connectd to the output of an amplifier. The input of the amplifier can then be used to monitor vertical excursions of ac signals or pulses during the selected period of time.

Biasing of a CRT with variable cathode bias is illustrated in Figure 13-37. The potentiometer controls the electron flow, or the intensity of the spot on the screen (brightness). The other elements are held constant at dc voltages. If $G2$ and $A1$ are combined, the electrode would probably be biased at 500 or 600 v.

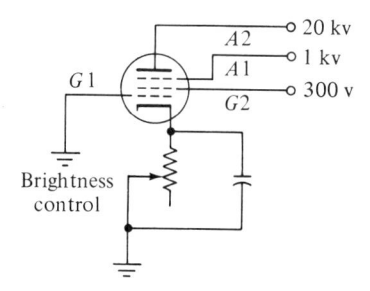

FIGURE 13-37 *CRT Bias Circuit*

Figure 13-38 shows a block diagram of the electrostatic deflection system. The amplifiers simply control the potential difference across the plates. As connected, the output of the amplifiers would have to swing both positively and negatively. Theoretically, the spot should be centered at zero difference between the plates. More often, however, both plates are connected into an amplifier at positive potentials that are balanced out as a net difference of zero. Then the amplifiers work on a pair of plates in push-pull fashion, where one increases while the other decreases.

Magnetic deflection is a technique used in television picture tubes. It entails vertical and horizontal deflection coils instead of plates. Both affect the beam in a similar manner. While the plates use voltage for deflection, the coils use current.

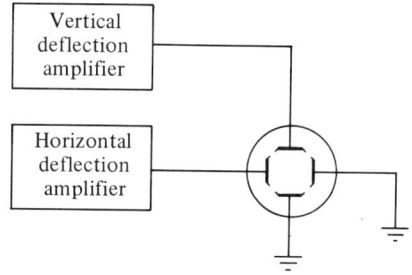

FIGURE 13-38 *CRT Deflection Diagram*

QUESTIONS

13-1   What are the electrodes called in a diode?

13-2   What is the purpose of a filament type heater?

13-3   What is the purpose of the cathode?

13-4   What is the purpose of the plate?

13-5   Assuming a constant diode plate voltage, what is the effect on current conduction through the tube when the cathode temperature is changed?

13-6   Explain how a diode rectifies ac signals.

13-7   Where is the control grid physically located in respect to the other electrodes in a triode?

13-8   What is the purpose of a control grid?

13-9   In Figure 13-14 what value of R_L is required for $E_{bb} = 160$ v, $E_b = 90$ v, and $I_b = 14$ ma?

13-10 In Figure 13-14 what value of E_{bb} is required for $E_b = 120$ v, $I_b = 5$ ma, and $R_L = 36$ k ohms?

13-11 What size cathode resistor is required for Question 13-9 if the required bias is $e_c = -1.5$ v?

13-12 What is the bias voltage for Question 13-10 if a 0.5-k ohm cathode resistor is used for the conditions listed?

13-13 What size cathode bypass capacitor is required for Question 13-12 if the frequency of the signal is to 0.4 MHz?

13-14 Refer to Figure 13-18. What are the upper and lower limits of E_b for $e_c = -8$ v?

13-15 Refer to Figure 13-18. What is the I_b for $e_c = -6$ v and $E_b = 175$ v?

13-16 What is the amplification factor in Figure 13-19 if the E_b scale is changed to 0, 40, 80, 120, 160, and the e_c scale is changed to 0, -1, -2, -3, -4?

13-17 For Question 13-16 what is the mutual transconductance?

13-18 For Question 13-16 what is the plate resistance?

13-19 Refer to Figures 13-22 and 13-23. If $R_L = 55$ k ohms, and $E_{bb} = 275$ v, what is the bias voltage for $E_b = 150$ v?

13-20 For Figure 13-23 if $E_{cc} = -14$ v, what are the values of E_b and I_b?

Vacuum Tubes

13-21 In Figure 13-14, if the amplification factor is 42, and the transconductance is 2100 μ mhos, what is the voltage gain?

13-22 What is the transconductance of a vacuum tube if the circuit voltage gain is 20, load resistance is 20 k ohms, and plate resistance is 5 k ohms?

13-23 When does the dip in plate current occur in a tetrode?

13-24 What is the purpose of a screen grid?

13-25 What currents make up the total cathode current in a tetrode or pentode?

13-26 What is the purpose of a suppressor grid in a pentode?

13-27 Why is the amplification factor for a pentode so much higher than for a triode?

13-28 Why is the plate resistance for a pentode so much higher than for a triode?

13-29 Refer to Figure 13-29. How much are E_b and I_b if e_c is changed to -3.0 v?

13-30 In Figure 13-29 if $E_{bb} = 300$ v, and $R_L = 60$ k ohms, what are the plate current and bias voltage for $E_b = 225$ v?

13-31 For Question 13-30, what is the screen voltage for $I_s = 1.5$ ma, and $R_s = 120$ k ohms?

13-32 What is the voltage gain for Question 13-30 if $\mu = 6000$, and $g_m = 4000\ \mu$ mhos?

13-33 When referring to tube specifications, what is meant by average or service ratings?

13-34 What is the voltage gain of a grounded grid amplifier if $\mu = 24$, $R_L = 30$ k ohms, $R_g = 400$ ohms, and $r_p = 6$ k ohms?

13-35 What size cathode resistor is needed in a grounded plate amplifier if $A_v = 0.91$, $r_p = 6$ k ohms, and $\mu = 25$?

13-36 What are the primary features of a beam power tube?

13-37 What are some of the features of a high power transmitter tube?

13-38 Briefly, what causes a gas tube to conduct?

13-39 What is the effect of illuminating the cathode of a phototube?

13-40 What is the function of the second grid in a CRT?

13-41 When the electron beam strikes the inside face of a CRT, how is the outside illuminated?

13-42 How is an electron beam deflected in an electrostatic CRT?

14 Semiconductors

The two general classes of semiconductors are diodes and transistors. Semiconductor diodes are miniature devices that are generally less expensive than vacuum tube diodes. They do not require filament supplies, a factor which reduces heat losses and results in the cooler operation of electronic equipment. Transistors are also miniature low cost devices which are designed to replace many vacuum tube applications. Figure 14–1 is a photograph of several types of transistors.

Low-current, low-voltage diodes are usually one-quarter inch long and one-eighth inch in diameter. Typical high-current diodes are the size of the end of a pencil, but can run much larger. The typical transistor is about the size of the eraser on a pencil. They, too, vary in accordance with current and voltage requirements.

FIGURE 14–1 *Typical Transistors*

14.1 Semiconductor Theory

Materials used in semiconductors have conduction properties that are in between those of a conductor and an insulator. For example, current flows in a conductor with very little to resist it. An insulator, on the other hand, presents a very high resistance to current flow. Germanium and silicon semiconductor materials have the crystalline structure best suited for diodes and transistors. A comparison of their resistivity to that of copper and mica is shown in Table 14–1.

TABLE 14–1 *Resistivity*

Material	Ohm–cm
Copper (Cu)	1.7×10^{-6}
Germanium (Ge)	4.6×10
Silicon (Si)	2.3×10^3
Mica	1.0×10^{17}

The atomic structures of the two semiconductors are unique in that their valence bands (outermost band) contain four electrons. The atom is in an unstable state until it has the maximum number of eight valence electrons. But with four, it can lose or gain electrons with equal ease. A germanium atom is diagrammed in Figure 14–2. A silicon atom was shown in Chapter 1. The germanium atom has thirty-two electrons, distributed in accordance with the rule of $2n^2$, where n is the number of the band. The silicon atom has fourteen electrons.

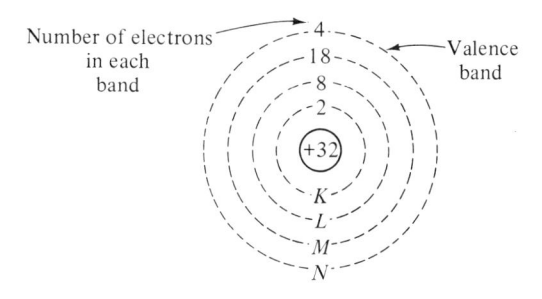

FIGURE 14–2 *Germanium Atom*

If two atoms of germanium were close together, the random orbital paths of the valence electrons would tend to be shared by each of the atoms. If many atoms were close together, they would tend to arrange themselves in a crystal-like structure. In this way the valence electrons complete their sharing process so that each atom in effect has a full valence band of eight electrons. The process bonds the atoms and is called **covalent bonding**. A two-dimensional crystal lattice is shown in Figure 14-3(a). Each of the germanium atoms is diagrammed with only the valence band. Each has four covalent bonds, one with each of four other atoms. The atoms remain in a balanced and stable state because they react as though they have a full valence band.

Figure 14–3(b) shows a **free electron** that has broken away from the bond. It leaves a **hole** behind it which is charged positively the same amount as

(a) Balanced covalent bonding   (b) Free electron

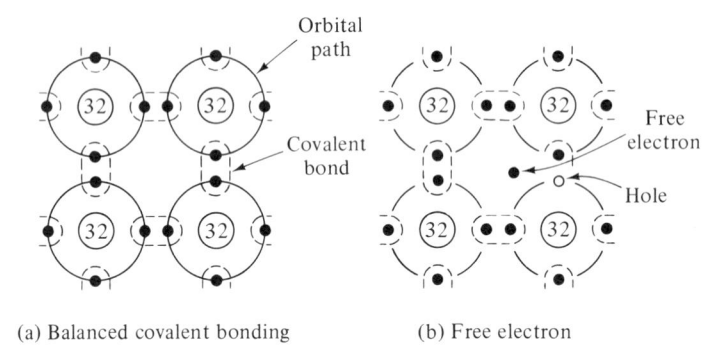

FIGURE 14–3 *Germanium Crystal Lattice (Two Dimension)*

the electron is charged negatively. A free electron drifts at random through a crystal until it finds a hole, and because of the positive charge of the hole, the electron is strongly attracted to fill the hole. Since electrons represent current, free electrons cause a drift current as they drift through the crystal. The current is normally not measurable because of the canceling effect of random motion.

When a voltage potential is connected across a crystal as in Figure 14–4, free electrons are attracted toward the positive terminal. At the same time, "apparent" hole flow is toward the negative terminal. It must be remem-

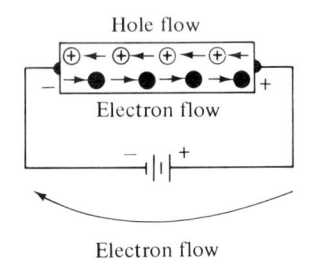

FIGURE 14–4 *Current Flow with Applied Potential*

bered that in a pure crystal considerable energy is required to free electrons from their stable valence bands. The diagram shows both holes and electrons in motion. **Conventional** current follows the direction of the holes, while electron flow goes the opposite direction. Advanced studies in semiconductors ordinarily use the conventional or hole flow direction for current flow.

One more step in describing hole flow is illustrated in Figure 14–5. Electron flow is toward the plus potential. In the first block a hole is shown

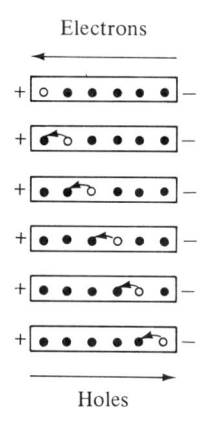

FIGURE 14–5 *Current Flow*

on the left end. Its positive charge, along with the influence of the external potential, attracts an electron. The second block has an electron moving over and filling the hole. That electron left a hole, which is then filled in the next block. The process continues with electron movement to the left while holes move to the right. Even though "apparent," there is a definite pattern to the movement of holes.

The process of electrons breaking free and then finding new holes to fill is called **recombination**. This process occurs on a continuing basis as long as free electrons are in motion, and it involves millions of electrons.

In a pure crystal, known as an **intrinsic crystal**, there is sufficient energy at room temperature to cause some motion of electrons. The motion of electrons increases as temperature increases because heat applies external energy to the crystal. If there were additional electrons or holes added from an impurity, it would take much less energy to cause electron or hole motion. There would be an abundance of charges which would reduce the requirement of electrons having to break free of their valence bands. The process of adding an impurity to a crystal is called **doping**. Effective doping is achieved with approximately one impurity atom for every 10^6 atoms in an intrinsic crystal. More than that would cause excessive motion in the crystal.

The criterion for creating extra electrons is to provide one additional electron in the valence band of each impurity atom so that even though an occasional electron is free to drift, covalent bonding is complete. Impurities with a valence of five fit the category. Extra holes, on the other hand, are created by doping with an impurity with a valence of three so that occasional atoms are missing an electron. Table 14–2 lists elements

TABLE 14–2 *Valence Band Electrons*

3		4		5	
Boron (B)	5	Silicon (Si)	14	Phosphorus (P)	15
Aluminum (Al)	13				
Gallium (Ga)	31	Germanium (Ge)	32	Arsenic (As)	33
Indium (In)	49			Antimony (Sb)	51

used in the manufacture of semiconductor materials. Both silicon and germanium are in the center column under a valance of four. The elements with a valence of five are used for doping when extra electrons are desired. Extra holes are produced by doping with an element in the column with a valence of three. When a crystal is doped, it is known as an **extrinsic crystal**.

Two-dimensional crystals are drawn in Figure 14–6 to illustrate a missing electron (a) and an extra electron (b). Notice that semiconductors are further classified as P-type or N-type. The P-type has extra holes so that holes are considered the **majority carriers** for current. N-type has extra electrons, so electrons are majority carriers for current. Table 14–2 is divided between silicon and germanium. The elements, boron, aluminum,

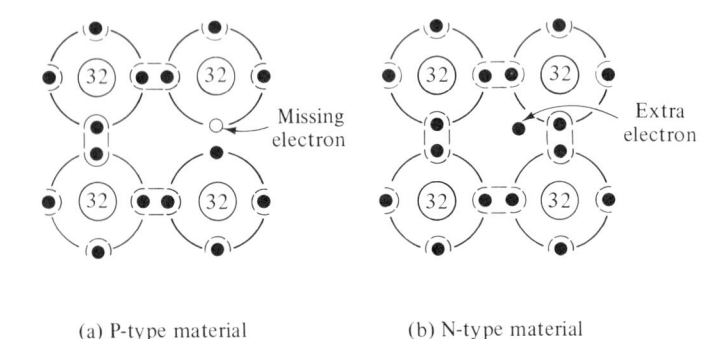

(a) P-type material   (b) N-type material

FIGURE 14–6 *Germanium Crystal with Doping*

and phosphorus, are normally used to dope silicon. The others are for germanium.

Since holes are majority carriers in P-type crystals, electrons are considered **minority carriers**. And in N-type crystals, electrons are majority carriers while holes are minority carriers. The minority carriers account for leakage curents which will be discussed later. A primary difference between holes and electrons is their *mobility*, which is the ease with which they move through a crystal. As shown in Table 14–3, electrons are twice as mobile

TABLE 14–3

Carrier	Mobility
Electron	3800 cm/sec
Hole	1900 cm/sec

as holes. This is expected since free electrons are outside the influence of the valence bands. Hole movement, though, requires a redistribution of charges as the holes progress through the crystal.

Increasing temperature produces thermal energy which generates many free electrons and holes. With more carriers in the crystal, the resistivity decreases. If the temperature of the crystal is raised high enough, about 85°C for germanium and 200°C for silicon, the number of thermally generated electrons and holes greatly exceeds the doping ratio. As a result, the extrinsic value of the semiconductor is lost because the number of electrons comes closer to equaling the number of holes. Also, the resistivity is reduced to an extremely low value.

14.2 P-N Junction

A P-N junction is formed by joining a P-type crystal to an N-type crystal. Alloying and diffusion are two common methods of forming P-N junctions. The process of alloying germanium, for example, begins by placing a minute piece of galium or indium on a wafer of N-type crystal. The wafer, including the impurity, is heated to about 450°C. The impurity melts so that some of its atoms penetrate the germanium. The result is a P-type layer on an N-type wafer which forms a P-N junction.

A P-N junction formed by diffusion results in a more precise junction, but the process is much slower. The crystal wafer is exposed to impurities that are being evaporated. The impurities diffuse into the wafer and form a P or N layer on the impurity, depending upon the impurity and type of crystal. The layer on the crystal forms the P-N junction. This process is widely used in producing the integrated circuits discussed in Chapter 15.

A charge potential must exist across the P-N junction because the P material has excess positive charges (holes), and the N material has excess negative charges (electrons). The distribution of charges across a P-N junction is shown in Figure 14-7. At the first instant of joining, excess electrons from the N material cross the junction toward the positive charges and diffuse into the area adjacent to the junction. At the same time, excess holes are attracted toward the N material and diffuse in that area adjacent to the junction. The diffusion of opposite charges creates an area called the **depletion region**. The negative charges on the P side of the junction finally increase to a point where they repel additional crossings of electrons. And the positive charges on the N side repel additional crossings of holes from the P material.

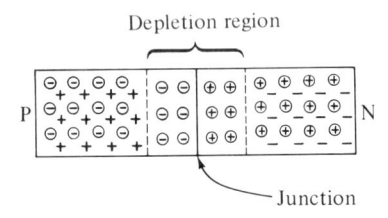

FIGURE 14-7 *P-N Junction Charge Distribution*

The potential across the junction becomes dependent upon the width of the depletion region. The density of the charges is spread due to the diffusion process. A wider region spreads the charges and results in a low charge density, while a narrower region condenses the charges and results in a high charge density. A high charge density allows freer flow of current across the junction. Conversely, a low charge density restricts the flow of current. This concept is important in the operation of P-N junctions with regard to bias potentials and current flow.

Biasing effects on P-N junctions are illustrated in Figure 14-8. The center drawing is without bias and is intended for reference. Forward bias occurs when a positive potential is applied to the P side while a negative potential is applied to the N side. The positive potential attracts electrons through the P material, and the depletion region narrows. Electrons are repelled through the N material by the negative terminal which further reduces the depletion region. As a result, current flows freely through the N material, across the junction, and through the P material. Hole flow would be simultaneous, but in the opposite direction.

The lower drawing illustrates reverse bias. On the P side, the negative potential attracts free holes, and this tends to widen the depletion region. The positive potential on the N side attracts free electrons, and this further

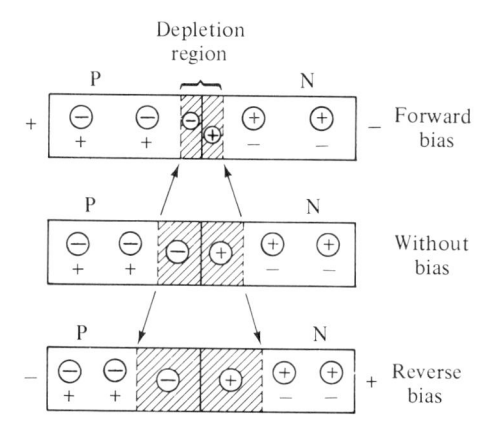

FIGURE 14-8 *Depletion Region*

widens the region. The biasing depletes the region of majority carriers so that current cannot flow. However, a slight amount of leakage current can flow as a result of minority carriers. Typical leakage current ranges from 1 to 10 μa for germanium and ranges from 0.01 to 0.1 μa for silicon. Junctions designed for high current, of course, have leakage paths that are correspondingly higher.

The two drawings in Figure 14-9 show the circuits used to achieve forward and reverse bias. In (a) the positive terminal of the battery is toward the P-type while the negative terminal is toward the N-type. Because forward bias results in low resistance through the device, usually less than 100 ohms, a resistor is used to limit the current flow. Excessive current, beyond the capability of the particular junction, would overheat and damage the device. The completed circuit provides a continuous supply of electrons into the junction. In (b) the junction is reverse-biased. A resistor is used in this case to limit the amount of leakage current in the circuit.

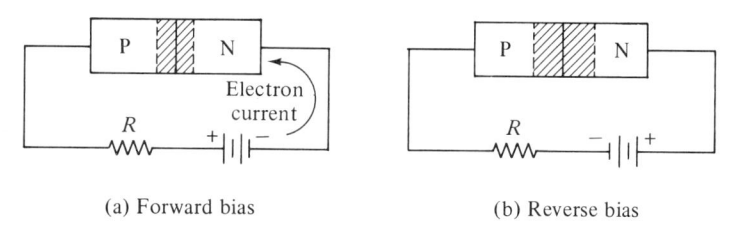

(a) Forward bias   (b) Reverse bias

FIGURE 14-9 *P-N Junction Bias*

14.3 Diodes

The forward and reverse bias characteristics of a P-N junction provide a natural means for rectification. Figure 14-10 shows the diode and its symbol as used in forward and reverse bias circuits. A *V-I* characteristic curve for a typical diode is shown in Figure 14-11. The forward bias area illustrates the non-linear characteristic of forward current (I_F) while increasing the voltage across the diode. Normal conduction of a diode occurs at a point where the curve begins to rise steeply (about 0.5 v in the plot). Diode drops vary from one type of device to another. In general, germanium is about 0.3 v, and silicon is about 0.6 v.

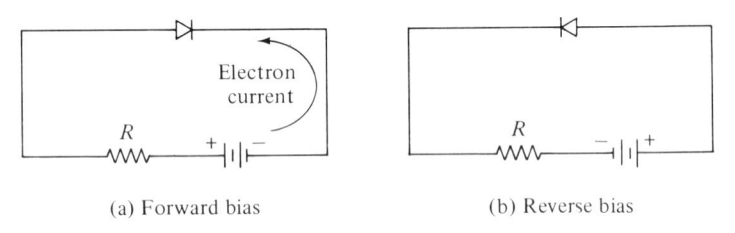

(a) Forward bias   (b) Reverse bias

FIGURE 14-10   *P-N Junction Diode Bias*

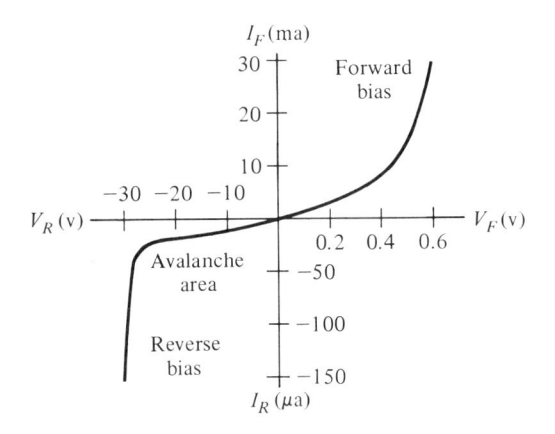

FIGURE 14-11   *Diode Characteristic Curve*

The reverse-bias area is on a reduced scale to illustrate leakage current. When a junction reaches its rated reverse voltage, it breaks down, and the current increases rapidly. This area is known as the **avalanche** or **zener** breakdown area.

Diode rectification is shown in Figure 14-12. Negative clipping occurs in (a). During positive excursions of the input square wave, the diode is forward-biased and conducts current. The output, then, reproduces the positive portion. During negative excursions, the diode is reverse-biased and cannot conduct. Thus, the output remains at zero, which is the ground reference level in the circuit. In (b) positive clipping occurs. The diode rejects positive pulses, but passes negative pulses.

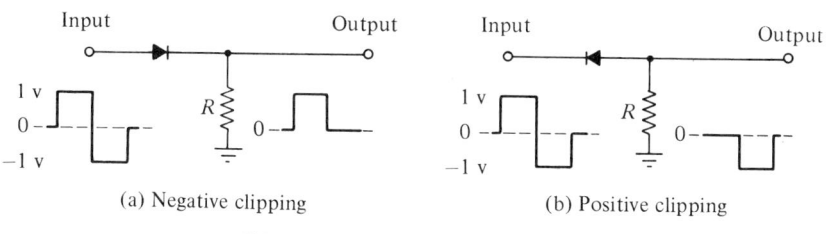

FIGURE 14-12 *Diode Rectification*

(a) Negative clipping (b) Positive clipping

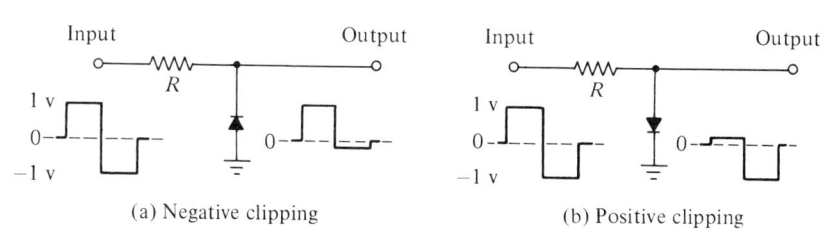

FIGURE 14-13 *Diode Clipping*

(a) Negative clipping (b) Positive clipping

The circuits in Figure 14-13 accomplish nearly the same result as the ones in the previous figure. In (a) the diode is essentially an open-circuit to positive pulses. Negative pulses forward bias the diode and conduct current. The diode resistance is very low compared to R, so that the resultant output is just slightly negative an amount equal to the voltage drop across the diode. The circuit in (b) is similar to the one in (a), except for positive clipping.

Diode action is applicable to sine waves in the same way it is to pulses or square waves. Other applications include blocking the paths of dc levels in special circuit configurations. Rectification in power supplies is an application that is described in Chapter 16.

14.4 PNP/NPN Junction

PNP and NPN junctions are an extension of P-N junctions. The junction theory is the same. In Figure 14-14(a), the charge distribution is shown for three crystals joined to form a PNP junction. The depletion regions are

(a) PNP junction (b) NPN junction

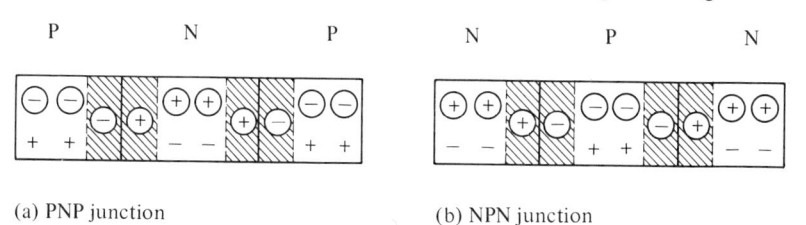

FIGURE 14-14 *Charge Distribution—Three Crystals*

like those for P-N junctions. Alloying and diffusion techniques apply for any number of junctions. The NPN junction in (b) is a mirror image of the PNP junction.

Biasing of PNP/NPN junctions is specifically arranged to provide **transistor action**, which is the amplification of current. To achieve transistor action, the base-emitter (*B-E*) junction is forward-biased, and the collector-base (*C-B*) junction is reverse-biased. The bias circuit is shown in Figure 14–15(a). Notice that the depletion regions are narrow at the *B-E* junction and wide at the *C-B* junction. As in *PN* diode circuits, the purpose of the series resistors is to limit the current. The names for each connection to the transistor are: **emitter, base**, and **collector**. The first letter of each word is its abbreviation.

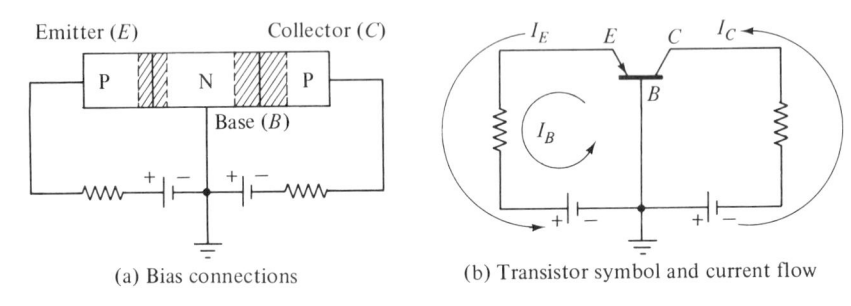

FIGURE 14–15 *PNP Transistor Junction Bias*

The symbol for a PNP transistor and the direction of current flow is shown in Figure 14–15(b). The forward-biased section, *B-E*, readily conducts current. Electron flow begins in the negative terminal of the battery, passes through the base-emitter junction, the series resistor, and back to the positive terminal. When forward-biased, the *B-E* junction provides an easy path for current carriers. Because holes are the majority carriers in P material, current flow passes initially from the emitter toward the base. The base is lightly doped and made very thin. The holes travel in large numbers toward the base, but very few holes recombine in the base because of the light doping and thin layer. Most of the holes are carried to the negative potential at the collector. Hole movement in one direction results in electron flow in the other; therefore, a large flow of holes to the collector results in an equally large flow of electrons to the emitter.

The transistor action of Figure 14–15 is a large current flow from the collector to the emitter with a small forward-bias current through the *B-E* junction. Typical transistors require from 1 to 5 percent of the total emitter current for forward bias, which means that about 95 to 99 percent of the emitter current is available as collector current. The dc current gain of a transistor is thus the ratio of collector current (I_C) to base current (I_B).

Current gain is called **beta**, abbreviated β. The dc current gain is calculated with the following formula.

$$\beta_{dc} = \frac{I_C}{I_B} \quad (14\text{-}1)$$

An NPN transistor is shown in Figure 14–16(a). The batteries are reversed from the previous circuit, but the same rules apply to bias. The *B-E* is forward-biased and the *C-B* is reverse-biased. The depletion regions are narrow at the *B-E* junction and wide at the *C-B* junction. In (b) the currents are for electron flow and are the opposite of those in Figure 14–15. In the emitter material, electrons are majority carriers, and they flow in large numbers toward the base. Very few recombine in the thin, lightly doped base. Therefore, the electrons continue in large numbers toward the positive potential on the collector. As in the PNP version, hole flow is opposite to electron flow, and a small current at the base region causes a large current in the collector region. Current gain [formula (14-1)] applies to both PNP and NPN transistors.

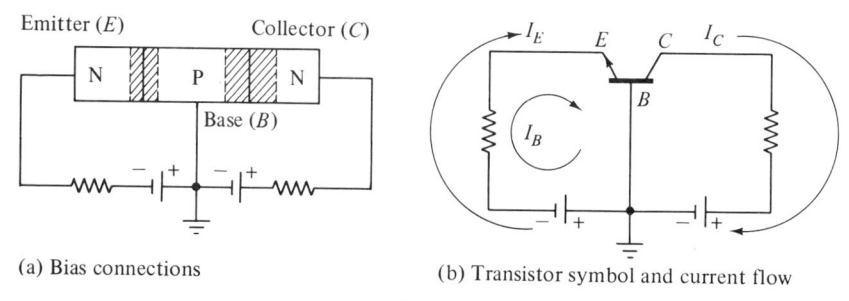

(a) Bias connections

(b) Transistor symbol and current flow

FIGURE 14–16 *NPN Transistor Junction Bias*

Notice that in Figures 14–15(b) and 14–16(b) the currents for the base and collector must flow through the emitter. Kirchhoff's law states that the sum of the currents entering a junction must equal the sum of the currents leaving the junction. Therefore, the sum of the base and collector currents is equal to the emitter current.

$$I_E = I_B + I_C \quad (14\text{-}2)$$

The formula can be rearranged to show

$$I_C = I_E - I_B$$
$$I_B = I_E - I_C$$

Another definition of transistor currents is the percentage of emitter current that is collector current. The percentage is presented in a decimal quantity and called **alpha**, abbreviated α.

$$\alpha_{dc} = \frac{I_C}{I_E} \quad (14\text{-}3)$$

As an example, if I_C is 95 percent of I_E, and I_E is 10 ma, I_C is calculated:

$$I_C = \alpha I_E = 0.95 \times 10 \times 10^{-3} = 9.5 \times 10^{-3} = 9.5 \text{ ma}$$

It should be obvious that the terms in the formula are completely interchangeable. For example, I_C can be replaced with $I_E - I_B$; I_E can be replaced with $I_B + I_C$; and I_B can be replaced with $I_E - I_C$. Values of beta and alpha are normally specified for the various devices depending upon their construction. However, the values are usually specified in terms of a minimum value or a range due to manufacturing tolerances.

Conversion of beta to alpha and vice-versa is accomplished by combining formulas (14–1) and (14–3).

$$\alpha = \frac{\beta}{\beta + 1} \tag{14-4}$$

$$\beta = \frac{\alpha}{1 - \alpha} \tag{14-5}$$

The conversion process is explained in the following examples.

$$\alpha = \frac{I_C}{I_E} = \frac{\beta I_B}{I_E} = \frac{\beta I_B}{\beta I_B + I_B} = \frac{\beta I_B}{I_B(\beta + 1)} = \frac{\beta}{\beta + 1}$$

$$\beta = \frac{I_C}{I_B} = \frac{\alpha I_E}{I_B} = \frac{\alpha I_E}{I_E - \alpha I_E} = \frac{\alpha I_E}{I_E(1 - \alpha)} = \frac{\alpha}{1 - \alpha}$$

Some examples of applying the current formulas, (14–1) through (14–5), are:

Example 14–1 If beta is 40, and I_B is 0.03 ma, how much are I_C and I_E?

Solution
$$I_C = \beta I_B = 40 \times 0.03 \times 10^{-3} = 1.2 \times 10^{-3}$$
$$= 1.2 \text{ ma}$$
$$I_E = I_B + I_C = 0.03 \text{ ma} + 1.2 \text{ ma} = 1.23 \text{ ma}$$

Example 14–2 If $I_E = 10$ ma, and $I_C = 9.8$ ma, how much are beta and I_B?

Solution
$$I_B = I_E - I_C = 10 \text{ ma} - 9.8 \text{ ma} = 0.2 \text{ ma}$$
$$\beta = \frac{I_C}{I_B} = \frac{9.8 \times 10^{-3}}{0.2 \times 10^{-3}} = 49$$

Example 14–3 For a beta of 100, and I_E of 1.01 ma, how much are I_C and I_B?

Solution
$$\beta = \frac{I_C}{I_B} = \frac{I_E - I_B}{I_B}$$
$$I_E = \beta I_B + I_B = I_B(\beta + 1)$$

$$I_B = \frac{I_E}{\beta + 1} = \frac{1.01 \times 10^{-3}}{100 + 1} = 0.01 \times 10^{-3} = 10\ \mu a$$

$$I_C = I_E - I_B = 1\ \text{ma} - 0.01\ \text{ma} = 0.99\ \text{ma}$$

Example 14-4 What is the value of alpha if I_C is 9.7 ma, and I_E is 10 ma?

Solution
$$\alpha = \frac{I_C}{I_E} = \frac{9.7\ \text{ma}}{10\ \text{ma}} = 0.97$$

Example 14-5 What are the values of I_E and I_C if alpha is 0.95 and I_B is 50 μa?

Solution
$$\alpha = \frac{I_C}{I_E} = \frac{I_E - I_B}{I_E}$$

$$\alpha I_E = I_E - I_B \qquad I_B = I_E - \alpha I_E = I_E(1-\alpha)$$

$$I_E = \frac{I_B}{1-\alpha} = \frac{50 \times 10^{-6}}{1.00 - 0.95} = \frac{50 \times 10^{-6}}{0.05}$$

$$= 1 \times 10^{-3} = 1\ \text{ma}$$

$$I_C = I_E - I_B = 1\ \text{ma} - 0.05\ \text{ma} = 0.95\ \text{ma}$$

Example 14-6 What is the value of alpha for Example 14-1?

Solution
$$\alpha = \frac{\beta}{\beta + 1} = \frac{40}{40 + 1} = 0.975$$

Example 14-7 What is the value of beta for Example 14-5?

Solution
$$\beta = \frac{\alpha}{1-\alpha} = \frac{0.95}{1.00 - 0.95} = 19$$

14.5 Bias Effects

For transistor action, which is the amplification of current, the *B-E* is forward-biased, and the *C-B* is reverse-biased. Examples in the last section demonstrated that the collector current is a function of beta and the input bias current I_B, which is the *B-E* forward-bias current. Assuming a constant collector bias voltage, changing the base current will change the collector current. There must be base current so that majority carriers are available for collector current. The curve in Figure 14-17 illustrates a non-linear rise in collector current as base current is increased. Base current can only flow when the junction is forward-biased. The base-emitter junction voltage follows the trend established in Figure 14-11 for diodes. If the junction is reverse-biased, collector current other than leakage current cannot flow.

There has to be a limit to how much collector current can flow when I_B is increased. That point is called **saturation** and occurs when further increases

342   Semiconductors   [14.5]

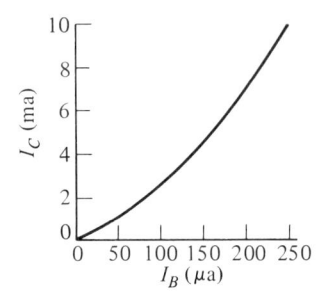

FIGURE 14-17 *Collector Characteristic*

in I_B produce no further increase in I_C. The saturation point depends upon the transistor characteristics and the supply voltages.

A summary of transistor symbols is contained in Table 14–4. The sym-

TABLE 14-4 *Transistor Symbols—dc*

Symbol	Description
V_{CC}	Collector supply voltage
V_C	Collector voltage
I_C	Collector current
R_C or R_L	Collector load resistor
V_{EE}	Emitter supply voltage
V_E	Emitter voltage
I_E	Emitter current
R_E	Emitter resistor
V_{BB}	Base supply voltage
V_B	Base voltage
I_B	Base current
R_B	Base resistor
β	Current gain—base to collector
α	Current gain—emitter to collector

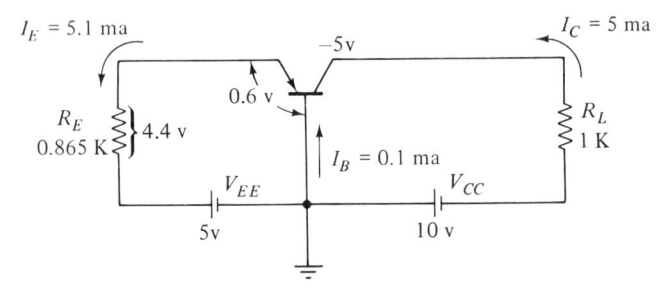

FIGURE 14-18 *Bias Effects for $B_{dc} = 50$*

bols are for dc applications; ac symbols are covered later. All voltage values are referenced to the common point in the circuit, usually ground. Notice that voltages carry a symbol of V instead of E to avoid confusion with the symbol for the emitter.

A bias arrangement controlled with emitter current is shown in Figure 14–18. The transistor is rated for a beta of 50. This means that I_B will be one-fiftieth of I_C. I_C, however, is dependent upon the amount of I_E, and I_E is controlled by V_{EE} and R_E, along with the voltage drop across the *B-E* junction (0.6 v for this example). The emitter current is calculated as follows.

$$I_E = \frac{V_{EE} - V_{BE}}{R_E} = \frac{5 - 0.6}{0.865 \times 10^3} = \frac{4.4}{0.865 \times 10^3}$$
$$= 5.1 \times 10^{-3} = 5.1 \text{ ma}$$

The base current is calculated from the formula derived in Example 14–3.

$$I_B = \frac{I_E}{\beta + 1} = \frac{5.1 \text{ ma}}{50 + 1} = 0.1 \text{ ma}$$

The collector current is

$$I_C = I_E - I_B = 5.1 \text{ ma} - 0.1 \text{ ma} = 5 \text{ ma}$$

The voltage on the collector is

$$V_C = V_{CC} - I_C R_L = -10 + (5 \times 10^{-3} \times 1 \times 10^3)$$
$$= -10 + 5 = -5 \text{ v}$$

Note that if I_E were increased sufficiently, an I_C of 10 ma would drop 10 v across R_L so that V_C would be 0. In this circuit, I_E controls the amount of I_C.

Example 14–8 In Figure 14–18, what are the values of I_C and I_B if R_E is 2.2 k ohms?

Solution
$$I_E = \frac{V_{EE} - V_{BE}}{R_E} = \frac{5 - 0.6}{2.2 \times 10^3} = \frac{4.4}{2.2 \times 10^3}$$
$$= 2 \text{ ma}$$

$$I_B = \frac{I_E}{\beta + 1} = \frac{2 \text{ ma}}{50 + 1} = 0.039 \text{ ma}$$

$$I_C = I_E - I_B = 2 \text{ ma} - 0.039 \text{ ma} = 1.961 \text{ ma}$$

Example 14–9 What is the collector voltage for Example 14–8?

Solution
$$V_C = V_{CC} - I_C R_L = -10 + (1.961 \times 10^{-3} \times 1 \times 10^3)$$
$$= -10 + 1.961 = -8.039 \text{ v}$$

Example 14-10 For Figure 14-18, what is the required R_E for an I_B of 0.15 ma?

Solution
$$I_E = I_B(\beta + 1) = 0.15 \text{ ma}(50 + 1) = 7.65 \text{ ma}$$
$$R_E = \frac{V_{EE} - V_{BE}}{I_E} = \frac{5 - 0.6}{7.65 \times 10^{-3}}$$
$$= \frac{4.4}{7.65 \times 10^{-3}} = 0.575 \text{ k}\Omega$$

Figure 14-19 illustrates a variation in bias arrangements. In this case, the emitter is connected to the ground reference. The collector of the PNP transistor is still reverse-biased because $V_B = -0.6$ v, and $V_C = -5$ v. The circuit is designed similarly to the previous one. When beta is 50 and I_B is 0.1 ma, I_C is calculated:

$$I_C = \beta I_B = 50 \times 0.1 \times 10^{-3} = 5 \text{ ma}$$

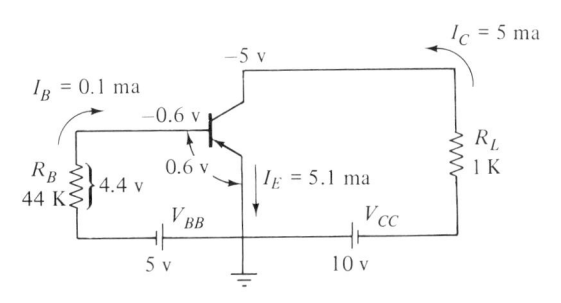

FIGURE 14-19 *Bias Effects with Base Current Control* ($B_{dc} = 50$)

Remaining calculations are

$$R_B = \frac{V_{BB} - V_{BE}}{I_B} = \frac{5 - 0.6}{0.1 \times 10^{-3}} = \frac{4.4}{0.1 \times 10^{-3}}$$
$$= 44 \times 10^3 = 44 \text{ k}\Omega$$
$$V_C = V_{CC} - I_C R_L = -10 + (5 \times 10^{-3} \times 1 \times 10^3)$$
$$= 5 \text{ v}$$

Example 14-11 In Figure 14-19, what are the collector current and voltage for a beta of 20?

Solution
$$I_C = \beta I_B = 20 \times 0.1 \times 10^{-3} = 2 \times 10^{-3} = 2 \text{ ma}$$
$$V_C = V_{CC} - I_C R_L = -10 + (2 \times 10^{-3} \times 1 \times 10^3)$$
$$= -8 \text{ v}$$

Example 14-12 For Figure 14-19, what values are required for R_L and R_B if beta is 30, V_C is -3 v, and I_C is 300 μa?

Solution

$$R_L = \frac{V_{CC} - V_C}{I_C} = \frac{10-3}{0.3 \times 10^{-3}} = 23.3 \times 10^3$$
$$= 23.3 \text{ k}\Omega$$

$$I_B = \frac{I_C}{\beta} = \frac{300 \ \mu a}{30} = 10 \ \mu a$$

$$R_B = \frac{V_{BB} - V_{B-E}}{I_B} = \frac{5 - 0.6}{10 \times 10^{-6}} = \frac{4.4}{10 \times 10^{-6}}$$
$$= 0.44 \text{ M}\Omega$$

14.6 Basic Transistor Configurations

There are three basic circuit configurations for operating transistors. Each configuration is defined in terms of which element of the transistor is *in common with both the input and output signals*. So far, biasing techniques have been described without input/output signals. Since dc current changes in the base or emitter circuits are transferred to the collector, it follows that by superposition, an ac signal can also cause current variations.

In Figure 14–20(a) the base is referenced to ground, which is the common connection for the input and output. Thus, the configuration is known as a **common-base circuit**. Notice that the input is into the emitter, and that changes in I_E cause slightly lesser changes in I_C. When I_C varies, the IR drop across R_L varies and provides an output signal voltage.

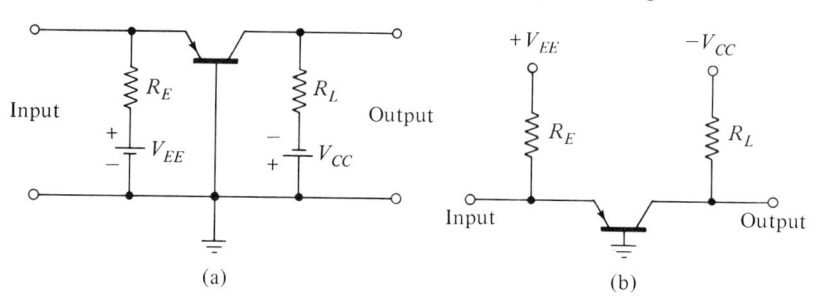

FIGURE 14–20 *Common-Base Configuration*

The biasing of the figure is the same as in Figures 14–15 and 14–18, which were also common-base circuits. An NPN transistor can be used instead if the two dc supplies are reversed. The diagram in Figure 14–20(b) is identical to (a), except the dc sources are assumed to return to ground. This drawing is a less complicated method of presentation, especially when many transistors may be used to construct a complete circuit.

The configuration in Figure 14–21(a) is a **common-emitter circuit**. Note that the emitter is common to both the input and output. Biasing is the same as described in Figure 14–19. This configuration has a definite current

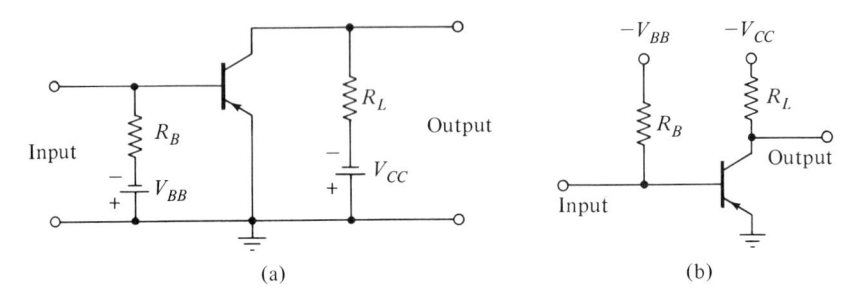

FIGURE 14-21 *Common-Emitter Configuration*

gain, which has the value of beta. DC beta has already been covered. Signal current gain is ac beta and is generally quite close to the dc beta value. The output is derived from the *IR* drop across R_L. If I_B varies, I_C varies and provides a varying voltage at the output. A simplified version of the common-emitter circuit is drawn in (b).

A **common-collector circuit** is shown in Figure 14–22(a). Biasing of this version has not been covered yet, though the rules are similar. The *B-E* junction is forward-biased, and the collector is reverse-biased. The input signal is into the base, like the common-emitter version. The output, though, is now taken from the emitter where an output voltage is developed across R_L. Current gain occurs since I_E is $I_B(\beta + 1)$.

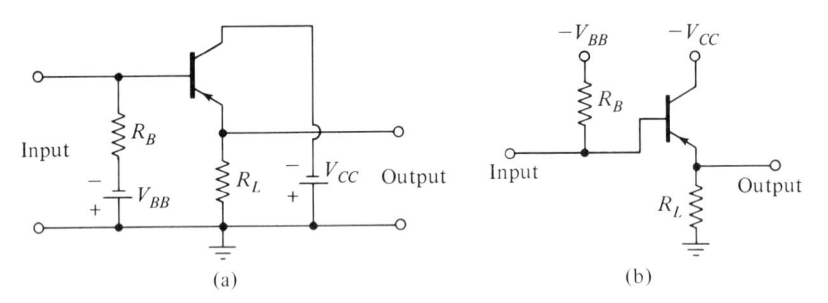

FIGURE 14-22 *Common-Collector Configuration*

Some typical characteristics of the configurations are listed in Table 14–5 for comparison. The common-base has a current gain just under unity because the signal is fed into the emitter. A very small voltage change at the emitter produces a large change in I_E and subsequently a large voltage change across the load resistor. Thus, it has the highest voltage gain. Power gain is the product of current and voltage gain.

The common-emitter has a current gain equivalent to the value of beta. Voltage gains are not ordinarily as high as the common-base; however,

TABLE 14-5 *Typical Characteristics*

Parameter	Common-base	Common-emitter	Common-collector
Current Gain	0.98	50	50
Voltage Gain	500	100	0.95
Power Gain	490	5000	47.5

the power gain is highest since this configuration has both current and voltage gain.

Although the common-collector has a current gain of approximately beta (actually $\beta + 1$), its voltage gain is less than unity. The *B-E* junction is forward-biased, so that the emitter voltage tends to follow the base. The signal is attenuated a slight amount due to the non-linear characteristics of a PN junction. The power gain is least of all the configurations.

14.7 Characteristic Curves

Characteristic curves for transistors are similar to vacuum tube curves. In general, the same rules apply in their analysis. And like vacuum tube curves, a set of transistor curves represents a typical device for the particular transistor type.

Figure 14-23 illustrates a set of curves for a common-base configuration. Voltage between the collector and base (V_{CB}) is plotted on the horizontal

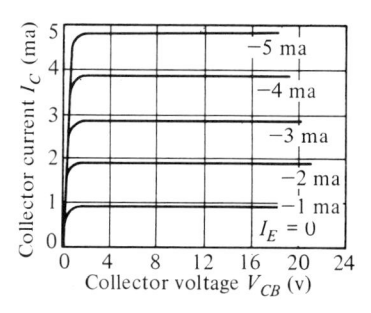

FIGURE 14-23 *Common-Base Characteristics*

axis, and collector current (I_C) is plotted on the vertical axis. A family of curves is plotted for 1 ma increments of I_E from 1 to 5 ma. Note that the I_E values are slightly higher than the I_C values, which is expected since $I_E = I_C + I_B$. For example, when $I_E = 5$ ma, I_C is about 4.8 ma. I_B, then, is calculated:

$$I_B = I_E - I_C = 5.0 \text{ ma} - 4.8 \text{ ma} = 0.2 \text{ ma}$$

In terms of dc alpha, using formula (14-3) gives

$$\alpha_{dc} = \frac{I_C}{I_E} = \frac{4.8 \text{ ma}}{5 \text{ ma}} = 0.96$$

A circuit for producing the common-base curves is shown in Figure 14-24. It is simply a minor rearrangement of Figure 14-20(b). The meters are inserted for measurements, and the dc supply voltages are variable. When V_{EE} is zero, I_E is zero. I_C is zero over the V_{CB} range of 0 to 24 v. Recall that I_C cannot flow unless the emitter-base junction is forward-biased.

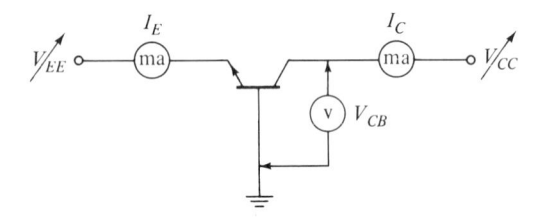

FIGURE 14-24 *Common-Base Circuit for Plotting Characteristic Curves*

The first curve is plotted by first adjusting V_{EE} so that $I_E = 1$ ma. Next, V_{CC} is increased in small increments while monitoring V_{CB} and I_C. Notice that when V_{CB} is hardly over zero, I_C rises sharply to an amount slightly under 1 ma. Since I_E is held at 1 ma, I_C cannot be any higher than αI_E. So increasing V_{CB} results in a "flat" curve at a constant value of αI_E.

Each of the other curves is plotted in the same way. I_E is adjusted to 2 ma, 3 ma, etc., and I_C is plotted for V_{CB} values of 0 to 24 v. Notice that all of the curves are flat and constant at I_C values just under the I_E values. Unfortunately, accurate values of alpha are impossible to calculate from common-base curves since I_c can only be approximated from the graph.

Figure 14-25 illustrates a set of curves for a common-emitter configuration. Collector voltage is plotted horizontally, and collector current is

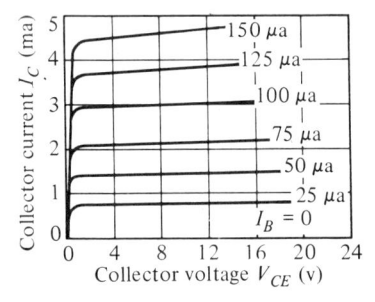

FIGURE 14-25 *Common-Emitter Characteristics*

plotted vertically. In this graph, though, I_C is plotted for constant values of base current.

A common-emitter circuit for plotting the curves is shown in Figure 14–26. The circuit is similar to the one in Figure 14–21(b). Ammeters monitor base and collector currents while a voltmeter monitors V_{CE}.

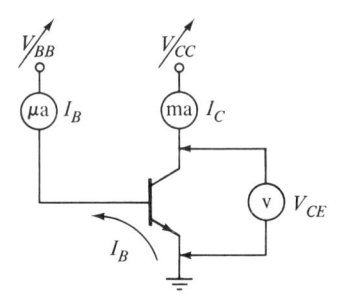

FIGURE 14–26 *Common-Emitter Circuit for Plotting Characteristic Curves*

When $I_B = 0$, the collector current must be zero. The first curve is plotted by first adjusting V_{BB} so that I_B is 25 μa. Then V_{CC} is increased starting from $V_{CC} = 0$. A very slight amount of collector voltage results in a sharp rise in collector current. Collector current rises to a value that is dc beta times the base current.

$$I_C = \beta_{dc} I_B$$

Each curve is plotted in the same way, by holding I_B constant and measuring I_C for varying V_{CE} values. The lower values of I_B produce fairly flat I_C curves, but higher I_B values show a slope or slight rise in I_C when V_{CE} is increased. This occurs because of internal leakage current which adds to the collector current.

Beta can be calculated directly from the curves. Curves for each of the base currents correspond to a certain amount of collector current. For example, formula (14–1) defined dc beta as

$$\beta_{dc} = \frac{I_C}{I_B}$$

The curve for $I_B = 25$ μa is at $I_C = 0.75$ ma.

$$\beta_{dc} = \frac{I_C}{I_B} = \frac{0.75 \times 10^{-3}}{25 \times 10^{-6}} = 30$$

At the upper end of the curves, a base current line of 150 μa provides an approximate collector current of 4.5 ma.

$$\beta_{dc} = \frac{I_C}{I_B} = \frac{4.5 \times 10^{-3}}{150 \times 10^{-6}} = 30$$

Due to the non-linear characteristics of transistors, beta is not necessarily the same for all values of collector current. The alpha of a transistor may also vary, depending on the device and the collector current.

The ac beta of a transistor is defined as a change in collector current divided by a change in base current. AC and dc beta values are often close, but not necessarily the same since dc represents static conditions, and ac represents dynamic conditions. The formula is

$$\beta_{ac} = \frac{\Delta I_C}{\Delta I_B} \tag{14-6}$$

Several examples for Figure 14–25 follow.

Example 14–13 What is the ac beta for an average V_{CE} of 8 v when I_B varies between 25 and 50 μa?

Solution At $V_{CE} = 8$ v, the two values of I_C that correspond to $I_B = 25$ and 50 μa are about 0.75 ma and 1.4 ma. Using formula (14–6) yields

$$\beta_{ac} = \frac{\Delta I_C}{\Delta I_B} = \frac{(1400 \times 10^{-6}) - (750 \times 10^{-6})}{(50 \times 10^{-6}) - (25 \times 10^{-6})}$$

$$= \frac{650}{25} = 26$$

Example 14–14 What is the ac beta for an average V_{CE} of 4 v when I_B varies between 4.45 ma and 3 ma?

Solution The two I_C points along the 4-v line intercept $I_B = 100$ μa and 150 μa.

$$\beta_{ac} = \frac{\Delta I_C}{\Delta I_B} = \frac{(4.45 \times 10^{-3}) - (3 \times 10^{-3})}{(150 \times 10^{-6}) - (100 \times 10^{-6})}$$

$$= \frac{1.45 \times 10^{-3}}{50 \times 10^{-6}} = 29$$

Notice the difference in the two betas of the examples. By shifting operational points, a variation from 26 to 29 occurred. Both values are relatively close to the dc beta of 30.

14.8 Specifications

This section discusses an introductory list of transistor specifications. The list covers the common specifications in general use.

Maximum ratings of transistors are given to indicate the highest voltages, currents, and power dissipation that the device will withstand without being damaged. Voltages and currents are normally listed for operation at room temperature, 25°C.

Collector-Base Voltage is the maximum reverse dc voltage between the collector and base.

Collector-Emitter Voltage is the maximum reverse dc voltage between the collector and emitter. For a PNP transistor, polarities are negative to collector and positive to emitter. The polarities are the opposite for an NPN.

Emitter-Base Voltage is the maximum reverse dc voltage between the emitter and base.

Collector Current is the maximum collector current.

Emitter Current is the maximum emitter current.

Device Dissipation is the maximum EI rating of the transistor. The collector and base circuits are summed so that the total power dissipation of the device can be calculated.

$$P_d = (V_{CE}I_C) + (V_{BE}I_B)$$

If the dissipation rating is given for room temperature (25°C) a derating factor must be used, which means the transistor is rated for a lower dissipation at higher temperatures. The derating factor is usually in the mw/°C rise, so that if the device is intended to operate at 50°C, for example, then 25 (50 − 25) times the derating factor (mw) is subtracted from the maximum rating. If the maximum rating at 25°C is 500 mw, and the derating factor is 3 mw/°C, the maximum device rating at 75°C is calculated

$$P_{max} = 500 \text{ mw} - \left(\frac{3 \text{ mw}}{°C}\right)(75°C - 25°C)$$
$$= 500 \text{ mw} - \frac{3 \text{ mw} \times 50°C}{°C}$$
$$= 500 \text{ mw} - 150 \text{ mw} = 350 \text{ mw}$$

Sometimes dissipation ratings are given for several temperatures so that a derating factor is not needed. For example:

Device Dissipation:	25°C	300 mw
	50°C	250 mw
	100°C	150 mw

h parameters are often used to specify certain transistor ratings. The parameters are intended to represent a typical operational characteristic for the type of device being specified. There are several different ways of expressing the subscripts for h parameters. For purposes of introduction, only one is explained here.

h_i is the input resistance of a transistor in ohms.

h_o is the output admittance or internal conductance in millimhos.

h_r is the feedback voltage ratio and is a number somewhere around 10^{-4} to 10^{-5}.

The h parameters are further defined with an additional subscript of b for common-base application or e for common-emitter application. For example, h_{ib}, h_{ob}, and h_{rb} apply to common-base, while h_{ie}, h_{oe}, and h_{re} apply to common-emitter.

Current transfer ratios are:

h_{fb} for common-base alpha

h_{fe} for common-emitter beta

Another parameter of interest is the frequency response of a transistor. The parameter defines the frequency cut-off point of the transistor, which occurs when the alpha or beta has decreased to 0.707 of the midband value. In other words, a value of alpha or beta is specified at a midband frequency of approximately 1 to 5 kHz. If the frequency were increased, a point would be reached where the value of alpha or beta would fall off to 70.7 percent of its midband value.

$h_{\alpha o}$ is the cutoff frequency for common-base alpha

$h_{\beta o}$ is the cutoff frequency for common-emitter beta

Other items included in transistor specifications are: (1) Type—PNP or NPN, germanium, or silicon (2) Outlines—physical dimensions (3) Characteristic curves—may include collector characteristics, junction characteristics, and input characteristics.

14.9 Basic Transistor Circuits

This section provides an introduction to basic circuits for the three configurations: common-base, common-emitter, and common-collector. All three are described for sinusoidal signal amplification. And the characteristic curves in Figures 14–23 and 14–25 are assumed applicable to these circuits.

Transistor circuits that amplify ac signals use notations that are similar to the dc symbols of Table 14–4. The primary symbol, however, is lower

TABLE 14–6 *Transistor Symbols—ac*

Symbol	Description
i_B	ac base current
i_C	ac collector current
i_E	ac emitter current
v_B	ac base voltage
v_C	ac collector voltage
v_E	ac emitter voltage
v_{in}	ac input voltage
v_{out}	ac output voltage

case for ac values. For instance, i is for ac current, and v is for ac voltage. The symbols are summarized in Table 14-6.

A common-base amplifier is drawn in Figure 14-27. The resistors, R_E and R_L, are used for obtaining proper dc bias for the circuit. The two capacitors couple ac signals in and out of the circuit. They also keep the dc levels from interfering with other circuits. Class A operation is desirable since the collector signal swings plus and minus around a reference point. To maintain the collector within the region of zero to $+5$ v, an operating point is selected midway. In the figure, $+2.5$ v is chosen for V_C. With an R_L of 2.5 k ohms,

$$I_C = \frac{V_{CC} - V_C}{R_L} = \frac{5\text{ v} - 2.5\text{ v}}{2.5\text{ k}\Omega} = 1\text{ ma}$$

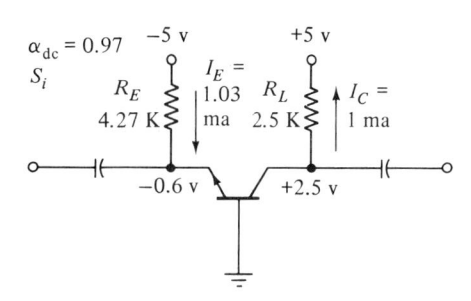

FIGURE 14-27 *Common-Base Amplifier*

DC alpha is approximately 0.97, as taken from the curves in Figure 14-23. The required emitter current to deliver an I_C of 1 ma is

$$I_E = \frac{I_C}{\alpha} = \frac{1\text{ ma}}{0.97} = 1.03\text{ ma}$$

The junction voltage for silicon, V_{BE}, is approximately 0.6 v. Thus, R_E has an IR drop of

$$V_{RE} = V_{EE} - V_{BE} = 5 - 0.6 = 4.4\text{ v}$$

The required emitter biasing resistor is then

$$R_E = \frac{V_{RE}}{I_E} = \frac{4.4\text{ v}}{1.03\text{ ma}} = 4.27\text{ k}\Omega$$

A small signal change at the emitter will result in a large signal change at the collector. Even though there is less than unity current gain, a small voltage change across the emitter-base junction causes a large voltage change across R_L. The input resistance of a common-base configuration is small; in the circuit described (Figure 14-27) it is about 25 ohms between the emitter and ground. A *p-p* signal of 10 mv develops an input current of approximately

$$i_E = \frac{v_{in}}{R_{in}} = \frac{10 \text{ mv}}{25 \Omega} = 0.4 \text{ ma}$$

The collector also sees a change of nearly 0.4 ma *p-p*. However, the change in current develops a larger signal across R_L.

$$v_{out} = i_C R_L \approx 0.4 \text{ ma} \times 2.5 \text{ k}\Omega \approx 1 \text{ v } p\text{-}p$$

The voltage gain is

$$A_v = \frac{v_{out}}{v_{in}} = \frac{1000 \text{ mv}}{10 \text{ mv}} = 100$$

Notice that the common-base circuit does not invert the signal. For example, when the input swings negatively, I_E is increased. Increasing I_E causes an increase in I_C, which increases the *IR* drop across R_L. When the *IR* drop increases, V_C moves further away from V_{CC} (+5 v) toward zero, which is in the negative direction, the same direction as the input.

A common-emitter amplifier is shown in Figure 14–28. An R_L of 2.5 k ohms was chosen, the same as used in the common-base amplifier of Figure

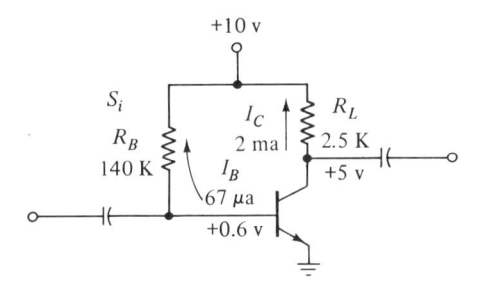

FIGURE 14–28 *Common-Emitter Amplifier*

14–27. A load line will be used in the analysis of this circuit. Figure 14–29 is an expanded version of Figure 14–25. The load line points are $V_{CE} = 10$ v for the open-circuit condition, and $I_C = 4$ ma for the short-circuit condition.

$$I_C = \frac{V_{CC}}{R_L} = \frac{10 \text{ v}}{2.5 \text{ k}\Omega} = 4 \text{ ma}$$

Assuming Class *A* operation, V_C is to be +5 v, which is midway on the load line. At $V_C = 5$ v, $I_C = 2$ ma and $I_B \approx 67$ μa. The proper value of R_B for the conditions described is

$$R_B = \frac{V_{BB} - V_{BE}}{I_B} = \frac{10 \text{ v} - 0.6 \text{ v}}{67 \text{ μa}} = \frac{9.4 \text{ v}}{67 \text{ μa}} = 140 \text{ k}\Omega$$

[14.9]   Semiconductors   355

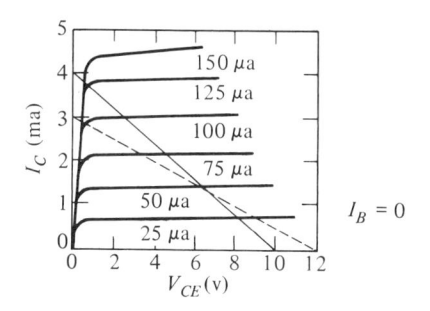

FIGURE 14-29 *Common-Emitter Amplifier Load Line Analysis*

Beta for the curve has previously been calculated at 30. Therefore, another way to calculate I_B when I_C is known is

$$I_B = \frac{I_C}{\beta} = \frac{2 \text{ ma}}{30} = 67 \text{ }\mu\text{a}$$

Notice that V_{BB} and V_{CC} are both connected to the same power supply (+10 v) in the illustration.

As in the common-base amplifier, the coupling capacitors allow ac signals to pass while blocking dc voltages. The common-emitter amplifier inverts the signal. For example, the positive portion of a cycle increases the base current, which increases the collector current. Higher I_C causes a higher drop across R_L which causes V_C to decrease in a negative direction higher degree. Negative portions of a cycle cause an opposite reaction.

The current gain of a common-emitter stage is a function of beta. DC beta is dc current gain, and ac beta is ac current gain. In Section 14.7, dc beta was 30 as calculated from Figure 14-25, while ac beta ranged from 26 to 29, depending upon the operating portion of the curves.

The voltage gain can be approximated by calculating the input and output currents and converting to ac signals. The input resistance of the particular circuit is approximately 0.8 k ohms (beta times the junction resistance). For an arbitrary ac signal of 10 mv, the input current is

$$i_B = \frac{v_{in}}{R_{in}} = \frac{10 \text{ mv}}{0.8 \text{ k}\Omega} \approx 12.5 \text{ }\mu\text{a}$$

If ac beta is approximately 30 (same as β_{dc}), the output current is

$$i_C = \beta_{ac} i_B = 30 \times 12.5 \text{ }\mu\text{a} \approx 0.375 \text{ ma}$$

The output signal, then, is calculated:

$$v_{out} \approx i_C R_L \approx 0.375 \text{ ma} \times 2.5 \text{ k}\Omega \approx 0.94 \text{ v}$$

And the voltage gain is

$$A_v = \frac{v_{out}}{v_{in}} = \frac{940 \text{ mv}}{10 \text{ mv}} = 94$$

A common-collector amplifier is shown in Figure 14–30. A dc bias point of $V_E = 3.1$ v was chosen to illustrate a variation from the usual mid-point. With $R_L = 1$ k ohm,

$$I_E = \frac{V_E}{R_L} = \frac{3.1 \text{ v}}{1 \text{ k}\Omega} = 3.1 \text{ ma}$$

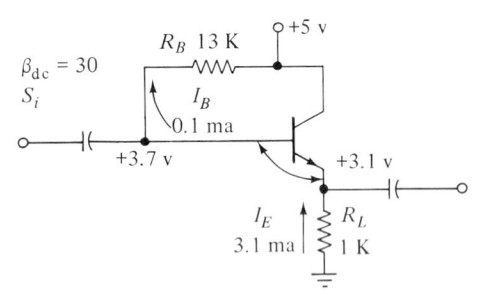

FIGURE 14–30 *Common-Collector Amplifiers*

The current transfer ratio of a common-collector configuration is called **gamma**, abbreviated γ. Gamma is related to beta in the following way.

$$\gamma = \beta + 1 \qquad (14\text{–}7)$$

The above formula is possible because the current transfer ratio is from base to emitter. However, gamma is seldom used in general practice since it is simpler to use values of beta. Base current, then, is related to emitter current in this way:

$$I_B = \frac{I_E}{\beta + 1} \qquad (14\text{–}8)$$

This does not change formula (14–1), $I_B = I_C/\beta$, since it is expected that I_E is higher than I_C. In Figure 14–30, base current is calculated from formula (14–8).

$$I_B = \frac{I_E}{\beta + 1} = \frac{3.1 \text{ ma}}{30 + 1} = 0.1 \text{ ma}$$

V_{BE} is 0.6 v, so that the dc level of the base is

$$V_B = V_E + V_{BE} = 3.1 \text{ v} + 0.6 \text{ v} = 3.7 \text{ v}$$

The base bias resistor is calculated:

$$R_B = \frac{V_{BB} - V_B}{I_B} = \frac{5 \text{ v} - 3.7 \text{ v}}{0.1 \text{ ma}} = \frac{1.3 \text{ v}}{0.1 \text{ ma}} = 13 \text{ k}\Omega$$

Inversion of the signal does not occur in a common-collector amplifier because the emitter follows the base level with only a dc offset of V_{BE} (0.6 v in the example). If the input swings positively, more base current flows which causes more emitter current to flow. Higher emitter current results in a higher IR drop across R_L, which causes V_E to swing positively.

A stabilized amplifier is illustrated in Figure 14–31. Semiconductor properties quite obviously change as a result of temperature. When they do, the dc operating points tend to drift. The amplifier in the figure coun-

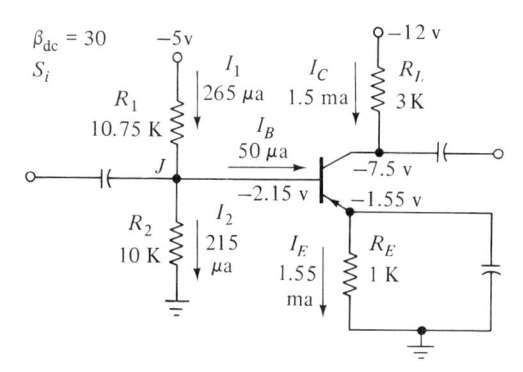

FIGURE 14–31 *Common-Emitter Stabilized Amplifier*

teracts dc drift by bringing the base closer to ground (lower bias resistors) and by bringing the emitter farther from ground, which provides dc feedback for compensation. DC operating points of the transistor can be established from the *dashed* load line in Figure 14–29. The first point is $V_{CE} = 12$ v, since V_{CC} in the circuit is -12 v. The second point is established with the transistor short-circuited.

$$I_C = \frac{V_{CC}}{R_L + R_E} = \frac{12 \text{ v}}{3 \text{ k}\Omega + 1 \text{ k}\Omega} = 3 \text{ ma}$$

Assuming a V_{CE} of about 6 v, the bias currents are $I_C = 1.5$ ma, and $I_B = 50$ μa. The emitter is below ground level due to R_E, so that V_C is more negative than -6 v.

$$V_C = V_{CC} - I_C R_L = -12 \text{ v} + (1.5 \text{ ma} \times 3 \text{ k}\Omega) = -7.5 \text{ v}$$

The emitter current is the sum of I_B and I_C.

$$I_E = I_B + I_C = 0.05 \text{ ma} + 1.5 \text{ ma} = 1.55 \text{ ma}$$

And V_E is calculated:

$$-V_E = I_E R_E = 1.55 \text{ ma} \times 1 \text{ k}\Omega = -1.55 \text{ v}$$

The drop across the transistor is

$$V_{CE} = V_C + V_E = -7.5 \text{ v} + 1.55 \text{ v} = -5.95 \text{ v}$$

which is close to the 6 v shown on the load line.

Input biasing of the circuit is simply a matter of determining the base voltage and keeping the currents organized. The base voltage is

$$V_B = V_E - V_{BE} = -1.55 \text{ v} - 0.6 \text{ v} = -2.15 \text{ v}$$

Recalling Kirchhoff's law, the currents into a junction must equal the currents out of the junction. In the circuit, I_1 flows from -5 v, through R_1, to the junction "J." The current then splits, with part of it entering the base, and the rest of it flowing through R_2 to ground. Since V_B is known (-2.15 v), the currents can be determined from Ohm's law, and a base current of 50 μa can be verified.

$$I_1 = \frac{-V_{CC} + V_B}{R_1} = \frac{5 \text{ v} - 2.15 \text{ v}}{10.75 \text{ k}\Omega} = 265 \text{ μa}$$

$$I_2 = \frac{-V_B}{R_2} = \frac{2.15 \text{ v}}{10 \text{ k}\Omega} = 215 \text{ μa}$$

$$I_B = I_1 - I_2 = 265 \text{ μa} - 215 \text{ μa} = 50 \text{ μa}$$

The two series capacitors in Figure 14-31, input and output, are used for the usual ac coupling and dc blocking. The capacitor across R_E holds the emitter at ac ground. Without the capacitor, the emitter would try to follow the input signal, and this would reduce the voltage gain an appreciable amount. Reducing the voltage gain as described is known as **degeneration**.

QUESTIONS

14-1 What atomic structure causes a material to be a good semiconductor?

14-2 How many electrons are there in a germanium atom? In a silicon atom?

14-3 What does covalent bonding mean?

14-4 What is a free electron? How does it relate to recombination?

14-5 What is the difference between intrinsic and extrinsic crystals?

14-6 What can be used to dope silicon to provide P-type material?

14-7 What can be used to dope germanium to provide N-type material?

14-8 What are the majority carriers in P- and N-type materials?

14-9 How is the term mobility related to semiconductors?

14-10 Define the term depletion region.

Semiconductors 359

14-11 How does bias affect the depletion region?

14-12 When a P-N junction is forward-biased, what are the directions of current flow?

14-13 What causes leakage current through a reverse-biased junction?

14-14 Refer to Figure 14-11. If the forward current is 15 ma, how much voltage is dropped across the diode?

14-15 How is diode rectification accomplished?

14-16 Suppose the output of Figure 14-12(b) is connected to the input of Figure 14-13(a). What is the expected output level of Figure 14-13(a)?

14-17 How are PNP/NPN junctions normally biased?

14-18 What is the meaning of transistor action?

14-19 Describe electron current flow through a PNP transistor.

14-20 Describe the terms beta and alpha. How are they related?

14-21 For a dc beta of 200, and I_E of 4.02 ma, how much are I_B and I_C?

14-22 How much are dc beta and I_B if $I_C = 1.5$ ma, and $I_E = 1.56$ ma?

14-23 What is the value of alpha for Question 14-21?

14-24 How much is beta for an alpha of 0.98?

14-25 Define the term saturation.

14-26 Refer to Figure 14-18. What are I_C and I_B for $R_E = 1.1$ k ohms?

14-27 What is the value of V_C for Question 14-26?

14-28 What value of R_E is required to produce $I_C = 1$ ma for Figure 14-18 if alpha is 0.95?

14-29 Refer to Figure 14-19. What is the value of I_C if dc beta changes from 50 to 75? What would be the value of V_C?

14-30 In Figure 14-19, there is a dc beta of 20. What values of R_B and R_L are required for $I_C = 1$ ma and $V_C = -6$ v?

14-31 Which transistor configuration is best suited for high voltage gain without inverting the input signal?

14-32 What advantage does a common-emitter amplifier have over the other configurations?

14-33 How are dc and ac alpha derived from Figure 14-23?

14-34 How are dc and ac beta derived from Figure 14-25?

14-35 Refer to Figure 14-29. What is the ac beta if the I_B curves are in increments of 10 μa, 20 μa, 30 μa, etc., and the I_C values are 0, 2 ma, 4 ma, etc? Use the dashed load line as it is and evaluate beta between limits of $V_{CE} = 3$ v and $V_{CE} = 8$ v.

14-36 In general, what do the maximum ratings of transistors mean?

14-37 In general, what is the purpose of h parameters?

14-38 Refer to Figure 14-27. If $R_E = 5.5$ k ohms, what are I_C and V_C?

14-39 Refer to Figure 14-28. If $R_B = 188$ k ohms, and dc beta is 40, what are the values of I_C and V_C?

14-40 Refer to Figure 14-30. For $V_E = 2$ v, what value of R_B is required?

15 Special Devices and Integrated Circuits

This chapter is a continuation of solid state technology. Several common devices are discussed along with field effect transistors and the theory of integrated circuitry.

15.1 Zener Diodes

A zener diode is constructed in much the same way as a P-N junction diode. The primary difference is that a zener is doped more heavily than a junction diode, and it is purposely operated with reverse bias. By heavily doping a diode, its junction breakdown potential is dramatically reduced. The zener is then operated with an applied potential that is beyond its breakdown rating. However, once the junction has broken down, very little change occurs in the voltage across the diode, even when the applied voltage varies. It is this feature that makes them important devices for regulating voltage.

Figure 15-1 illustrates the effect of doping on the depletion region. Drawing (a) represents a distribution for normal P-N doping. Heavy doping which reduces the width of the region is shown in (b). A narrower region reduces the amount of voltage required to cause breakdown.

Two types of breakdown can occur at a junction. Breakdown below about five volts is **zener breakdown,** which is caused by the electric field that

362 Special Devices and Integrated Circuits [15.1]

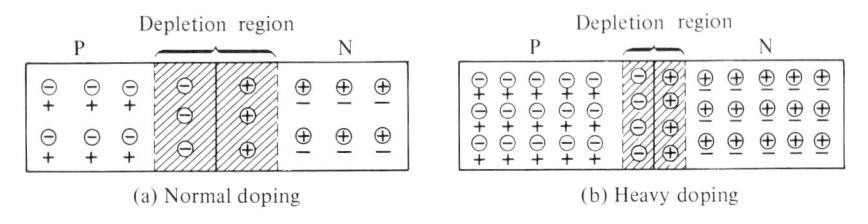

(a) Normal doping

(b) Heavy doping

FIGURE 15-1 *Doping Effect on Depletion Region*

is created by the reverse voltage. The electric field dislodges electrons from the outer bands. When the intensity of the field is high enough, the reverse current increases quite rapidly. When breakdown voltages are higher than about five volts, it is the result of **avalanche breakdown**. Since minority carriers flow during reverse bias, they are able to reach high velocities. The high velocity carriers dislodge electrons from their outer bands. The dislodged electrons in turn release other electrons. The process is cumulative, so that at the breakdown point high current can flow through the reverse-biased junction.

A basic zener diode test circuit is drawn in Figure 15-2. Notice that the diode is reverse-biased. The symbol for a zener is similar to a regular diode

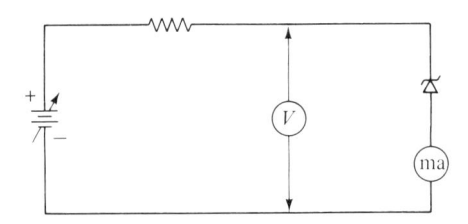

FIGURE 15-2 *Zener Diode Test Circuit*

except for the angles at the cathode end. The battery is variable, and meters are inserted to measure the voltage across the diode and current through the diode. The voltage drop across the milliammeter is considered negligible. In cases where the drop may be significant, the voltmeter is connected directly across the diode.

By varying the dc voltage, a curve can be plotted as shown in Figure 15-3. The forward characteristic is similar to a normal junction diode. The reverse characteristic, however, is a low leakage current until the breakdown value occurs (at -15 v on the graph). Then a sharp bend occurs as the reverse current increases very rapidly. The basic characteristic of a zener diode is a nearly square response curve in the reverse direction. And the breakdown voltage (V_z) is fairly stable for large variations in reverse current (I_z).

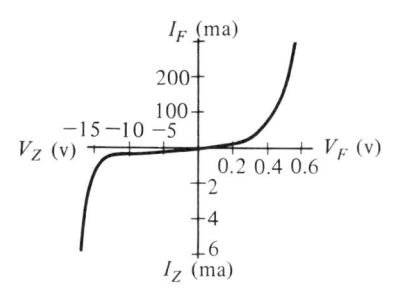

[15.1] Special Devices and Integrated Circuits 363

FIGURE 15-3 *Zener Diode Characteristic Curve*

Typical zener diodes are designed to breakdown at voltages from 2 to 200 v. The specific point is a matter of doping and construction of the diode. Manufacturers specify the breakdown point, along with a tolerance. For example, a diode might be specified at -15 v $\pm$ 10 percent, which means the diode would breakdown and operate within the limits of -13.5 v to -16.5 v. Zeners are available at closer or wider tolerances. Also, operation in the breakdown region will not damage the diode as long as it is operated within the current limits specified for the device.

Figure 15-4 illustrates an analysis of a zener diode for operation as a reference supply. The curve in (a) is a duplication of the reverse curve of Figure 15-3. The V_z and I_z portions are extended to allow load line plotting. The load line is plotted from the circuit in (b). The V_z point is found by opening the diode, which becomes the V_{in} value of 24 v. Although V_{in} is positive, it reverse-biases the diode, so V_{in} becomes negative for purposes of plotting on the graph. The I_z point is found by shorting the diode.

$$I_z = \frac{V_{in}}{R_S} = \frac{24 \text{ v}}{3 \text{ k}\Omega} = 8 \text{ ma}$$

The load line is drawn between the points $V_z = -24$ v and $I_z = 8$ ma, which represent the extremes at which the diode can operate. The load line

(a) Load line   (b) Reference voltage

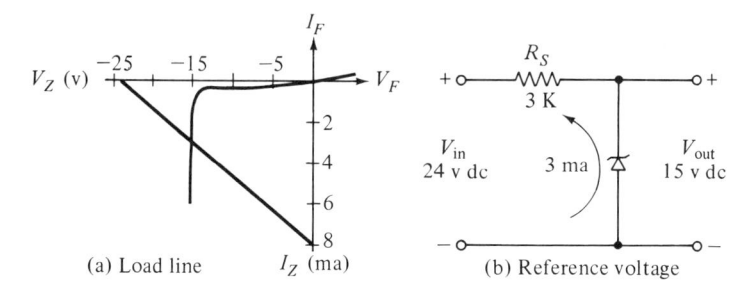

FIGURE 15-4 *Zener Diode Analysis*

crosses the characteristic curve at $V_z = -15$ v and $I_z = 3$ ma. This means that the diode will drop 15 v and conduct 3 ma of current in the circuit, assuming a no-load output condition. The analysis can be verified by Ohm's law. The voltages are

$$V_{RS} = I_z R_S = 3 \text{ ma} \times 3 \text{ k}\Omega = 9 \text{ v}$$
$$V_z = V_{in} - V_{RS} = 24 \text{ v} - 9 \text{ v} = 15 \text{ v}$$

Example 15-1 What is the operating current of the zener in Figure 15-4 if V_{in} is 20 v, and R_S is 2.5 k ohms?

Solution For the open-circuit zener, $V_z = V_{in} = 20$ v. For the short-circuit zener,

$$I_z = \frac{V_{in}}{R_S} = \frac{20 \text{ v}}{2.5 \text{ k}\Omega} = 8 \text{ ma}$$

A load line between $V_z = -20$ v, and $I_z = 8$ ma crosses the curve at $I_z = 2$ ma, which is then the operating current of the zener.

Figure 15-5 illustrates the zener effects of a varying load resistance. A reverse-biased curve is plotted in (a) for a zener voltage of 6 v. The circuit in (b) is similar to the one in Figure 15-4(b). R_S is arbitrarily given a value of 1.2 k ohms so that with 6 v across the zener, R_S will also drop 6 v. By Ohm's law, the current through R_S is

$$I_{RS} = \frac{V_{RS}}{R_S} = \frac{6 \text{ v}}{1.2 \text{ k}\Omega} = 5 \text{ ma}$$

By Kirchhoff's law, the 5 ma must be distributed between the zener and R_L. The size of R_L will decide how the current is distributed. Load lines for each extreme of R_L are plotted in (a). First, the absolute maximum current (I_z) is found by shorting the zener.

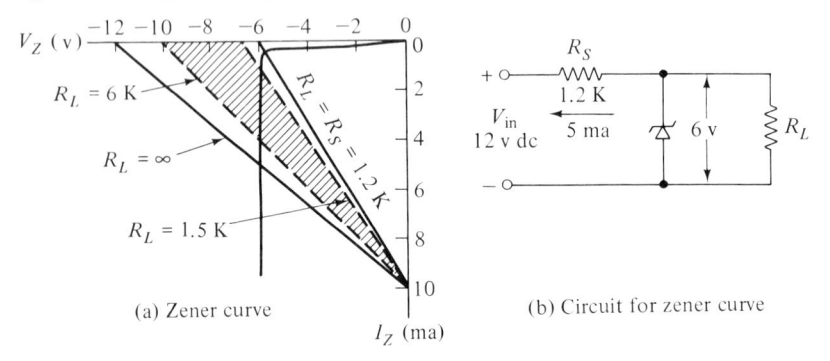

(a) Zener curve

(b) Circuit for zener curve

FIGURE 15-5 *Load Line Analysis for Varying Load Resistance*

[15.1] Special Devices and Integrated Circuits 365

$$I_z \text{(max)} = \frac{V_{in}}{R_S} = \frac{12 \text{ v}}{1.2 \text{ k}\Omega} = 10 \text{ ma}$$

The 10-ma point is then the focal point for varying R_L values because regardless of R_L, I_z (max) is unchanged when the zener is short-circuited. If R_L is equal to R_S, half of the input voltage (V_{in}) is dropped across R_L and the zener. Since breakdown is at 6 v, the zener cannot draw appreciable current because R_L is drawing 5 ma which is the maximum the circuit can deliver when $R_L = R_S$.

$$I_{max} = \frac{V_{in}}{R_S + R_L} = \frac{12 \text{ v}}{1.2 \text{ k}\Omega + 1.2 \text{ k}\Omega} = 5 \text{ ma}$$

Any value of R_L that is less than R_S simply divides the voltage so that less than 6 v is dropped across the zener, in which case the zener is like an open circuit. The other extreme occurs when an R_L is infinite, or open-circuited. V_z is -12 v in the open zener case, and the line crosses the curve at $I_z = 5$ ma. If the zener is to regulate the output voltage, an R_L that is greater than R_S and less than infinite is required. In the practical sense, the zener should handle at least 10 percent of the open-load value, which is $10\% \times 5$ ma $= 0.5$ ma, in the example. However, the load line in Figure 15-5(a) is nonlinear at $I_z = 0.5$ ma because the knee of the curve is at that point. A more practical low limit for this example is $I_z = 1$ ma. With $I_z = 1$ ma, I_{RL} must be 5 ma $-$ 1 ma $=$ 4 ma. The load line (dashed) is pivoted to the left, so it will cross the curve at the $I_z = 1$ ma point. It shows an open-zener voltage of 6.67 v. The low limit of R_L is then

$$R_L \text{(min)} = \frac{V_{zen}}{I_{RL}} = \frac{6 \text{ v}}{4 \text{ ma}} = 1.5 \text{ k}\Omega$$

For purposes of demonstration, a value of $R_L = 6$ k ohms was chosen for the upper limit. Its open circuit (V_z) is

$$V_z \text{(open zener)} = \frac{R_L}{R_L + R_S} V_{in} = \frac{6 \text{ k}\Omega}{6 \text{ k}\Omega + 1.2 \text{ k}\Omega} \times 12 \text{ v} = 10 \text{ v}$$

Notice that the current through the zener swings from a low of 1 ma to a high of 4 ma for loads that vary between 1.5 k and 6 k ohms. V_z is relatively constant at -6 v for the variations.

Example 15-2 Referring to Figure 15-5, what current flows through the zener if R_L is 6 k ohms, and V_{in} drops to 9 v?

Solution When the zener is shorted, the I_z point is

$$I_z = \frac{V_{in}}{R_S} = \frac{9 \text{ v}}{1.2 \text{ k}\Omega} = 7.5 \text{ ma}$$

The V_z point is found in the following manner.

$$V_z = \frac{R_L}{R_L + R_S} V_{in} = \frac{6 \text{ k}\Omega}{6 \text{ k}\Omega + 1.2 \text{ k}\Omega} \times 9 \text{ v} = 7.5 \text{ v}$$

A load line between the points $I_z = 7.5$ ma and $V_z = 7.5$ v crosses the curve at $I_z = 1.5$ ma, which is the operating current of the zener.

Example 15–3 What current flows through R_S and R_L for Example 15–2?

Solution
$$I_{RS} = \frac{V_{in} - V_z}{R_S} = \frac{9 \text{ v} - 6 \text{ v}}{1.2 \text{ k}\Omega} = 2.5 \text{ ma}$$

$$I_{RL} = I_{RS} - I_z = 2.5 \text{ ma} - 1.5 \text{ ma} = 1 \text{ ma}$$

The above can be cross-checked.

$$I_{RL} = \frac{V_z}{R_L} = \frac{6 \text{ v}}{6 \text{ k}\Omega} = 1 \text{ ma}$$

In Figure 15-6 load lines are drawn for variations of input voltage to the circuit in Figure 15-5(b). However, the graph is drawn for an open load

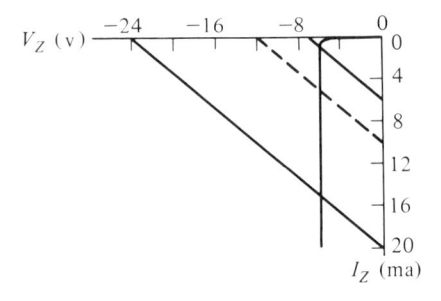

FIGURE 15–6 *Load Line Analysis for Varying Input Voltage*

resistance. By combining the analyses from Figure 15–5(a), the analogy can be proven valid for various loads. If V_{in} increases to as high as 24 v, the short-circuit I_z is 20 ma. On the other hand, a decrease in V_{in} to 7.2 v results in $I_z = 6$ ma. Each extreme crosses the curve at a constant $V_z = -6$ v. However, the zener current varies from 1 ma to 15 ma. The dashed line represents the design center for $V_{in} = 12$ v. For $R_S = 1.2$ k ohms, the 20-ma line represents the maximum current (15 ma) the zener would have to handle. Loading the circuit would of course divide the current so that R_L would use some, and the zener would not be required to handle the full 15 ma.

In plotting and using zeners, it is important to operate within the current limits specified by the manufacturer. Selection of the proper value of R_S

guarantees that the circuit will not exceed the pre-selected value of I_z.

15.2 Photo Devices

A solid state photo device is constructed so its characteristics change when illuminated by external light. Some devices are made to change resistivity and therefore cause a resistance change. Other devices are made to conduct current. The characteristics of a photo device are normally specified for a specific amount of light. The usual unit of illumination is called the **foot-candle**. So a photo device might be specified in terms of resistance at one foot-candle of illumination. On the other hand, a current device might be specified in terms of microamperes or milliamperes at one foot-candle.

Materials that are sensitive to light include germanium and silicon. (For germanium or silicon transistors, the cases are normally light-tight so that light cannot interfere with their operation as current amplifiers.) Other materials are selenium and several compounds of cadmium. Whatever the material, doping is kept low in order to minimize thermal agitation of the atomic structure. Ideally, a photo device would be an insulator when dark, and a conductor when illuminated. In the practical sense, though, most devices are less than an insulator when dark, and their conduction simply increases with illumination.

Light produces energy in units called **photons**. Photon energy varies with the frequency of light. Recall from basic physics that light occupies certain bands of frequencies, depending on its color. The lower frequencies containing violet and blue produce higher energy than the higher frequencies containing red. Photomaterials are therefore sensitive to light in terms of the photon energy of the light. In addition, some materials are light frequency sensitive. Silicon, for instance, is especially sensitive to infrared light.

Photoresistors are constructed of materials which exhibit low energy levels in their dark state. Typical dark resistances of photoresistors run from 10^5 to 10^7 ohms. When illuminated, resistance ranges from 10^2 to 10^3. Typical dark-to-light ratios are at least 100:1. The physical construction of a photoresistor is similar to that shown in Figure 15–7(a). The

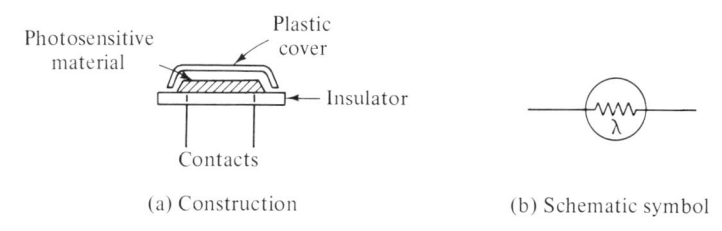

FIGURE 15–7 *Photoresistor*

photosensitive material is placed behind a transparent material so that light can enter the device. The transparent material is usually plastic due to its low cost. Some devices use covers that are formed into a lens which concentrates the light onto the sensing material. Lenses, of course, increase the sensitivity of the device. The schematic symbol for a photoresistor is shown in drawing (b).

An application for a photoresistor is shown in Figure 15–8. The relay requires 20 ma to operate and close its contacts. With a 30-v suppply, the total circuit resistance for 20 ma is found:

$$R_T = \frac{E_T}{I_T} = \frac{20 \text{ v}}{20 \text{ ma}} = 1.5 \text{ k}\Omega$$

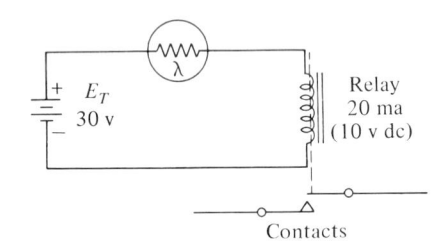

FIGURE 15–8 *Photoresistor and Relay*

The relay is specified for operation at 10 v. Its resistance is

$$R_{rel} = \frac{E_{rel}}{I_{rel}} = \frac{10 \text{ v}}{20 \text{ ma}} = 0.5 \text{ k}\Omega$$

So the photoresistor must be

$$R_P = R_T - R_{rel} = 1.5 \text{ k}\Omega - 0.5 \text{ k}\Omega = 1 \text{ k}\Omega$$

A photo device must be selected in conjunction with a light source which exhibits 1 k ohm or less when illuminated. If the dark resistance is 100 times as much (100 k ohms), the circuit current is

$$I_{dark} = \frac{E_T}{R_P + R_{rel}} = \frac{30 \text{ v}}{100 \text{ k}\Omega + 0.5 \text{ k}\Omega} = 0.3 \text{ ma}$$

which is considerably less than the amount required to operate the relay.

Example 15–4 If a photoresistor operates at 500 ohms when illuminated and at 1 M ohm when dark, how large a voltage source is required to operate a 15-v, 10-ma relay?

Solution The relay coil resistance is

$$R_{rel} = \frac{V_{rel}}{I_{rel}} = \frac{15 \text{ v}}{10 \text{ ma}} = 1.5 \text{ k}\Omega$$

$$E_T = I_{rel}(R_{rel} + R_P)$$
$$= 10 \times 10^{-3}(1.5 \times 10^3 + 0.5 \times 10^3)$$
$$= 10 \times 10^{-3} \times 2 \times 10^3 = 20 \text{ v}$$

A photoresistor can also be used to drive a transistor. In Figure 15-9, the relay requires 100 ma at 5 v to operate. A practical photoresistor would require a very high voltage to operate the relay. The device would have to be designed to carry 100 ma of current, which is considered quite high and expensive. A simple solution is to use an inexpensive photoresistor (rated at 1 k ohm when illuminated) to drive a transistor.

In Figure 15-9, if the base is driven hard enough, the transistor becomes saturated, and the voltage drop across it is nearly zero. With a 100-ma requirement, the 50-ohm relay coil requires

$$V_{rel} = IR = 100 \times 10^{-3} \times 50 = 5 \text{ v}$$

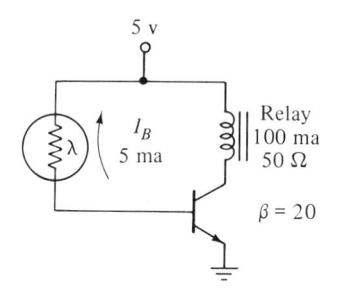

FIGURE 15-9 *Photoresistor Amplifier*

Beta for the device is 20, so that the required base current is

$$I_B = \frac{I_C}{\beta} = \frac{100 \text{ ma}}{20} = 5 \text{ ma}$$

If the base-emitter drop is assumed to be negligible, which is reasonable for a germanium device, the photoresistance must be

$$R_P = \frac{V_{BB}}{I_B} = \frac{5 \text{ v}}{5 \text{ ma}} = 1 \text{ k}\Omega$$

And since the photoresistor was specified as 1 k ohm when illuminated, the circuit is proven to be functional. Two items of interest can occur. First, the resistance of the photo device might decrease for some reason—extra light, for instance. In that case, the base receives more current than required to drive the collector at 100 ma. However, the collector cannot carry additional current since the coil resistance limits I_C. Thus, the circuit continues to function even if R_P decreases. If R_P increases, the circuit becomes marginal and can fail to function if the collector current drops below the required value to operate the relay.

If the beta increases, the base drive increases, and the same effects are true as with a lesser photo device resistance. It is therefore important that the circuit be designed for a photo device resistance of a value equal to or less than the absolute requirement. The circuit must also be designed for a minimum beta, so that reasonable changes in resistance or beta will not seriously affect operation of the circuit. If resistance decreases, and beta increases, the circuit will be overdriven, which should not cause any harm as long as the components do not exceed their specified limits. A minimumly specified beta means that the transistor is at least as good as the minimum. For example, if beta minimum were 20, typical devices might be 30 or more. Therefore, designing to a beta of 20 allows degradation of the device to the minimum beta without upsetting the circuit.

Phototransistors operate in much the same way as conventional transistors, except that the base of the phototransistor is driven with photons of light rather than current. The base region is made very thin and is designed for high sensitivity to light energy. A lens arrangement of some type is normally used to concentrate light directly onto the photosensitive base material. Plastic is frequently used because of its low cost. Phototransistors are designed to deliver microamperes to milliamperes. Typical devices handle as much as 20 ma without further amplification, so many types of relays can be operated directly from a phototransistor.

A simple phototransistor circuit is shown in Figure 15–10. The device is an NPN type, and the collector is reverse-biased in respect to the emitter. In the example, the phototransistor is designed to deliver 5 ma of I_C when illuminated. A 5-ma, 1-k ohm relay requires a source voltage of 5 v.

$$V_{CC} = I_C R_{rel} = 5 \text{ ma} \times 1 \text{ k}\Omega = 5 \text{ v}$$

For reliable relay operation, sufficient light must be guaranteed to ensure an I_C of 5 ma, which means the phototransistor must saturate. Typical

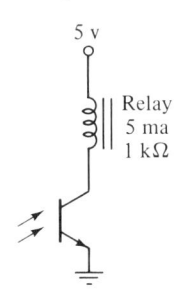

FIGURE 15–10 *Phototransistor Relay Circuit*

[15.2]   Special Devices and Integrated Circuits   371

leakage currents, without light, are less than one microampere, which is low enough to prevent relay operation under dark conditions.

Example 15-5   What phototransistor sensitivity is required to operate a 10-v relay with a coil resistance of 4 k ohms?

Solution
$$I_c \text{ (min)} = \frac{E_{rel}}{R_{rel}} = \frac{10 \text{ v}}{4 \text{ k}\Omega} = 2.5 \text{ ma}$$

The circuit in Figure 15-11 illustrates the use of a phototransistor as a driver for a transistor relay circuit. The transistor must saturate in order to operate the relay since the relay requires 20 v for its operation.

$$V_{rel} = I_C R_{rel} = 50 \text{ ma} \times 0.4 \text{ k}\Omega = 20 \text{ v}$$

With a minimum beta of 50, the required base drive is

$$I_B = \frac{I_C}{\beta} = \frac{50 \text{ ma}}{50} = 1 \text{ ma}$$

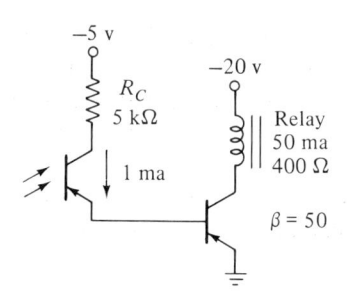

FIGURE 15-11   *Phototransistor Driver Circuit*

A phototransistor with a minimum output of 1 ma is required for the base circuit. If V_{BE} is neglected, R_C is used to limit the base current to 1 ma when the phototransistor saturates.

Example 15-6   In Figure 15-11, what size R_C is required if a 100-ma relay is used, and the minimum beta is 25?

Solution
$$I_B = \frac{I_C}{\beta} = \frac{100 \text{ ma}}{25} = 4 \text{ ma}$$

$$R_C \approx \frac{V_{BB}}{I_B} = \frac{5 \text{ v}}{4 \text{ ma}} = 1.25 \text{ k}\Omega$$

Several important parameters are used to specify phototransistors. A summary of them follows.

I_C Collector current for a specified illumination. Minimum, average, and maximum values of I_C may be specified.

V_C	The maximum reverse bias for the collector.
V_E	The maximum reverse bias for the emitter.
P_d	The maximum rated power dissipation of the device.
Response	The point in the frequency spectrum of peak sensitivity for the device, usually expressed in **angstrom units** (Å) or microns (μ). Angstroms and microns are wavelengths of light.

$$1 \text{ Å} = 10^{-10} \text{ meters}$$
$$1 \text{ } \mu = 10^{-6} \text{ meters}$$

Photosensitive devices are used in a number of everyday applications: light meters or automatic eyes for cameras, burglar sensing alarms, dusk-to-dawn switches for yard lights, and timing devices for mechanical components.

15.3 Field Effect Transistors

Field effect transistors, abbreviated FET, are *voltage* amplifying devices; whereas, junction transistors are *current* amplifying devices. An important feature of the FET is its very high input resistance, which is around 10^5 ohms. In that respect, it is very similar to a vacuum tube—another voltage amplifying device.

The basic structure of a FET is shown in Figure 15-12. The example is made up of a P-type channel with two very thin pieces of N-type contacts.

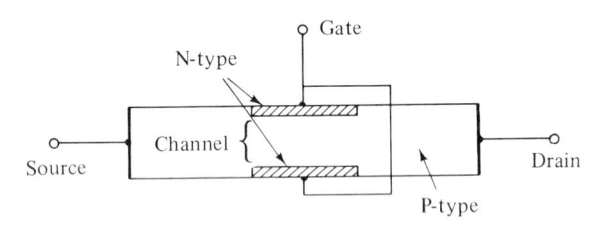

FIGURE 15-12 *Basic FET Construction*

The two contacts are connected together and are called the *gate*. The channel has a contact at each end, and they are called the **source** and **drain**. If the source and gate were both held at ground potential, the drain would be biased with a negative supply voltage. Then the gate-drain junction would be reverse-biased. The gate of a FET is normally reverse-biased with respect to the source. That is, the source is at ground, the N-type gate is slightly positive, and the drain is negative. The gate does not draw current because of its reverse-biased operation.

An N-channel FET is constructed in the same way as a P-channel. The channel, however, is N-type, and the gates are thin contacts of P-type material. Biasing is slightly negative on the gate and positive on the drain, which is the opposite of the P-channel in the figure.

Schematic symbols are drawn in Figure 15-13 for the two types of FET's. The drain, gate, and source are abbreviated D, G, and S respectively. In (a) the P-channel FET is designated with the arrow pointing outward, while the N-channel in (b) has the arrow pointing inward.

(a) P-channel   (b) N-channel

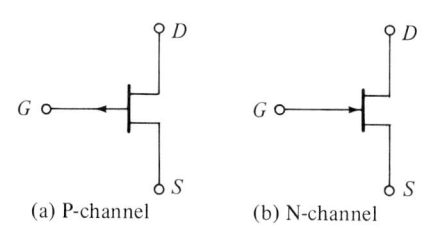

FIGURE 15-13 *FET Schematic Symbols*

(a) P-channel   (b) N-channel

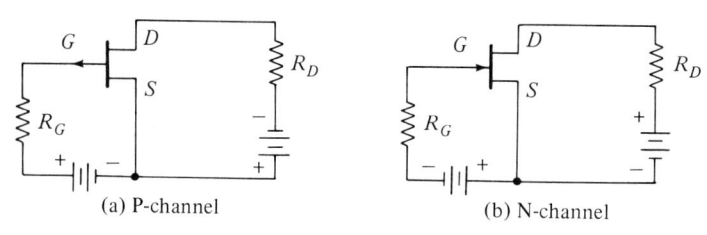

FIGURE 15-14 *FET Bias Circuits*

Biasing for the two types is illustrated in Figure 15-14. A P-channel is shown in drawing (a). As previously discussed, the drain is reverse-biased with a negative potential. The gate is also reverse-biased, but with a positive potential. Notice that P-channel FET is similar to a PNP transistor, except that the gate (base) is reverse-biased. The N-channel, similar to an NPN, is shown in (b). Its drain is reverse-biased with a positive potential, and its gate is reverse-biased with a negative potential.

The depletion region of a FET is somewhat similar to a P-N junction because reverse bias increases the width of the region. A FET, though, has two junctions and consequently two depletion regions, one on each side of the channel. The shaded areas in Figure 15-15 represent depletion regions. In (a) current flows freely through the channel between the two regions.

When the regions are separated, the device is said to be **below pinch-off**, because when they close, a phenomenon known as **pinch-off** occurs, which restricts a change in current. Note that electron current flows from source to drain, toward the positive drain potential. If it is assumed that both the

(a) Below pinch-off (b) At pinch-off (c) Above pinch-off

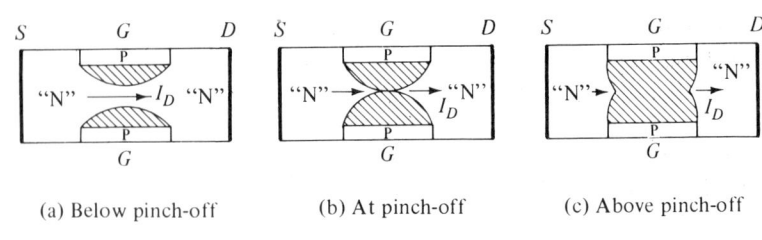

FIGURE 15-15 *Depletion Region (Shaded) Effect on Pinch-Off*

source and gate are at ground potential, an increase in drain potential increases the reverse bias, and the depletion regions increase as shown in (b). When the two regions just meet, the FET is at pinch-off. At pinch-off, abbreviated V_{po}, the drain current stops increasing and remains relatively constant.

The condition above pinch-off, when the depletion regions merge, further restricts a change in current flow. However, the drain voltage is increasing, which tends to increase current flow. But the increasing drain bias increases the depletion region which tends to restrict current flow. The two are compensating so that the net effect is only a small change in drain current.

A voltage-current curve is plotted in Figure 15-16 which illustrates the nature of current flow through a FET. To plot the curve, both the source and gate are held at ground potential. The drain voltage is increased slowly while the drain current is measured. Initially, I_D increases rapidly since the channel is open. As the channel starts to close, the rate of change for I_D decreases. At pinch-off (V_{po}), pinch-off current (I_{po}) is reached. The I_{po} point is maintained fairly constant for further increases in V_{DS}. In practical applications I_D continues to rise slowly. Breakdown of the device occurs at some point designated BV_{DSS} by the manufacturer. Normal operation for an amplifier would be in the "flat" section of the curve between V_{po} and BV_{DSS}, otherwise, slight changes in V_{DS} would cause significant changes in I_D.

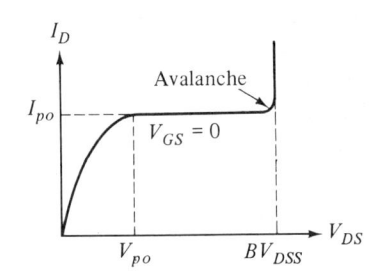

FIGURE 15-16 *FET VI Characteristics*

Figure 15–17 illustrates a set of characteristic curves for a FET. As with vacuum tube characteristics, a curve is plotted for a number of gate-to-source bias voltages, V_{GS}. Notice that higher values of reverse bias (V_{GS}) reduce the drain current. The curves are similar to the VI curve in Figure 15–16, except that they show the normal increase in I_D between V_{po} and BV_{DSS}, and the curves are plotted by holding V_{GS} constant while increasing

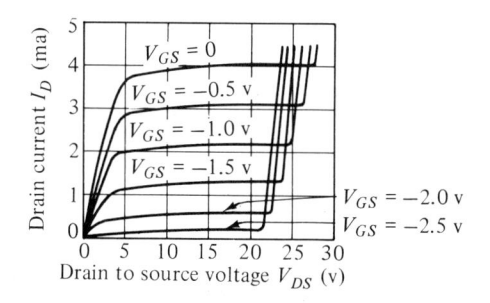

FIGURE 15–17 *FET Characteristic Curves*

V_{DS}. Notice the similarity to pentode and common-base curves. A V_{GS} of about -3.0 v would result in zero I_D. These curves are drawn for an N-channel device. This explains the positive values of V_{DS} and the negative values of V_{GS}. P-channel curves are similar, but the polarities are reversed.

Symbols for FET circuits are summarized in Table 15–1. The letters are capitalized to indicate dc values. Capacitors C_D and C_G are usually bypass

TABLE 15–1 *FET Circuit Symbols*

Symbol	Description
V_{DD}	dc supply voltage for drain
V_{DS}	dc voltage, drain to source
I_D	dc drain current
R_D	Drain resistor
C_D	Drain capacitor
V_{GG}	dc supply voltage for gate
V_{GS}	dc voltage, gate to source
I_G	dc gate current
R_G	Gate bias resistor
C_G	Gate capacitor
V_{SS}	dc supply voltage for source
V_S	dc voltage at source with respect to ground
I_S	dc source current
R_S	Source bias resistor
C_S	Source bypass capacitor

capacitors, but are sometimes used as coupling capacitors. Lower case letters, such as v_{in} and v_{out}, refer to ac signal levels.

An amplifier circuit for a P-channel device is shown in Figure 15–18. C_G and C_D are coupling capacitors for ac signals. R_G is simply a high resistance path for leakage current through the back-biased junction. It is

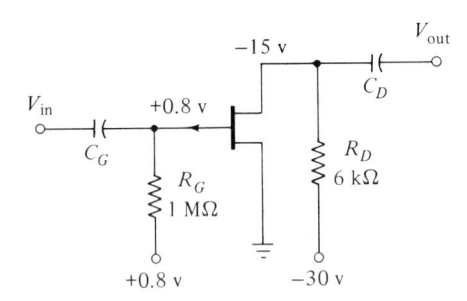

FIGURE 15–18 *P-Channel Amplifier*

made large so it will not seriously affect the input resistance of the device, which is around 100 k ohms. A load line for the circuit is plotted in Figure 15–19. To determine the load line, the FET is short-circuited, and the maximum I_D point is calculated.

$$I_D \text{ (max)} = \frac{V_{DD}}{R_D} = \frac{30 \text{ v}}{6 \text{ k}\Omega} = 5 \text{ ma}$$

The V_{DS} point is equal to V_{DD} (-30 v) when the FET is open-circuited. Load line points, then, are $I_D = 5$ ma and $V_{DS} = -30$ v. If the amplifier is to operate class A, assumed to be $V_{DS} = -15$ v, the proper V_{GS} bias is approximately $+0.8$ v. The drain current at this point is $I_D = 2.5$ ma. DC values of V_{DS} and I_D for the circuit can be established at any point along the load line. For example, if $V_{GS} = 0$ v, then V_{DS} is -7 v, and I_D is 3.8 ma.

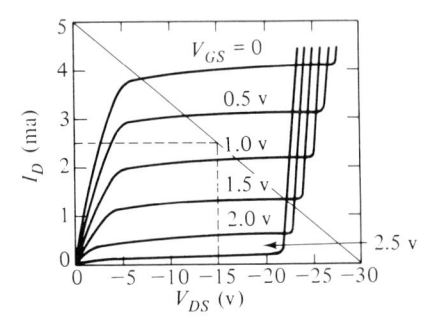

FIGURE 15–19 *Load Line for FET*

Example 15-7 What bias, V_{GS}, is required for Figures 15-18 and 15-19, if $V_{DD} = 20$ v, $R_D = 10$ k ohms, and V_{DS} is to be -7.5 v?

Solution The open-circuit V_{DS} point is -20 v. The short-circuit I_D point is

$$I_D \text{ (max)} = \frac{V_{DD}}{R_D} = \frac{20 \text{ v}}{10 \text{ k}\Omega} = 2 \text{ ma}$$

A load line is drawn between $V_{DS} = -20$ v and $I_D = 2$ ma. A vertical line upward from $V_{DS} = -7.5$ v intercepts the load line at $V_{GS} = 1.5$ v. Therefore, the bias value is 1.5 v.

Example 15-8 What are the load line points for $V_{GS} = 1.5$ v, $V_{DS} = -7.5$ v, and $R_D = 2$ k ohms? Refer to Figures 15-18 and 15-19.

Solution I_D is 1.25 ma for the intersection of $V_{GS} = 1.5$ v and $V_{DS} = -7.5$ v.
The IR drop across R_D is

$$E_{RD} = I_D R_D = 1.25 \text{ ma} \times 2 \text{ k}\Omega = 2.5 \text{ v}$$

The V_{DS} point of the load line is the open-circuit V_{DD} value.

$$V_{DD} = V_{DS} - E_{RD} = -7.5 \text{ v} - 2.5 \text{ v} = -10 \text{ v}$$

And the I_D point is simply

$$I_D \text{ (max)} = \frac{V_{DD}}{R_D} = \frac{10 \text{ v}}{2 \text{ k}\Omega} = 5 \text{ ma}$$

A self-biased amplifier is illustrated in Figure 15-20 for an N-channel FET. The curves of Figure 15-19 are used for this version, but with the polarities of V_{GS} and V_{DS} reversed. The coupling capacitors are the same as for the P-channel version. R_G is used to bleed off any leakage current that might occur, which is the only current that can flow through the gate.

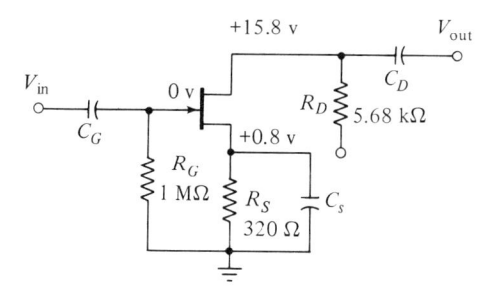

FIGURE 15-20 *N-Channel Amplifier*

Because current does not flow through the gate, it follows that I_D must be the same as I_S. If a resistor, R_S, is placed in series with the source, the current through it causes a voltage drop that reverse-biases the gate-source junction. To illustrate the point, notice that electron flow is from ground, through R_S, through the FET source to drain, and through R_D to the positive potential of $V_{DD} = +30$ v. The polarity at the top of R_S must be positive, and because the gate is zero, the G-S junction is reverse-biased. The bypass capacitor, C_S, is intended to keep the source at ac ground during signal conditions.

The load line is the same as that plotted for Figure 15-18. The open-circuit voltage is $V_{DS} = +30$ v (-30 v on the graph). Short-circuiting the FET gives

$$I_D \text{ (max)} = \frac{V_{DD}}{R_D + R_S} = \frac{30 \text{ v}}{5.68 \text{ k}\Omega + 0.32 \text{ k}\Omega} = 5 \text{ ma}$$

The required bias of 0.8 v is unchanged for $V_{DS} = 15$ v. Under this condition, $I_S = I_D = 1.25$ ma. Proper bias for the FET is verified in the following manner.

$$R_S = \frac{V_{GS}}{I_S} = \frac{0.8 \text{ v}}{1.25 \text{ ma}} = 0.32 \text{ k}\Omega$$

Notice that V_{DS} is still 15 v.

$$V_{DS} = V_D - V_S = 15.8 \text{ v} - 0.8 \text{ v} = 15 \text{ v}$$

The actual voltage on the drain is higher by an amount equal to the V_{GS} (+0.8 v), since the drain resistance was reduced to maintain the same load line as illustrated.

Example 15-9 What size R_D and R_S are required to operate the amplifier in Figure 15-20 at $V_{DS} = 12$ v? Use the load line drawn in Figure 15-19.

Solution When $V_{DS} = 12$ v, $V_{GS} \approx 0.5$ v, and $I_D \approx 3.05$ ma. R_S is

$$R_S = \frac{V_{GS}}{I_D} = \frac{0.5 \text{ v}}{3.05 \text{ ma}} \approx 164 \text{ }\Omega$$

Since the load was initially calculated for $R_D = 6$ k ohms,

$$R_D = 6 \text{ k}\Omega - 0.164 \text{ k}\Omega = 5.836 \text{ k}\Omega$$

Verifying the result, the IR drop across R_D should be

$$V_{RD} = V_{DD} - V_{DS} - V_S = 30 \text{ v} - 12 \text{ v} - 0.5 \text{ v}$$
$$= 17.5 \text{ v}$$

$$V_{RD} = I_D R_D = 3.05 \text{ ma} \times 5.836 \text{ k}\Omega \approx 17.5 \text{ v}$$

FET's can be connected in several ways, such as common-source (the circuits discussed), common-gate, and common-drain. The same general philosophy used with vacuum tubes and transistors can be applied. However, because this section is intended as an introduction to the devices, additional studies are left to the student.

15.4 Integrated Circuits

The name, integrated circuits, is given to special processes used for constructing and manufacturing miniature solid state circuits. Integration is the technology for combining components into "standard" circuit packages that are as small or smaller than the packages used for individual transistors. The ideal method of manufacture would consist of processing the components for the circuit simultaneously and from similar materials. For example, diffusion techniques are common in the fabrication of individual transistors. Rather than diffusing a single chip for one transistor, many transistors can be diffused simultaneously from the same chip with only a slight modification in the process. These same transistors can be connected as transistors and/or diodes. The junction capacitances can be utilized for small capacitors, or the resistivity of the material can be used for resistors.

In forming circuit components on a single chip, interconnections can be made during the process which result in short lead lengths. The total circuit can then be packaged in a small transistor-type can or some other suitable housing. The photograph in Figure 15–21 shows several integrated circuit

FIGURE 15–21 *Typical Integrated Circuit Packages*

packages. Each package contains many components and several circuit stages.

Advantages of integrated circuits are summarized as follows.

1. **Savings in weight and size** is an important consideration in the aerospace programs. It is also a factor in large electronic system installations. Some computers back in the fifties required very large rooms to house the various units and devices. They were heavy and bulky. Today an equivalent computer might fit on a desk top.
2. **Cost savings** are made because the components are manufactured in unison. The interconnections are part of the process, and less materials are needed for packaging the product.
3. **Increased reliability** is due to better interconnections between components. These interconnections eliminate much of the wiring and printed circuit board connections. Improved manufacturing and testing techniques are more practical during mass production.
4. **Improved frequency response** is due to shorter lead lengths and smaller components. Integration reduces leakage capacitance which would tend to shunt high frequency signals.
5. **Low power** results in cooler operation. The use of low signal levels is also feasible since extraneous noise pulses have less effect on the circuits; their size and short lead lengths are less susceptible to noise pickup.

There are several techniques for constructing integrated circuits: **monolithic, thinfilm,** and **hybrid**. Monolithic techniques involve diffusion processes where common materials are used for all components. Thin-film is a process where passive components are deposited on insulated substrate materials. Passive components include resistors, capacitors, and interconnections. Active components are transistors and diodes, which are added separately. The hybrid technique combines the methods of both monolithic and thin-film.

In comparison to monolithic techniques, thin-film and hybrid techniques require more space due to discrete components, are more costly, and are less reliable due to the additional component connections. For those reasons, monolithic integrated circuits are preferred when mass quantities of similar circuitry are required. Digital computer systems are cases in which many logic circuits are used repeatedly. A specific circuit package might be used over one hundred times in the same system. Eight to ten specific types might well account for over half of the total circuitry in a computer.

[15.4] Special Devices and Integrated Circuits 381

Construction of an *IC* transistor begins with a small wafer of substrate material, which is approximately 0.006 inch thick. P-type silicon (S_i) is used in the drawing in Figure 15–22. The substrate material is actually made larger than required for diffusion and electrical properties to provide physical strength for the device. The silicon-oxidized coating (S_iO_2) is provided by heating the substrate to a temperature in excess of 1000°C.

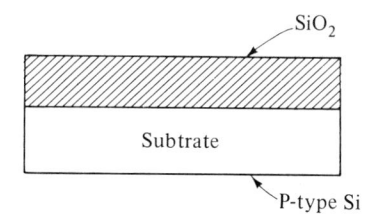

FIGURE 15–22 *Oxidized Substrate*

In Figure 15–23 a photosensitive emulsion is used to coat the top of the material. An etching mask (photographic negative) is placed over the emulsion. The mask has transparent and opaque areas, so that when exposed to light, the materials beneath the opaqued portions can be

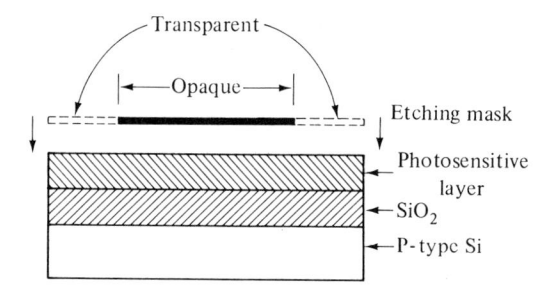

FIGURE 15–23 *Masking Process*

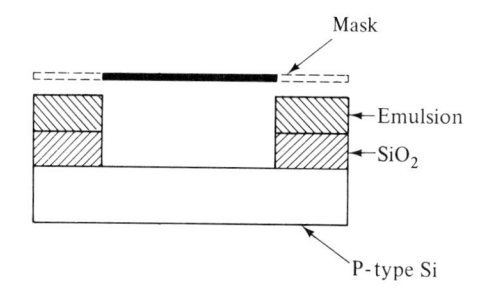

FIGURE 15–24 *Result of Photoetching*

removed down to the substrate. Figure 15–24 illustrates the result of etching, which is accomplished with special solvents.

Two diffusion processes are shown in Figure 15–25. First, a heavily doped $n+$ layer is diffused into the substrate by exposing the material to an N-type dopant such as arsenic. The diffusion occurs when the material is placed in a high temperature oven. While in the oven, the $n+$ area is exposed to gas containing N-type impurities, which create a lightly doped N region. The N region is the layer that will become the collector for a transistor. This layer is often called the **epataxial layer**.

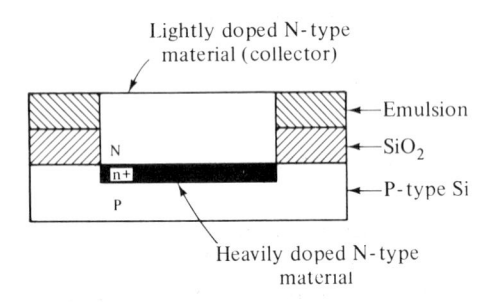

FIGURE 15–25 *Collector Diffusion*

P-type and N-type impurities are then diffused into the epataxial layer as shown in Figure 15-26. The P-layer becomes the base while the N-type becomes the emitter. Note that the process has produced an NPN transistor. The finished process is shown in Figure 15–27. Terminals are con-

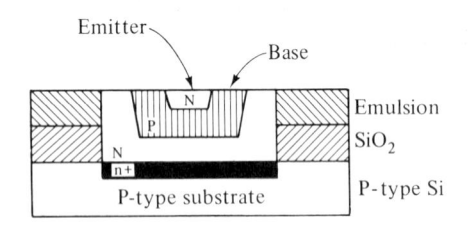

FIGURE 15–26 *Base and Emitter Diffusions*

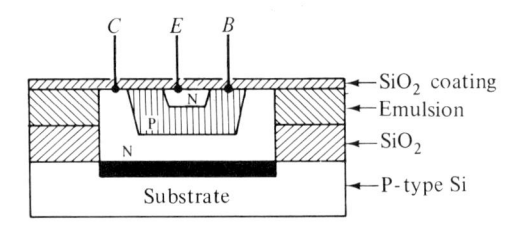

FIGURE 15–27 *Coating and Connections*

nected to the layers, and a protective coating of silicon oxide (S_iO_2) is applied across the top of the transistor.

Figure 15-28 illustrates the principles of producing other components by the same monolithic process. For example, (a) and (b) are transistors that can be used as diodes. The emitter-base junction is used in (a). The base-collector is shorted in (b) and, with the emitter, provides a diode. The second method, (b), provides greater current handling capability than that of the base junction alone. In (c) a resistor is formed by using the resistivity of the base region. In this case, the emitter diffusion step is omitted. The value of resistance can be varied by the spacing of the terminals or by varying the length of the P-region. Capacitance is achieved by utilizing the back-biased junction capacitance of a transistor. This is illustrated in (d). Capacitance values are limited, however, to values below 20 pf, though somewhat higher values are possible with selective diffusion methods.

Two transistors that are diffused on a single substrate are shown in Figure 15-29(a). The diffusion process is identical to that used for single transistors. Any number of transistors can be manufactured on one substrate in the same way. The P-type substrate is connected to the lowest (most negative) potential in the circuit. In that way, isolation is provided

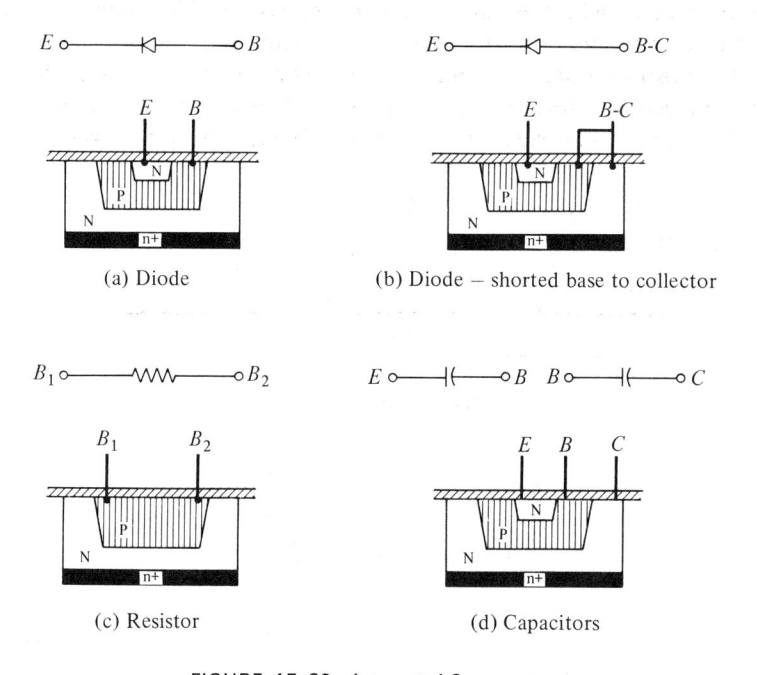

(a) Diode

(b) Diode — shorted base to collector

(c) Resistor

(d) Capacitors

FIGURE 15-28 *Integrated Components*

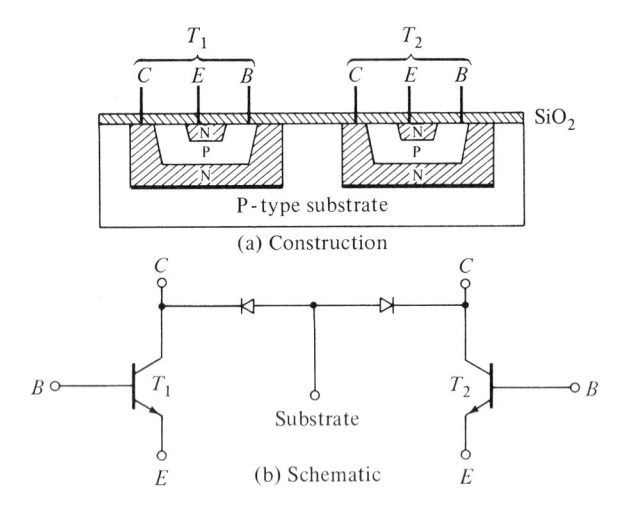

FIGURE 15–29 *Two Transistors on One Substrate*

by the reverse-biased junction of the collector-to-substrate. The circuit schematic is drawn in (b). The collector-to-substrate diodes are shown back-to-back. When the anode (substrate) is negative, the diodes are reverse-biased, which isolates the two transistors. The same principle applies to multiple transistors on the same substrate.

The substrate wafer in Figure 15–30(a) contains three devices that are interconnected to form a transistor circuit. The components are formed by the techniques already discussed. For reference, the resistor was demon-

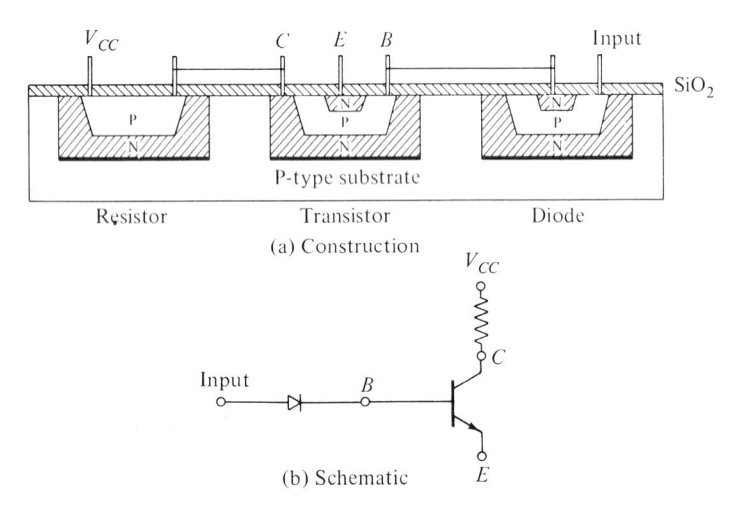

FIGURE 15–30 *Component Circuit on One Substrate*

strated in Figure 15–28(c), the transistor in Figures 15–27 and 15–29, and the diode in Figure 15–28(a). The connections between devices are arranged so that none of them cross. In this way, printed circuit patterns can be produced photographically for interconnecting components. The equivalent circuit is shown in (b). Additional components could have been added quite conveniently. A resistor divider could be connected to the base for biasing the transistor, or coupling capacitors could be added for the base and collector.

Construction details for two types of integrated circuit packages are shown in Figure 15–31. The device in (a) is known as a plastic dual in-line package. This type typically contains four to six logic circuits. The device in (b) is a 36-lead memory package. Both types are typical of the packages used in modern digital computer systems.

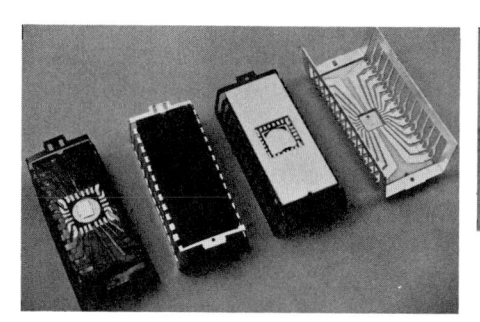

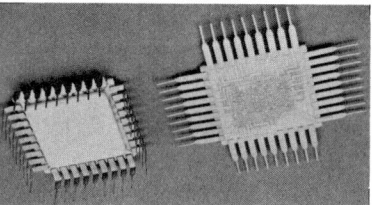

FIGURE 15–31 *Integrated Circuit Construction (Photographs Courtesy of Fairchild Semiconductor)*

Miniaturization of electronic devices is possible to the extent that an entire radio can be made smaller than a pencil eraser. In large systems, the various substrate wafers can be stacked to form integrated circuit modules that contain thousands of circuits. Once manufactured and tested, the modules should be very reliable if they are operated within their ratings. In a failure does occur, their low cost justifies replacing the entire module. Service time is therefore minimized along with down-time or off-the-air time.

Some of the specific applications for integrated circuits are radios, televisions, high fidelity amplifiers, transmitters, radars, and computer systems. Miniaturization also means more functions in a smaller space. Integrated circuit families that offer a wide choice of functions and operating conditions are supplied by a significant number of manufacturers. Frequently a family from one manufacturer can be interfaced or matched to that of another manufacturer.

QUESTIONS

15-1 Why is a zener diode more heavily doped than a regular junction diode?

15-2 What happens to the depletion region when the doping of a diode is increased?

15-3 What is the difference between zener and avalanche breakdown?

15-4 In what portion of the curve in Figure 15-3 does a zener operate?

15-5 What are the load line points and operating current for Figure 15-4 if $R_S = 4$ k ohms, and $V_{in} = 20$ v?

15-6 In Figure 15-5, what is the operating current of the zener and the value of R_L if the V_z point of the load line is pivoted to -8 v?

15-7 Refer to Figure 15-5. How much is the zener current for $R_S = 2$ k ohms, and $R_L = 6$ k ohms?

15-8 What is the zener current for Question 15-7 if V_{in} is 16 v?

15-9 How does a zener diode regulate voltage?

15-10 In general, what principles cause a phototransistor to function as an electronic device?

15-11 In Figure 15-8, how much does the resistance of the photoresistor have to be to activate the relay if $E_T = 40$ v?

15-12 Refer to Figure 15-9. What minimum value of beta is required to operate a 40-ma relay if the resistance of the photoresistor drops to 10 k ohms when illuminated?

15-13 If beta in Figure 15-9 decreases to 10, and a 20-ohm relay is used, what does the photoresistance have to be for operation of the circuit?

15-14 What are key differences between a phototransistor and a conventional junction transistor?

15-15 Assuming that the phototransistor in Figure 15-11 saturates with 0.4 ma, what value of R_C and what relay parameters are required to saturate the transistor?

15-16 In Figure 15-11, what size R_C is required if minimum beta is 40, and the resistance of the relay coil is 200 ohms?

15-17 How many angstrom units represent one micron?

Special Devices and Integrated Circuits   387

15-18 What phenomenon controls current flow through a FET?

15-19 What bias polarities are normally applied to a P-channel FET?

15-20 What bias polarities are normally applied to an N-channel FET?

15-21 What does the term **pinch-off** mean?

15-22 Why are the curves in Figure 15-17 relatively flat above the pinch-off potential?

15-23 Referring to the FET curves in Figure 15-17, what value of drain current is expected for $V_{GS} = -2.0$ v, and $V_{DS} = 15$ v?

15-24 In Figure 15-18, what value of R_D provides $V_{DS} = -15$ v with a V_{GS} of 2.0 v?

15-25 For Figure 15-18, assume $V_{DD} = -24$ v, $I_D = 2$ ma, and $R_D = 8$ k ohms. What are the values of V_{DS} and V_{GS}?

15-26 What are the load line points for Figure 15-18 if $V_{DS} = -10$ v, $V_{GS} = 1.3$ v, and $R_D = 10$ k ohms?

15-27 For Figure 15-20, $R_S = 0.5$ kΩ, $V_{DD} = -24$ v, and I_D (max) = 3 ma (which is the I_D load line point). What are the values of I_D, V_{GS}, V_{DS}, and R_D?

15-28 For the conditions in Question 15-24, what are the values of R_D and R_S for Figure 15-20?

15-29 What are the advantages of integrated circuits over discrete circuits?

15-30 Why are monolithic *IC*'s more economical to manufacture than thin-film or hybrid techniques?

15-31 Briefly explain how monolithic techniques provide components such as diodes, resistors, and capacitors when actually fabricated for transistors.

15-32 When more than one transistor is diffused onto a single substrate wafer, what prevents them from shorting to one another?

16 Power Supplies

The many types of electronic circuits studied in this text have used or assumed a dc source voltage from a battery. Batteries, of course, have their application for automobile radios and portable equipment, such as radios, transmitters, tape recorders, amplifiers, and television receivers. A disadvantage of batteries is that they require a means of recharging, or else they must be replaced periodically. They are impractical for high power equipment because the size necessary for a battery to deliver sufficient voltage-current would have to be very large. For those reasons, electronic power supplies are normally used whenever ac line voltage is convenient.

This chapter introduces and discusses various types of power supplies, utilizing either vacuum tube rectifiers or solid state rectifying diodes. In addition, regulating circuits are discussed which regulate and maintain a constant voltage or current with changing load conditions.

16.1 Vacuum Tube Power Supplies

A typical vacuum tube power supply is diagrammed in Figure 16-1. The circuit is called a **half-wave-supply** because it rectifies only one-half of the input waveform. An input of 117 v ac is used to drive the primary of the transformer; this is a typical value obtained from a regular household outlet. However, it should be noted that power is also delivered in terms of

[16.1]   Power Supplies

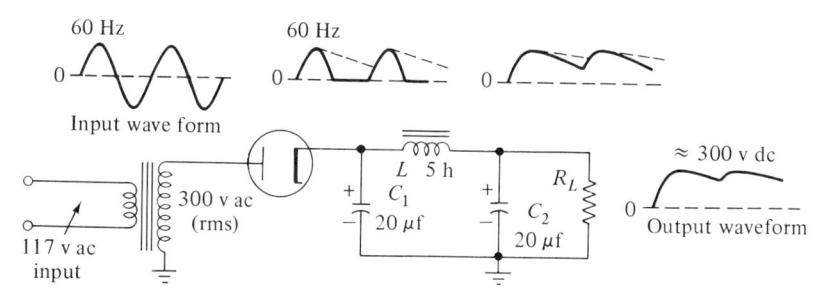

FIGURE 16-1 *Half-Wave* VT *Power Supply*

220 v ac and 440 v ac, particularly for commercial and industrial applications. Primary windings may be obtained for these voltages, too.

The secondary windings may be selected for about any voltage that satisfies the requirement for a dc output voltage. The figure shows a value of 300 v ac. In power supply work, it is customary to show ac voltages in rms values. As a very rough approximation, the dc output voltage across R_L is 300 v dc. The exact value of dc output depends on the type of filter and the loading (R_L) on the supply.

The vacuum tube rectifier is shown with its plate facing toward the secondary of the transformer; ground is used as a reference for the supply. In this position, the tube can pass only the positive excursions of the waveform. The plate must be positive in respect to the cathode in order to pass current (see Chapter 13). During negative portions of the plate voltage, the rectifier is reverse-biased. The waveforms are shown above the schematic. Two cycles are illustrated at the plate. Since the positive cycles pass, the waveform at the cathode shows two positive cycles with voids where the negative cycles cut off the tube. Notice that the frequency is 60 Hz at both sides of the rectifier. At this point, the filter takes over to smooth out the pulses causing a dc level across the load.

The filter for the power supply in Figure 16-1 consists of C_1, C_2, and L. The filter capacitors are polarized. Filter capacitors are normally designed for breakdown and leakage ratings in one direction. When the first pulse rises at the cathode, C_1 charges to the *peak* value of the waveform. The choke represents a high reactance to changes in the voltage. The high reactance, then, slows down the discharging of C_1 during negative cycles. The dashed lines show the discharge level which is also the dc level of the voltage at that point.

The waveform across C_2 (and R_L) is shown above the components (see Figure 16-1). The capacitor maintains a minimum charge equal to the lowest point of the dc level (shown by solid lines). However, the positive ripple portions add additional charge to C_2 which discharges a lesser amount. The result is an output waveform that has had the filtering effects of both capacitors, the choke, and the load resistance. Even though the capacitors charge to peak values of the ac voltage, these peaks are lower than at the plate because of the voltage drop across the tube. The output level is lower yet due to the voltage drop across the choke. These IR drops are highly dependent upon the current drawn by the load.

A **full-wave rectifying circuit** is shown in Figure 16-2. The plate circuit of each rectifier is referenced to ground via the center tap of the secondary.

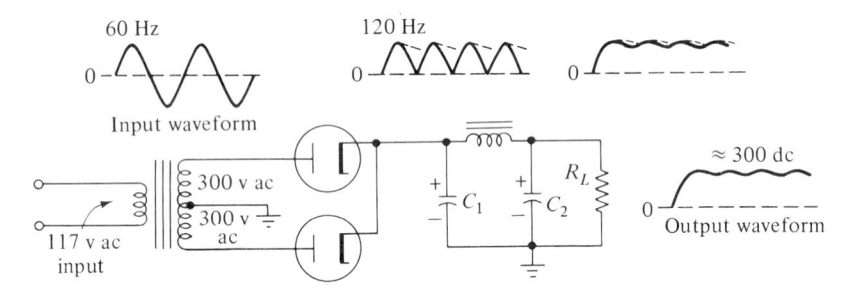

FIGURE 16-2 *Full-Wave* VT *Power Supply*

Each half of the secondary is rated at 300 v ac making a total secondary voltage of 600 v ac. However, the output level is still about 300 v dc even though the ac input is doubled. During positive cycles the upper diode conducts so that the cathode supports the first and third pulse. During negative half cycles the plate of the lower rectifier sees a signal that is actually positive with respect to ground. The upper diode is reverse-biased while the lower diode is forward-biased. This cathode now supports negative cycles as shown by the second and fourth pulses. As a result of conduction from both phases of a cycle, the waveform at the cathodes is 120 Hz, while the frequency of the waveform at the plates is 60 Hz.

Filtering of the full-wave version is similar to that used for the half-wave version. C_1 charges to the peak value, and its discharge is impeded by the choke. It discharges a lesser amount because the succeeding pulses are closer together. When the filters and load are identical, the resulting waveform is smoother than the waveform for a half-wave circuit.

There are several significant differences between half- and full-wave power supplies. Half-wave power supplies are generally less expensive because they need only one rectifier and less secondary voltage from a transformer. However, the filtering may be more extensive due to higher capacitance and inductance. A full-wave power supply is more efficient because it uses both phases of the input, and filtering is more efficient due to less time between peaks.

[16.1] Power Supplies 391

A **full-wave bridge rectifier** combines operating advantages of both the half- and full-wave circuits. In Figure 16-3, the secondary is 300 v ac. The dc output is approximately 300 v dc. Thus, full-wave rectification is possible even when a low secondary voltage is used. A disadvantage, of course, is that four rectifiers are required. Note that the secondary is not referenced directly to ground. The dc return to ground must always pass through a rectifier, V_1 or V_2. During positive cycles V_3 conducts positive voltage to the input of the π filter. Since the bottom of the secondary is negative with

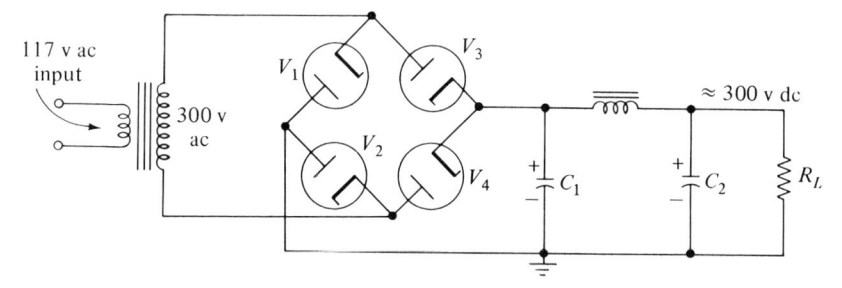

FIGURE 16-3 *Full-Wave Bridge Power Supply*

respect to the top, V_2 simultaneously conducts a negative voltage to the ground reference. During the first half cycle, the 300-v ac signal is impressed across C_1. The voltage peak will actually be less than 300 v because the *IR* drops across the tubes must be subtracted from the 300-v secondary. During the negative half-cycle V_1 conducts negatively to ground, and V_4 conducts positively to the top of C_1. It is important to remember that the secondary is not referenced directly to ground. Therefore, a maximum swing in secondary voltage is referenced plus at the top for positive pulses. Negative pulses cause the bottom to appear plus while the top appears minus.

In Figure 16-4, the secondary is connected to a dual rectifier tube. The principle of full-wave rectification used here is similar to the one applied

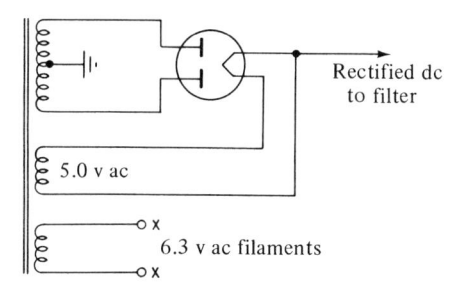

FIGURE 16-4 *Full-Wave Dual Rectifier*

in Figure 16–2. In this one, though, the filament and cathode are the same element. This fact places the dc voltage on the filament supply making it necessary to connect the filament to a separate transformer winding. Common filament supplies are rated at 6.3 v ac. The breakdown voltage rating of most tubes, filament to cathode, is not always high enough to withstand the full dc potential of a power supply. Notice that both filaments are shown in the drawing. The 5-v winding is for the rectifier and is typical for rectifiers with a common filament-cathode.

16.2 Power Supply Filters

A power supply filter is designed to smooth the pulsating pulses from a rectifying circuit and to establish a dc level. Filters shown in the last section were *LC* type filters. An *RC* type filter, where a resistor is used in placed of an inductor, is less expensive but also less effective. Both are discussed in this section.

A **choke input filter**, also known as an **L filter**, is shown in Figure 16–5(a). The input of the filter, E_{in}, is connected to the output of a rectifier circuit. The reactance of the choke impedes fluctuations or changes in its input voltage. The capacitor, of course, charges to the peak value impressed across it. A slow discharge between pulses smooths the ac component. A bleeder resistance, R_b, is shown for the first time in Figure 16–5. In cases

(a) Choke input filter

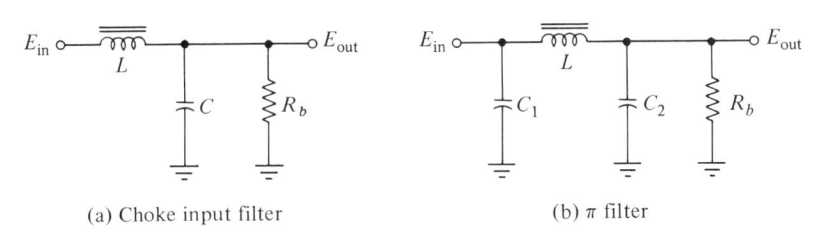

(b) π filter

FIGURE 16–5 *Choke-Type Filters*

where the load may change to a small value, or open entirely, a bleeder is necessary to maintain some reasonable amount of regulation. Without it, and if the load were opened, the filter capacitor would charge to the peak value at the output of the rectifier. Current could no longer flow through the rectifier, and the output voltage would be about $\sqrt{2}$ times the loaded voltage. If the load were occasionally reduced during normal operation, electronic circuitry would be exposed to a considerable amount of voltage variation. For example, if the full-load E_{out} is 100 v, the no-load E_{out} is approximately $\sqrt{2} \times 100 \text{ v} = 141 \text{ v}$. A very small load would see variations from the 100 v full-load to as much as 141 v when not loaded.

The purpose of a bleeder is to ensure that the rectifiers conduct current at all times and that the full-load to no-load variations are kept to a reasonable minimum. A rule of thumb is to use an R_b that is ten times the resistance of the full-load value. For example, if R_L is 1000 ohms, R_b should be about 10,000 ohms.

Typical values for a choke input filter are $L \approx 5$ to 10 h and $C \approx 10$ to 20 µf. Inductance values are primarily dependent upon the load resistance. The inductance of the choke in henries is approximately

$$L \approx \frac{R_L}{1000} \qquad (16\text{--}1)$$

whereas a load resistance of 10 k ohms would require a choke of about 10 h.

A value of capacitance is dependent upon the inductance of the choke, the frequency, and the amount of smoothing required. For example, the C (in farads) for a full-wave rectifier (120 Hz) is

$$C \approx \frac{100 \times 10^{-6}}{L} \qquad (16\text{--}2)$$

The capacitance for a half-wave rectifier would be twice as much since the pulses occur every other half cycle instead of every half cycle.

Example 16–1 Assume a 60-Hz signal that is full-wave rectified. If R_L is 4.5 k ohms, what are the approximate values of L and C for an L filter?

Solution
$$L \approx \frac{R_L}{1000} \approx \frac{4.5 \text{ k}\Omega}{1 \text{ k}\Omega} \approx 4.5 \text{ h}$$
$$C \approx \frac{100 \times 10^{-6}}{L} \approx \frac{100 \times 10^{-6}}{4.5} \approx 22 \text{ µf}$$

A **π-type filter** is shown in Figure 16–5(b). The difference between it and the L filter in (a) is that an additional capacitor (C_1) is shunted across the input. The π filter is more efficient in reducing ac components than the L filter because the input capacitor absorbs the initial fluctuations. The charge and discharge action of the filter was described in Section 16–1. Very approximate values of L and C can be calculated using the same formulas [(16–1) and (16–2)] used for L filters.

Filter design techniques are considerably more involved than depicted in this text. Sophisticated design is a complex subject reserved for more advanced studies in electronics.

RC filters find widespread application due to their lower cost. A simple π-type RC filter is illustrated in Figure 16–6(a). This filter and a π-type choke filter operate in a similar manner. But the resistance is not as effective as a choke in resisting a change in voltage since the reactance of a choke increases with frequency. For example, practical values of R are 1.5 k ohms and less. The reactance of a 10-h choke in a full-wave power supply application is

$$X_L = 2\pi f L = 2\pi \times 120 \times 10 = 7.52 \text{ k}\Omega$$

which is 5 or 10 times as high as a typical RC filter resistance. RC filters of this type, however, find widespread application in small ac/dc vacuum tube radios and amplifiers.

In Figure 16-6(b), a slight variation of (a) is drawn. The key addition is the 22-ohm series resistor. It serves to absorb initial current surges when C_1 charges, thus offering added protection for the cathode of a rectifier tube. The values are typical for ac/dc radio applications. E_{in} is the output of a 117-v ac rectifier and is about 130 v dc. E_{out} normally runs at 110 v dc. The dc values will vary, of course, depending upon the size of R_1, R_2, and the load. Capacitor values typically run from 20 to 50 μf.

(a) Simple π-type RC filter

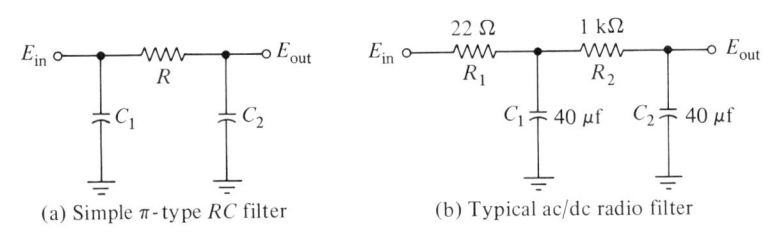

(b) Typical ac/dc radio filter

FIGURE 16-6 RC *Filters*

16.3 Solid State Power Supplies

Solid state power supplies are similar to vacuum tube supplies. The primary difference is in the use of semiconductor rectifiers instead of vacuum tubes. Using these rectifiers has distinct advantages. Filament supplies are not required, and cooler operation occurs since heating is not necessary. Semiconductor diodes are also more compact than vacuum tube rectifiers.

Rectifying diodes for power supplies are generally made from selenium or silicon. Silicon is more popular because it has higher voltage-current ratings for its physical size. Silicon can carry large amounts of current per unit area. Also, the depletion regions in silicon are wider, and this results in higher reverse breakdown voltages.

A half-wave rectifier is shown in Figure 16-7(a). Notice its similarity to Figure 16-1. A full-wave rectifier is shown in (b), and it operates as the vacuum tube version in Figure 16-2 does. In both circuits the anodes are connected to the transformer secondary, and the cathodes are connected to the filter side. These connections cause the filters to receive a positive potential. In many applications, a negative supply voltage is required.

(a) Half-wave   (b) Full-wave

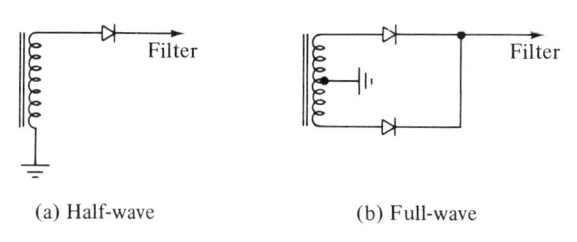

FIGURE 16-7 *Diode-Type Power Supplies*

Solid state rectifiers are easy to convert since the diodes need only be turned around. Although many vacuum tube rectifiers can be reversed, the one in Figure 16-4 cannot since the filament-cathode is a single element. Separate cathodes are required for full-wave rectifiers if they are to be reversed.

A solid state version of the bridge rectifying circuit of Figure 16-3 is drawn in Figure 16-8. This power supply can be utilized for plus or minus supply voltages by selecting the proper connections between the rectified dc and the filter. In other words, if the ground reference is connected to the negative terminal, it becomes a positive supply. Connecting the ground reference to the positive terminal makes it a negative supply.

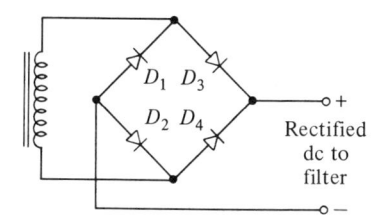

FIGURE 16-8 *Full-Wave Bridge Rectifier*

The diodes in the full-wave bridge rectifier are biased in a way similar to that used for the vacuum tube counterpart. When the top of the secondary is positive, D_3 and D_2 are forward-biased. D_4 and D_1 are forward-biased when the bottom of the secondary is positive. In either case, the output is polarized.

A voltage doubling power supply that uses solid state devices is shown in Figure 16-9. The line voltage in this circuit is connected to the circuit without a transformer. The doubling principle is the same when a step-up transformer is used. The resistor, R_s, is in series to limit the charging current through C_1 and C_2. A value of 30 ohms is typical. The junction of the rectifiers alternates between positive and negative during each cycle of the input voltage. Positive half-cycles allow D_1 to conduct current. The current

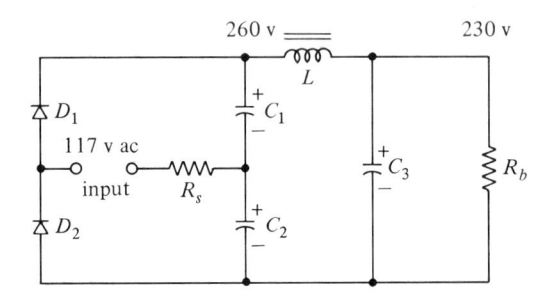

FIGURE 16-9 *Voltage Doubler Power Supply*

through D_1 charges C_1 to the peak value of the voltage at the cathode of D_1. Since C_1 is referenced to the other side of the ac input line, it charges to a value of approximately 130 v dc. During negative half-cycles, D_2 conducts current and charges C_2 to about 130 v dc. The polarities of the charged capacitors are shown in the diagram. The voltage across both of the capacitors is the sum of the individual potentials, or 260 v dc. The total of 260 v is applied to the choke input filter. The *IR* drop across L leaves about 230 v dc across R_b, the bleeder resistance.

Supplies of the type illustrated in Figure 16-9 are very common in television receivers because of their low cost. Since a power transformer is not required, the tube filaments would probably be connected in series.

A vacuum tube version of this supply would be identical except that the diodes would be replaced with rectifier tubes. Doublers are used whenever additional voltage is required from a given ac supply, especially in low-current applications in which the cost of components can be kept to a minimum. For example, a television receiver application requires rectifiers that are rated at 300 to 400 ma each. C_1 and C_2 are typically 150 μf each.

16.4 Voltage Regulation

A voltage regulator is a circuit designed to provide a constant output voltage (1) from a changing input voltage and (2) with changing load conditions. Variations in a supply voltage can easily change the operating characteristics of circuits, especially those requiring close bias conditions. Line voltage (117 v ac) can vary as much as ± 10 percent. This variance can cause a change of approximately the same ratio at the output of the supply.

The input voltage and the output load can vary the output of a power supply. A voltage regulator can be used in this case because it maintains a more consistent load on the supply. When a reference element such as a zener or gas tube is used, a regulator can compensate for input variations.

Many of the regulator designs provide a feedback path which corrects for changes in the output.

The two basic circuits in Figure 16-10 have been described in previous sections. In (a), the zener diode is rated for a given voltage. If the load decreases, more output current is shunted through the diode. If the load increases, more current is used, and less current flows through the diode. R_S is a series IR dropping resistor that allows E_{in} to be higher than E_{out}. Changes in the input voltage are absorbed by R_S so that the zener voltage remains constant over its operating range. Zeners are available for voltages from approximately 2 v to 200 v.

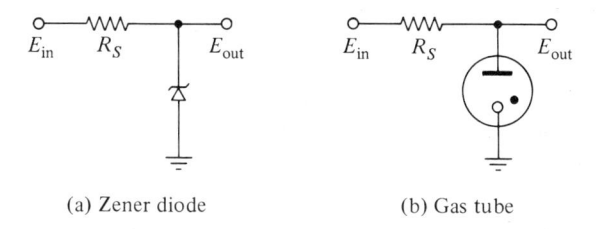

(a) Zener diode   (b) Gas tube

FIGURE 16-10 *Basic Regulator Circuits*

In Figure 16-10(b), the VR tube maintains a constant voltage at the output. Typical values of regulated voltage for the output are 105 v and 150 v. The VR tube compensates for varying loads, and R_S absorbs changes in the input voltage. Both circuits are known as **shunt regulators** because they shunt the load.

Two improved shunt regulators are shown in Figure 16-11. Both use a transistor to shunt variations of the input voltage or output load. In (a), the output voltage is held constant at a level determined by

$$E_{out} = V_z + V_{BE} \qquad (16\text{-}3)$$

Electron current flow for the biasing is from emitter to base (V_{BE}), through the zener, and back through R_S to the positive supply. R_B is provided for leakage current.

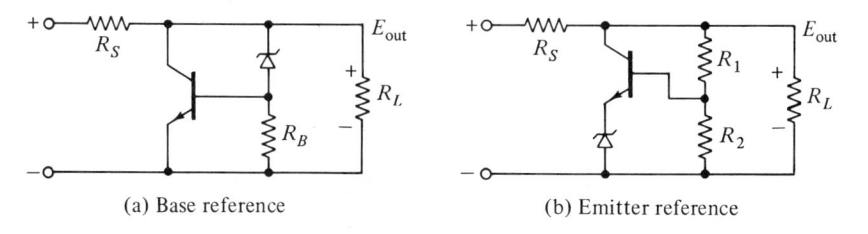

(a) Base reference   (b) Emitter reference

FIGURE 16-11 *Shunt Regulators*

Example 16–2 Assume $V_{BE} = 0.7$ v, and $V_z = 5.0$ v $\pm$ 10%. What is the voltage range across R_L?

Solution
$$V_z(+10\%) = 5.0 + (5.0 \times 0.1) = 5.5 \text{ v}$$
$$E_{\text{out}} = V_z + V_{BE} = 5.5 + 0.7 = 6.2 \text{ v (high value)}$$
$$V_z(-10\%) = 5.0 - (5.0 \times 0.1) = 4.5 \text{ v}$$
$$E_{\text{out}} = V_z + V_{BE} = 4.5 + 0.7 = 5.2 \text{ v (low value)}$$

Because V_z is relatively constant, increases in E_{out} will increase I_B and subsequently I_C. These increases will cause the transistor to shunt more current, drop more voltage across R_S, and return E_{out} to the reference level. Decreases in E_{out} will decrease I_B and subsequently I_C. These decreases will cause the transistor to shunt less current, drop less voltage across R_S, and return E_{out} to the reference level. If any ripple voltage (unfiltered portions of the power supply peaks) is present, and if the recovery time of the circuit is fast enough, the regulator will smooth the output level.

The shunt regulator in Figure 16–11(b) is regulated by the voltage divider, R_1 and R_2, V_{BE}, and V_z. The base bias is stabilized by the voltage drops of V_{BE} and V_z. R_1 and R_2 provide either more or less base current drive, depending upon the output voltage. For example, if E_{out} tends to rise, I_B will rise and increase I_C. The increased I_C will shunt more current and tend to reduce E_{out} to the established level of the voltage divider. However, if E_{out} tends to decrease, I_B and therefore I_C will decrease. These decreases cause less current to be shunted and avails more current to the load. This causes E_{out} to rise to the established level.

$$E_{\text{out}} = V_z + V_{BE} + V_{R1} \tag{16–4}$$

Example 16–3 In Figure 16–11(b), E_{out} is 10 v, $V_z = 8.3$ v, $V_{BE} = 0.7$ v, $I_C = 10$ ma, beta is 50, and $I_B = I_{R2}$. What are the values of R_1 and R_2?

Solution
$$V_B = V_z + V_{BE} = 8.3 + 0.7 = 9.0 \text{ v}$$
$$I_B = \frac{I_C}{B} = \frac{10 \text{ ma}}{50} = 0.2 \text{ ma}$$
$$I_{R2} = I_B = 0.2 \text{ ma}$$
$$I_{R1} = I_B + I_{R1} = 0.2 + 0.2 = 0.4 \text{ ma}$$
$$V_{R1} = E_{\text{out}} - V_B = 10 - 9 = 1 \text{ v}$$
$$R_1 = \frac{V_{R1}}{I_{R1}} = \frac{1 \text{ v}}{0.4 \text{ ma}} = 2.5 \text{ k}\Omega$$
$$R_2 = \frac{V_B}{I_{R2}} = \frac{9 \text{ v}}{0.2 \text{ ma}} = 45 \text{ k}\Omega$$

In comparing the circuits of Figure 16–11, (a) has an advantage because the zener only has to carry base current. The zener in (b) must carry the full emitter current of the transistor which requires a zener current rating of $(\beta + 1)$ times that of the base reference. An interesting feature of (b), though, is that a potentiometer can be placed in the base divider to provide a means for varying the regulated output voltage. A potentiometer can be placed in the emitter of (a), but the regulated output can only be adjusted upward, unless the zener reference voltage is decreased.

Series voltage regulators are illustrated in Figure 16–12. They waste less current than shunt types because they regulate by varying the voltage drop

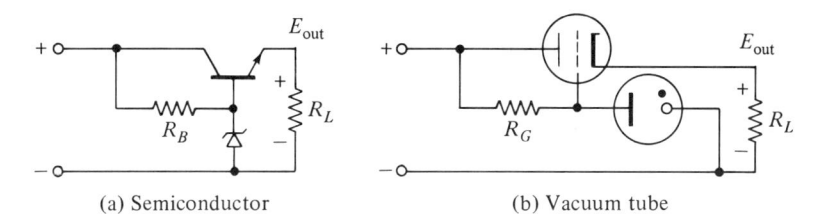

FIGURE 16–12 *Series Regulators*

across the device rather than by bleeding off a significant portion of the output current. In (a), E_{out} is maintained at a constant level by the voltage drops across the zener and base-emitter junction.

$$E_{\text{out}} = V_z - V_{BE}$$

R_B ensures forward bias of the transistor and establishes a reference level at the cathode of the zener diode. Changes in the input (the collector voltage) have little or no effect on the collector current, which is established by the emitter current. Therefore, input variations are stabilized, and E_{out} remains constant.

Changes in the load do not have any effect on the emitter voltage. Less load causes V_E to increase, but this increases the bias so that the transistor conducts less and causes V_E to return to its reference level. More load has the opposite effect.

The vacuum tube version of a series voltage regulator is shown in Figure 16–12(b). Because the VR tube is biased via the resistor R_G, the grid is held at a constant voltage level. With a constant bias supply, the cathode is maintained at a constant value for variations in the input. When the load resistance increases [which allows E_{out} (V_K) to increase], the bias is increased, and the current through the tube is reduced. This provides compensation so that E_{out} decreases to the reference level. Opposite compensation occurs if the load resistance decreases.

An improved series regulator is shown in Figure 16–13. Series regulation is provided by Q_1 in the usual way. R_S provides current limiting, and R_C causes Q_1 to be forward-biased and provides collector voltage to Q_2. Collector voltage for Q_2 is $E_{out} + V_{BE}(Q_1)$. R_S is usually 10 to 50 ohms for regulators in the range of 6 to 24 v. The output level, E_{out}, is established by the zener, V_{BE} of Q_2, and R_B.

$$E_{out} = V_z + V_{BE} + V_{RB} \qquad (16\text{–}5)$$

The series transistor is regulated in a shunt fashion similar to that used in Figure 16–11(b). The advantage of a second transistor in a regulator is that it adds sensitivity to changes in the input voltage or load. Any variations whatever are amplified by Q_2, which provides closer control of E_{out}.

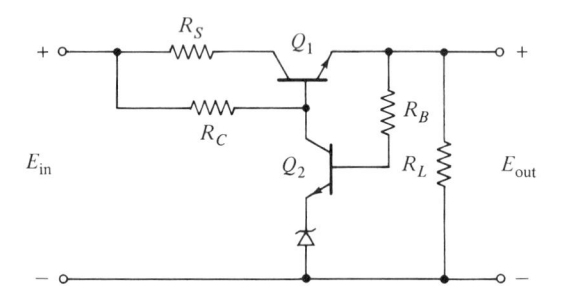

FIGURE 16–13 *Two-Transistor Series Voltage Regulator*

All the regulators in this section have been illustrated for positive voltages. The solid state versions can be adapted to negative supplies by reversing the zener diodes and replacing NPN transistors with PNP transistors. Figure 16–13 is redrawn in Figure 16–14 for a negative power supply. Variable bias for Q_2 is shown in Figure 16–14 (R_1, R_2, and R_3) to illustrate a technique for obtaining an adjustable output voltage.

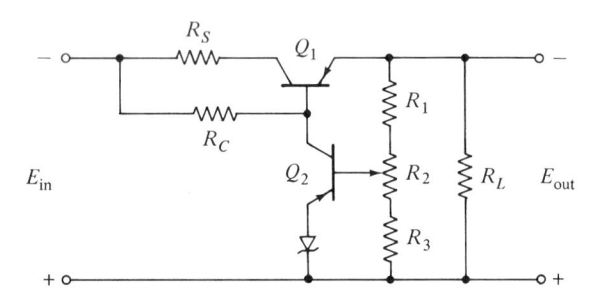

FIGURE 16–14 *Series Regulator for Negative Power Supply*

16.5 Current Regulation

Regulated current, or **constant current generators**, have applications in several areas. Many device parameters need to be explored and plotted with a constant current. Characteristic curves for transistors are plotted for various values of a constant base current. Precision magnetic recording devices require a constant current through an inductor.

The basic principle of current regulation is illustrated in Figure 16–15. The intent is to hold the emitter current constant so that the collector current is constant, since I_C is a function of the formula

$$I_C = \alpha I_E \tag{16-6}$$

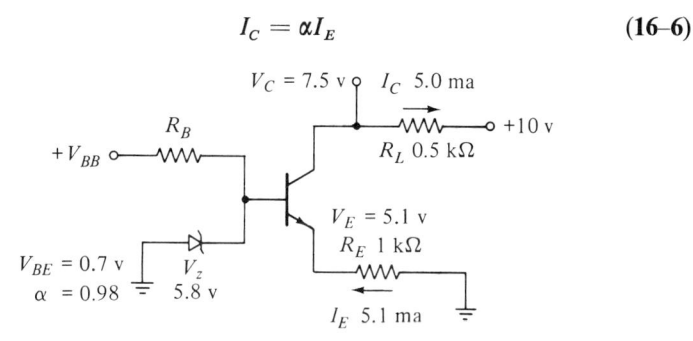

FIGURE 16–15 *Basic Current Regulation*

If I_E is constant, I_C is controlled by alpha, without regard for the collector supply voltage. Regarding the circuit, V_B is held at a constant 5.8 v by the 5.8-v zener diode. Bias for the zener is supplied via R_B and V_{BB}. Since the junction drop is specified as $V_{BE} = 0.7$ v, the emitter is 0.7 v less positive than the base.

$$V_E = V_B - V_{BE} = 5.8 \text{ v} - 0.7 \text{ v} = 5.1 \text{ v}$$

Emitter current is simply

$$I_E = \frac{V_E}{R_L} = \frac{5.1 \text{ v}}{1 \text{ k}\Omega} = 5.1 \text{ ma}$$

And with an alpha of 0.98,

$$I_C = \alpha I_E = 0.98 \times 5.1 \text{ ma} = 5.0 \text{ ma}$$

So that the collector voltage is

$$V_C = V_{CC} - I_C R_L = 10 - (5 \times 10^{-3} \times 0.5 \times 10^3)$$
$$= 10 - 2.5 = 7.5 \text{ v}$$

If the collector supply voltage varies, V_C will vary accordingly. But V_E, I_E and I_C will remain constant as long as V_z, V_{BE} and R_E do not change.

The circuit in Figure 16–16 more closely represents a circuit for operation from a regulated supply. Note that V_{EE} must be regulated to ensure a constant IR drop across R_E. The zener is biased with R_B so that V_B is held at the V_z reference value. I_C is a function of αI_E, and I_C is controlled by the parameters of formula (16–6).

Figure 16–17 is a rearranged version of the circuit in Figure 16–16. Regulated voltage is supplied at the input to R_S. The zener references the base to a predetermined value that, with V_{BE}, will establish a constant voltage at the emitter. This voltage, along with V_{reg}, sets I_E at a value dependent upon R_S. Changes in the load will change the output voltage of the circuit (IR drop across R_L), but the current will remain constant through R_L.

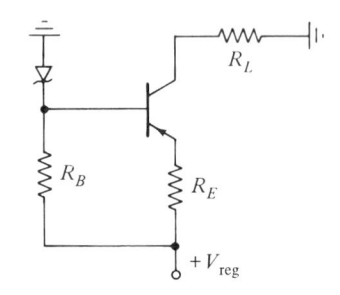

FIGURE 16–16 *Basic Current Regulator for Voltage Regulator Output*

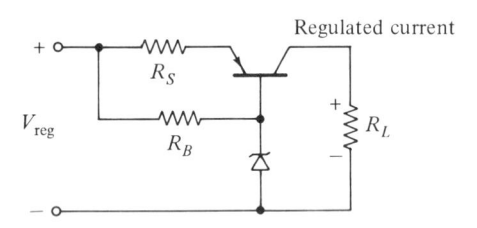

FIGURE 16–17 *Current Regulator*

In discussing voltage regulators, recall that according to Ohm's law load changes had to vary the output current. Load variations in a constant current supply must then vary the output voltage. Ohm's law does not permit both a constant voltage and current for changing load resistances.

The drawing in Figure 16–18 is a complete power supply, voltage regulator, and current regulator. The power supply is the customary full-wave rectifier type, filtered with a π-type filter, L, C_1, and C_2. Unregulated voltage is applied across R_1, the bleeder resistor for the supply. In the series regulator portion, Q_1 and D_3 are the key elements that provide regulated dc to the input (emitter circuit) of Q_2. D_4 is the reference device for current regulation. The entire circuit could fit an application where unregulated and regulated voltages are required, along with a constant current source. Any or all of the functions can be tapped and used.

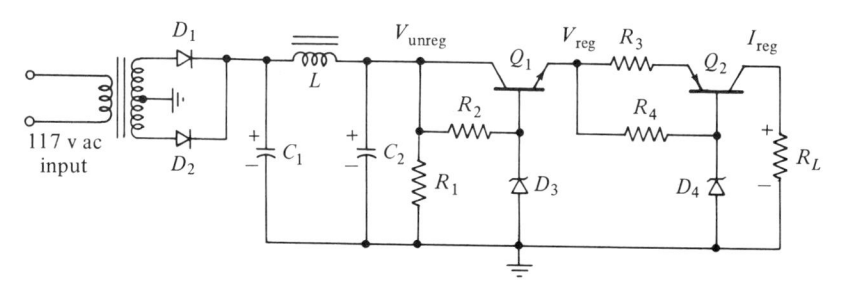

FIGURE 16–18 *Regulated Voltage and Current Supply*

QUESTIONS

16–1 What is the difference between half- and full-wave rectification?

16–2 Briefly explain the filtering action used in Figure 16–1.

16–3 If the input to a half-wave rectifier is 300 v rms, and the capacitors in the filter charge to peak values, why is the dc output voltage not equal to the peak value of 300 v rms?

16–4 If the same components are used in both filters, why does the one in Figure 16–2 provide a smoother output than the one in Figure 16–1?

16–5 What are the advantages of full-wave over half-wave rectification?

16–6 What are the advantages of half-wave rectification compared to full-wave rectification?

16–7 What are the key features of a full-wave bridge rectifier?

16–8 What is the difference between an L and a π filter?

16–9 What are the approximate values for a full-wave L-type filter if the load resistance is 12 k ohms? Assume a line frequency of 60 Hz.

16–10 If the capacitance in a half-wave L-type filter is 100 μf, what approximate load resistance is required for the filter to be effective? Assume a 60-Hz line frequency.

16-11 Why is an *LC* filter generally more effective than an *RC* filter?

16-12 What are the advantages of semiconductors over vacuum tubes for use in power supplies?

16-13 Why is silicon preferred over selenium for rectifier applications?

16-14 In Figure 16-9, which components comprise the filter?

16-15 What are the advantages of a voltage doubler?

16-16 What are the key differences between a shunt and a series voltage regulator?

16-17 In Figure 16-11(a), what is the range of E_{out} if $V_{BE} = 0.6$ v, and $V_z = 14.0$ v $\pm 10\%$?

16-18 In Figure 16-11(b), assume that R_2 is not connected. What value of V_z is required for $E_{out} = 20$ v if $R_1 = 70$ k ohms, $V_{BE} = 0.6$ v, and $I_B = 0.1$ ma?

16-19 In Figure 16-11, which circuit has an advantage over the other and why?

16-20 What is the value of the E_{out} in Figure 16-12(a) if $V_z = 7.0$ v, and $V_{BE} = 0.6$ v?

16-21 Refer to Figure 16-13. $V_z = 9.4$ v, $V_{BE} = 0.6$ v, beta $= 50$ for both transistors, $E_{out} = 14.0$ v, and $I_{out} = 51$ ma. What is the value of R_B?

16-22 What is the purpose of current regulation?

16-23 How would the circuit in Figure 16-17 be connected to regulate current from a negative power supply?

16-24 Refer to Figure 16-17. What is the value of regulated current if $V_{reg} = 30.0$ v, $V_z = 19.4$ v, $V_{BE} = 0.6$ v, $R_S = 0.4$ k ohms, and alpha $= 0.96$?

APPENDIX A

Electronic Units and Symbols

Electronic Quantities and Units

Quantity	Symbol	Unit
Capacitance	C	farad
Charge	Q or q	coulomb
Conductance	G	mho
Current	I or i	ampere
Electromotive force	E or e	volt
Frequency	F or f	hertz (Hz)
Impedance	Z	ohm
Inductance	L	henry
Period of time	T or t	second
Power	P	watt
Reactance	X	ohm
Resistance	R	ohm
Voltage	V or v	volt

Resistor Color Coding

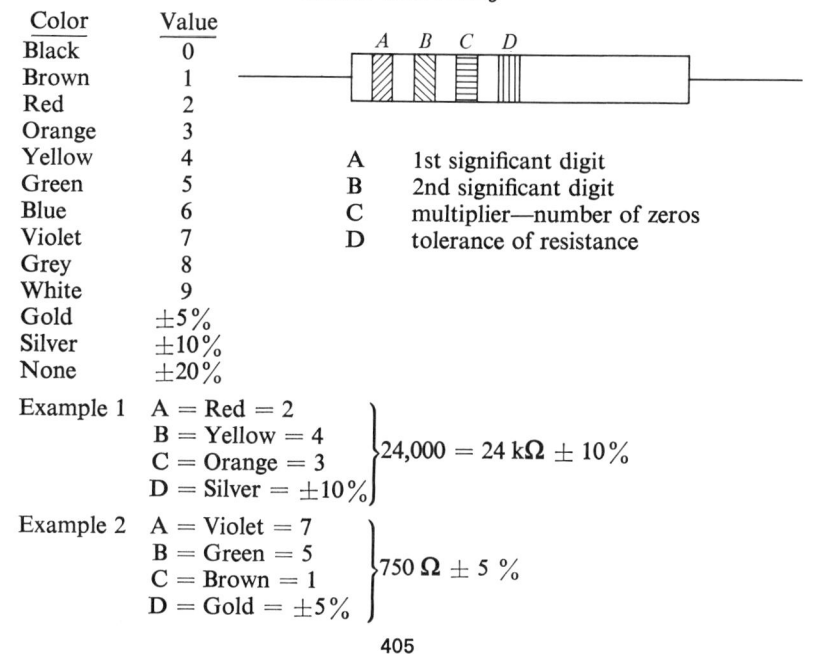

Color	Value
Black	0
Brown	1
Red	2
Orange	3
Yellow	4
Green	5
Blue	6
Violet	7
Grey	8
White	9
Gold	$\pm 5\%$
Silver	$\pm 10\%$
None	$\pm 20\%$

A 1st significant digit
B 2nd significant digit
C multiplier—number of zeros
D tolerance of resistance

Example 1 A = Red = 2
B = Yellow = 4
C = Orange = 3
D = Silver = $\pm 10\%$
$24,000 = 24$ k$\Omega \pm 10\%$

Example 2 A = Violet = 7
B = Green = 5
C = Brown = 1
D = Gold = $\pm 5\%$
$750\ \Omega \pm 5\%$

ELECTRONIC SYMBOLS

Device		Symbol
AC source voltage		
Antenna		
Batteries		
Capacitors	fixed	
	variable	
Coils, chokes	fixed	
	variable	
Crystal		
Fuse		
Grounds	dc or signal	
	chassis	
Meters		

[Appendix A] Electronic Units and Symbols 407

Device	Symbol
Relay and contacts	
Resistors fixed	
variable	
Solenoid	
Speaker	
Thermistor	
Transformers air core	
iron core	
autotransformer	

Electronic Units and Symbols [Appendix A]

Device		Symbol
Transistors	NPN	(NPN transistor symbol with B, C, E)
	PNP	(PNP transistor symbol with B, C, E)
Transistors-FET	N-channel	(N-channel FET symbol with G, D, S)
	P-channel	(P-channel FET symbol with G, D, S)
Diodes	P-N junction	(diode symbol)
	zener	(zener diode symbol)
	photo	(photodiode symbol)
Photoresistor		(photoresistor symbol with λ)
Phototransistors	N-type	(N-type phototransistor symbol)
	P-type	(P-type phototransistor symbol)

[Appendix A] Electronic Units and Symbols 409

Device		Symbol
Vacuum tubes	diode	
	triode	
	tetrode	
	pentode	
	gas filled diode	
	phototube	

APPENDIX B

Mathematic Functions and Tables

Constants

$\pi = 3.14$ $1/\pi = 0.318$
$2\pi = 6.28$ $1/2\pi = 0.159$
$(2\pi)^2 = 39.5$ $\sqrt{\pi} = 1.77$
$4\pi = 12.6$ $1/\sqrt{\pi} = 0.564$
$\pi^2 = 9.87$ $\sqrt{2} = 1.41$
$\pi/2 = 1.57$ $1/\sqrt{2} = 0.707$

Trigonometric Functions

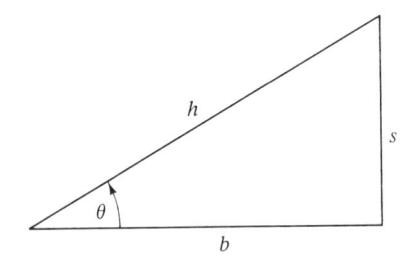

h = hypotenuse
s = side
b = base

$$\sin\theta = \frac{s}{h} \quad \cos\theta = \frac{b}{h}$$

$$\tan\theta = \frac{s}{b} \quad \cot\theta = \frac{b}{s}$$

Squares, Cubes, and Roots

n	n^2	n^3	$\sqrt{n}$	$\sqrt[3]{n}$	n
1	1	1	1.000000	1.000000	1
2	4	8	1.414214	1.259921	2
3	9	27	1.732051	1.442250	3
4	16	64	2.000000	1.587401	4
5	25	125	2.236068	1.709976	5
6	36	216	2.449490	1.817121	6
7	49	343	2.645751	1.912931	7
8	64	512	2.828427	2.000000	8
9	81	729	3.000000	2.080084	9
10	100	1,000	3.162278	2.154435	10
11	121	1,331	3.316625	2.223980	11
12	144	1,728	3.464102	2.289428	12
13	169	2,197	3.605551	2.351335	13
14	196	2,744	3.741657	2.410142	14
15	225	3,375	3.872983	2.466212	15
16	256	4,096	4.000000	2.519842	16
17	289	4,913	4.123106	2.571282	17
18	324	5,832	4.242641	2.620741	18
19	361	6,859	4.358899	2.668402	19
20	400	8,000	4.472136	2.714418	20
21	441	9,261	4.582576	2.758924	21
22	484	10,648	4.690416	2.802039	22
23	529	12,167	4.795832	2.843867	23
24	576	13,824	4.898979	2.884499	24
25	625	15,625	5.000000	2.924018	25
26	676	17,576	5.099020	2.962496	26
27	729	19,683	5.196152	3.000000	27
28	784	21,952	5.291503	3.036589	28
29	841	24,389	5.385165	3.072317	29
30	900	27,000	5.477226	3.107233	30
31	961	29,791	5.567764	3.141381	31
32	1,024	32,768	5.656854	3.174802	32
33	1,089	35,937	5.744563	3.207534	33
34	1,156	39,304	5.830952	3.239612	34
35	1,225	42,875	5.916080	3.271066	35
36	1,296	46,656	6.000000	3.301927	36
37	1,369	50,653	6.082763	3.332222	37
38	1,444	54,872	6.164414	3.361975	38
39	1,521	59,319	6.244998	3.391211	39
40	1,600	64,000	6.324555	3.419952	40

Squares, Cubes, and Roots

n	n^2	n^3	$\sqrt{n}$	$\sqrt[3]{n}$	n
41	1,681	68,921	6.403124	3.448217	41
42	1,764	74,088	6.480741	3.476027	42
43	1,849	79,507	6.557439	3.503398	43
44	1.936	85,184	6.633250	3.530348	44
45	2,025	91,125	6.708204	3.556893	45
46	2,116	97,336	6.782330	3.583048	46
47	2,209	103,823	6.855655	3.608826	47
48	2,304	110,592	6.928203	3.634241	48
49	2,401	117,649	7.000000	3.659306	49
50	2,500	125,000	7.071068	3.684031	50
51	2,601	132,651	7.141428	3.708430	51
52	2,704	140,608	7.211103	3.732511	52
53	2,809	148,877	7.280110	3.756286	53
54	2,916	157,464	7.348469	3.779763	54
55	3,025	166,375	7.416198	3.802952	55
56	3,136	175,616	7.483315	3.825862	56
57	3,249	185,193	7.549834	3.848501	57
58	3,364	195,112	7.615773	3.870877	58
59	3,481	205,379	7.681146	3.892996	59
60	3,600	216,000	7.745967	3.914868	60
61	3,721	226,981	7.810250	3.936497	61
62	3,844	238,328	7.874008	3.957892	62
63	3,969	250,047	7.937254	3.979057	63
64	4,096	262,144	8.000000	4.000000	64
65	4,225	274,625	8.062258	4.020726	65
66	4,356	287,496	8.124038	4.041240	66
67	4,489	300,763	8.185353	4.061548	67
68	4,624	314,432	8.246211	4.081655	68
69	4,761	328,509	8.306624	4.101566	69
70	4,900	343,000	8.366600	4.121285	70
71	5,041	357,911	8.426150	4.140818	71
72	5,184	373,248	8.485281	4.160168	72
73	5,329	389,017	8.544004	4.179339	73
74	5,476	405,224	8.602325	4.198336	74
75	5,625	421,875	8.660254	4.217163	75
76	5,776	438,976	8.717798	4.235824	76
77	5,929	456,533	8.774964	4.254321	77
78	6,084	474,552	8.831761	4.272659	78
79	6,241	493,039	8.888194	4.290840	79
80	6,400	512,000	8.944272	4.308869	80

[Appendix B]   Mathematic Functions and Tables   413

Squares, Cubes, and Roots

n	n^2	n^3	$\sqrt{n}$	$\sqrt[3]{n}$	n
81	6,561	531,441	9.000000	4.326749	81
82	6,724	551,368	9.055385	4.344481	82
83	6,889	571,787	9.110434	4.362071	83
84	7,056	592,704	9.165151	4.379519	84
85	7,225	614,125	9.219544	4.396830	85
86	7,396	636,056	9.273618	4.414005	86
87	7,569	658,503	9.327379	4.431048	87
88	7,744	681,472	9.380832	4.447960	88
89	7,921	704,969	9.433981	4.464745	89
90	8,100	729,000	9.486833	4.481405	90
91	8,281	753,571	9.539392	4.497941	91
92	8,464	778,688	9.591663	4.514357	92
93	8,649	804,357	9.643651	4.530655	93
94	8,836	830,584	9.695360	4.546836	94
95	9,025	857,375	9.746794	4.562903	95
96	9,216	884,736	9.797959	4.578857	96
97	9,409	912,673	9.848858	4.594701	97
98	9,604	941,192	9.899495	4.610436	98
99	9,801	970,299	9.949874	4.626065	99
100	10,000	1,000,000	10.000000	4.641589	100

Trigonometric Functions

Degrees	Sine	Cosine	Tangent	Cotangent	
0	.0000	1.0000	.0000	......	90
1	.0175	.9998	.0175	57.29	89
2	.0349	.9994	.0349	28.636	88
3	.0523	.9986	.0524	19.081	87
4	.0698	.9976	.0699	14.301	86
5	.0872	.9962	.0875	11.430	85
6	.1045	.9945	.1051	9.5144	84
7	.1219	.9925	.1228	8.1443	83
8	.1392	.9903	.1405	7.1154	82
9	.1564	.9877	.1584	6.3138	81
10	.1736	.9848	.1763	5.6713	80
11	.1908	.9816	.1944	5.1446	79
12	.2079	.9781	.2126	4.7046	78
13	.2250	.9744	.2309	4.3315	77
14	.2419	.9703	.2493	4.0108	76
15	.2588	.9659	.2679	3.7321	75
16	.2756	.9613	.2867	3.4874	74
17	.2924	.9563	.3057	3.2709	73
18	.3090	.9511	.3249	3.0777	72
19	.3256	.9455	.3443	2.9042	71
20	.3420	.9397	.3640	2.7475	70
21	.3584	.9336	.3639	2.6051	69
22	.3746	.9272	.4040	2.4751	68
23	.3907	.9205	.4245	2.3559	67
24	.4067	.9135	.4452	2.2460	66
25	.4226	.9063	.4663	2.1445	65
26	.4384	.8988	.4877	2.0503	64
27	.4540	.8910	.5095	1.9626	63
28	.4695	.8829	.5317	1.8807	62
29	.4848	.8746	.5543	1.8040	61
30	.5000	.8660	.5774	1.7321	60
31	.5150	.8572	.6009	1.6643	59
32	.5299	.8480	.6249	1.6003	58
33	.5446	.8387	.6494	1.5399	57
34	.5592	.8290	.6745	1.4826	56
35	.5736	.8192	.7002	1.4281	55
36	.5878	.8090	.7265	1.3764	54
37	.6018	.7986	.7536	1.3270	53
38	.6157	.7880	.7813	1.2799	52
39	.6293	.7771	.8098	1.2349	51
40	.6428	.7660	.8391	1.1918	50
41	.6561	.7547	.8693	1.1504	49
42	.6691	.7431	.9004	1.1106	48
43	.6820	.7314	.9325	1.0724	47
44	.6947	.7193	.9657	1.0355	46
45	.7071	.7071	1.0000	1.0000	45
	Cosine	Sine	Cotangent	Tangent	Degrees

APPENDIX C

Conversion Tables and Graphs

Length, Area, and Volume

To Convert	Into	Multiply By
centimeters	feet	3.281×10^{-2}
	inches	0.3837
	meters	0.01
	millimeters	10.0
cubic centimeters	cu feet	3.531×10^{-5}
	cu inches	0.06102
	cu meters	10^{-6}
cubic feet	cu cm	28,320.0
	cu inches	1728.0
	cu meters	0.02832
cubic inches	cu cm	16.39
	cu feet	5.787×10^{-4}
	cu meters	1.639×10^{-5}
cubic meters	cu cm	10^{6}
	cu feet	35.31
	cu inches	61,023.0
feet	centimeters	30.48
	meters	0.3048
	millimeters	304.8
inches	centimeters	2.540
	feet	8.333×10^{-2}
	meters	2.540×10^{-2}
	millimeters	25.40
meters	centimeters	10^{2}
	feet	3.281
	inches	39.37
	millimeters	10^{3}
millimeters	centimeters	0.1
	feet	3.281×10^{-3}
	inches	0.03937
	meters	0.001
square centimeters	sq feet	1.076×10^{-3}
	sq inches	0.1550
	sq meters	10^{-4}
	sq millimeters	10^{2}

square feet	sq cm	929.0
	sq inches	144.0
	sq meters	0.09290
	sq millimeters	9.290×10^4
square inches	sq cm	6.452
	sq feet	6.944×10^{-3}
	sq millimeters	645.2
square meters	sq cm	10^4
	sq feet	10.76
	sq inches	1,550
	sq millimeters	10^6
square millimeters	sq cm	10^{-2}
	sq feet	1.076×10^{-5}
	sq inches	1.550×10^{-3}

Reactance Chart

Inductive Reactance $X_L = 2\pi f L$

    To determine X_L locate intersection of f and L
                                      X_L is horizontally to left

    To determine f locate intersection of X_L and L
                                      f is vertically at bottom

    To determine L locate intersection of X_L and f
                                      L is diagonally up to right

Capacitive Reactance $X_C = \dfrac{1}{2\pi f C}$

    To determine X_C locate intersection of f and C
                                      X_C is horizontally to left

    To determine f locate intersection of X_C and C
                                      f is vertically at bottom

    To determine C locate intersection of X_C and f
                                      C is diagonally down to right

Resonant Frequency $f_r = \dfrac{1}{2\pi\sqrt{LC}}$

    To determine f_r locate intersection of L and C
                                      f_r is vertically at bottom

    To determine L locate intersection of f_r and C
                                      L is diagonally up to right

    To determine C locate intersection of f_r and L
                                      C is diagonally down to right

[Appendix C] Conversion Tables and Graphs 417

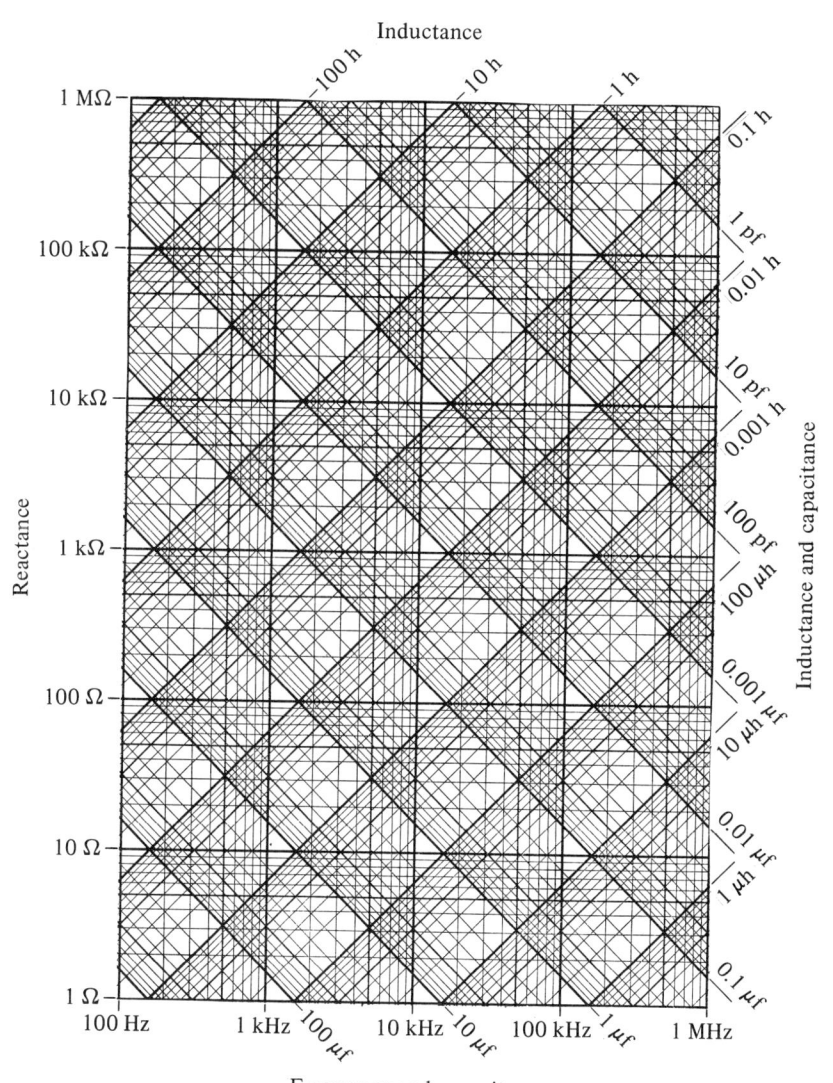

REACTANCE CHART

Universal Chart For:
Resistance, Inductance, Reactance, and Capacitance

The following formulas are for parallel resistance (R_T), parallel inductance (L_T), parallel reactance (X_{LT} and X_{CT}), and series capacitance (C_T).

$$R_T = \frac{R_1 R_2}{R_1 + R_2} \qquad L_T = \frac{L_1 L_2}{L_1 + L_2} \qquad X_{LT} = \frac{X_{L1} X_{L2}}{X_{L1} + X_{L2}}$$

$$X_{CT} = \frac{X_{C1} X_{C2}}{X_{C1} + X_{C2}} \qquad C_T = \frac{C_1 C_2}{C_1 + C_2}$$

To determine component values:
1. Locate and mark total units (R_T or L_T, etc.) on the vertical axis-left side.
2. Draw a horizontal line from the total units mark to intersect the diagonal line.
3. Use a straight edge to pivot around the intersection at the diagonal line.
4. Values of R_1, R_2 or L_1, L_2 or X_{L1}, X_{L2} or X_{C1}, X_{C2} or C_1, C_2 are points where a straight edge crosses vertical and horizontal axes and still passes through the intersection of diagonal line.

Note: Vertical and horizontal unts must be in the same order. For example, both scales are multiplied by 1, 10, 10^2, 10^3, etc. or 10^{-1}, 10^{-2}, 10^{-3}, etc.

[Appendix C] Conversion Tables and Graphs 419

UNIVERSAL CHART

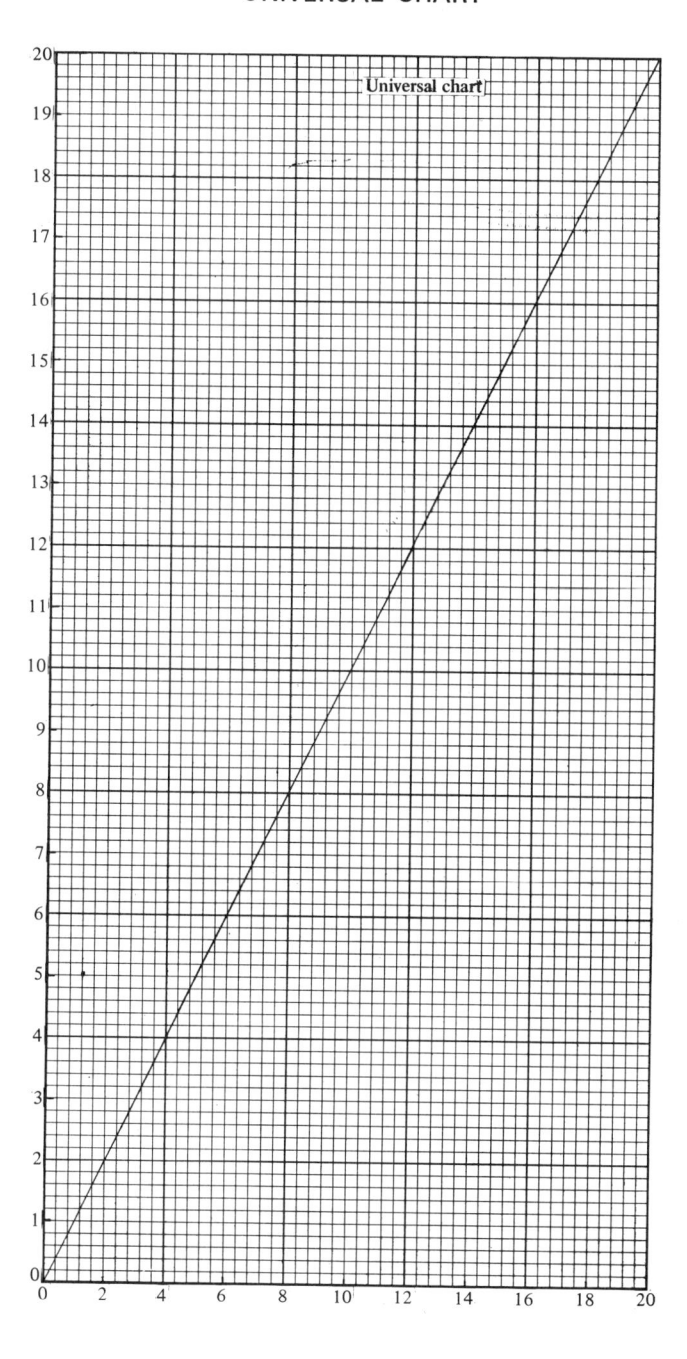

APPENDIX D

References

Electronics

Grob, Bernard. *Basic Electronics*. New York: McGraw-Hill Book Company, 1965.

Romanowitz, H. Alex and Russell E. Puckett. *Introduction to Electronics*. New York: John Wiley & Sons, Inc., 1968.

Ryder, J. D. *Electronic Fundamentals and Applications*. Englewood Cliffs, N. J.: Prentice-Hall, Inc., 1964.

Terman, Frederick E. *Electronics and Radio Engineering*. New York: McGraw-Hill Book Company, 1955.

Semiconductor Devices and Circuits

Schilling, D. L. *Electronic Circuits: Discrete and Integrated*. New York: McGraw-Hill Book Company, 1968.

Tocci, Ronald J. *Fundamentals of Electronic Devices*. Columbus, Ohio: Charles E. Merrill Publishing Company, 1970.

Turner, James F. *Digital Computer Analysis*. Columbus, Ohio: Charles E. Merrill Publishing Company, 1968.

Transistors

Kiver, Milton S. *Transistors*. New York: McGraw-Hill Book Company, 1962.

Malvino, A. P. *Transistor Circuit Approximations*. New York: McGraw-Hill Book Company, 1968.

Pierce, J. F. *Transistor Circuit Theory and Design*. Columbus, Ohio: Charles E. Merrill Publishing Company, 1965.

Riddle, Robert L. and Marlin P. Ristenbatt. *Transistor Physics and Circuits*. Englewood Cliffs, N.J.: Prentice-Hall, Inc., 1966.

Siedman, A. M. and L. S. Marshall. *Semiconductor Fundamentals*. New York: John Wiley & Sons, Inc., 1963.

Vacuum Tubes

GE Tube Manual. Syracuse, N.Y.: General Electric Company.

Krackhardt, Russell H. *Vacuum Tube Electronics*. Columbus, Ohio: Charles E. Merrill Publishing Company, 1966.

RCA Tube Manual. Harrison, N.J.: Radio Corporation of America.

Sylvania Tube Manual. Emporium, Pa.: Sylvania Electric Products.

APPENDIX E

Answers to the Odd Numbered Questions

Chapter 1

1-1 22

1-3 The approximate atomic weight of an atom is the sum of the protons and neutrons.

1-5 Covalent bonding occurs when atoms share their valence electrons. For example, a semiconductor has four valence electrons. By sharing electron pairs, each atom fills its valence band with the maximum of eight electrons.

1-7 Heat and light

1-9 2.5 coulombs

1-11 200 msec

1-13 −9 coulombs

1-15 Electrons are attracted toward the positive terminal, repelled from the negative terminal, and those in between are pushed along by those coming from behind.

1-17 1500 cm

1-19 2.5×10^3 ohms

Chapter 2

2-1 Stroking the material with another magnetic or by electromagnetic induction (current through a coil).

2-3 Field is strongest inside of a magnet. Lines of force are south to north inside and north to south outside of a magnet.

2-5 Electrons spinning on their own axis; because orbits are random and tend to cancel each other.

2-7 9500 maxwells

2-9 3 cm²

2-11 200 gauss

2-13 A small gap concentrates the magnetic lines of force so more lines can be directed into another material.

2-15 Ceramic materials having ferromagnetic properties of iron; very high permeability; insulating quality keeps eddy current losses to a minimum.

2-17 Grasp the solenoid in left hand, thumb pointing in the direction of north pole—fingers point in the direction of current flow.
2-19 7 amp
2-21 0.628 cm
2-23 300
2-25 6.28 gilberts per maxwell
2-27 Saturation is the point at which an increasing field intensity produces no significant increase in flux density.
2-29 When a field is removed, the material settles down at a lower flux density (B_r). In a varying current situation, B_r is the flux density when the field is at zero.
2-31 The conductor moves across the external field. Lenz's law explains that forces oppose. Thus, the circular field around the conductor increases on the side leading through the field while canceling on the trailing edge. The magnetic induction follows Flemming's left-hand rule.
2-33 50 μ v

Chapter 3

3-1 33 v
3-3 50 μa
3-5 14.7 w
3-7 33.55 v
3-9 108 mw
3-11 9 w
3-13 2.4 w
3-15 143 ma; 147 Ω

Chapter 4

4-1 100 Ω
4-3 0.1 A
4-5 2.5 ma
4-7 −60 v
4-9 $R_2 = 0.3$ kΩ; $R_4 = 0.25$ kΩ; $E_{R1} = 2.4$ v; $E_{R3} = 1.8$ v; $E_{R4} = 3$ v; $E_{R5} = 1.2$ v
4-11 $E_{P1} = 4$ v; $E_{P2} = 6$ v; $E_{P3} = 15$ v; $E_{P4} = 20$ v
4-13 $R_1 = 3$ kΩ; $R_3 = 5$ kΩ; $E_{P1} = -7$ v; $E_{P3} = -15$ v
4-15 240 mw; 600 Ω
4-17 10 w

Chapter 5

5-1 20 v
5-3 10 A

5–5	2.86 kΩ
5–7	13.3 kΩ
5–9	10 Ω
5–11	60 v
5–13	70 w
5–15	80 μw
5–17	0 v; 2 ma; 15 mw

Chapter 6

6–1	0.6 A
6–3	24 v
6–5	$E_1 = 9$ v; $E_2 = 6$ v; $E_3 = 10$ v; $E_4 = 5$ v
6–7	-25 v
6–9	$I_T = 5.3$ ma; $I_2 = 3.18$ ma
6–11	$R_1 = 1.56\ \Omega$; $R_2 = 2.4\ \Omega$; $R_3 = 1.5\ \Omega$
6–13	$P_4 = 5.75$ mw; $P_5 = 3.83$ mw

Chapter 7

7–1	36°
7–3	0.4 A
7–5	14.16 mv
7–7	7.2 v; -14.8 v
7–9	140°
7–11	5 MHz
7–13	0.2×10^4 cm
7–15	0.487 ft
7–17	8 nsec
7–19	0.1 μsec

Chapter 8

8–1	75 μ coulombs
8–3	5 μf
8–5	120 microcoulombs
8–7	6
8–9	Clockwise
8–11	0.05 μf
8–13	80 pf
8–15	$E_1 = 5.33$ v; $E_2 = 10.67$ v; $E_3 = 64$ v
8–17	$I_1 = 8\ \mu a$; $I_2 = 5\ \mu a$; $I_3 = 2\ \mu a$; $I_T = 15\ \mu a$
8–19	$E_1 = 4$ v; $E_2 = 2$ v; $E_3 = E_4 = 12$ v; $Q_1 = Q_2 = 1800 \times 10^{-12}$ coulombs; $Q_3 = 720 \times 10^{-12}$ coulombs; $Q_4 = 1080 \times 10^{-12}$ coulombs

8–21 4 MHz
8–23 0.64 kΩ
8–25 93.6 μa

Chapter 9

9–1 A changing current creates a changing magnetic field. The changing magnetic lines of flux cut through the conductor and induce the voltage.
9–3 7.2 v
9–5 6.4 mh
9–7 2 cm^2
9–9 10 mh
9–11 0.8
9–13 $N_{S1} = 250$ turns; $N_{S2} = 50$ turns
9–15 300 w
9–17 12 v; 36 turns
9–19 100 v ac
9–21 Laminated cores reduce eddy current losses.
9–23 4 h
9–25 24 mh
9–27 $e_1 = 0.08$ v; $e_2 = 0.14$ v
9–29 $i_1 = 64$ ma; $i_2 = 32$ ma
9–31 6.28 kΩ
9–33 $X_{L1} = 64\ \Omega$; $X_{L2} = 250\ \Omega$; $X_{LT} = 314\ \Omega$
9–35 3.2 Ω
9–37 80 mv

Chapter 10

10–1 81 Ω
10–3 −21.8°
10–5 10 Ω
10–7 21.36 v
10–9 21.8°
10–11 14 μa
10–13 4 msec
10–15 10 MΩ
10–17 19.18 v; 33°
10–19 138.7 ma; 72 Ω; 36.8°
10–21 −63.5°
10–23 328 mh
10–25 50 Ω; 25 v
10–27 1.05 kΩ

[Appendix E] Answers to the Odd Numbered Questions 425

10-29 1.697 v
10-31 25 ma; 0.4 kΩ
10-33 $R = 65\ \Omega$ and $X_C = 31.6\ \Omega$
10-35 10.5 Ω

Chapter 11

11-1 X_L leads by 90°, and X_C lags by 90°; X_L and X_C are 180° out of phase with each other.
11-3 Resonant frequency is the point at which $X_L = X_C$.
11-5 398 kHz
11-7 9.4 pf
11-9 $C = 40$ pf; $e_L = 99.5$ v; $e_C = 99.5$ v; $i_T = 25$ ma
11-11 At resonance, X_L equals X_C. The 180° phase difference cancels the actual inductive and capacitive currents except for the current produced by the small internal ac resistance of the coil. The remaining current is approximately $R_C/X_L i_L$, which flows through the source.
11-13 160
11-15 5 v
11-17 1.2 kΩ
11-19 5 Ω
11-21 16.9 Hz
11-23 1 kΩ; 0.008 μf
11-25 $X_L = 2.18$ kΩ; $X_C = 5.3$ kΩ; $L = 11.6$ mh
11-27 Band-pass filters pass a band of frequencies while attenuating those frequencies on either side. Band-stop filters reject or attenuate a selected band of frequencies.
11-29 The effect is a band-pass filter. At resonance, the impedance is maximum and has very little effect on the loading. On either side of resonance, the impedance is minimal and essentially shorts those frequencies to ground.
11-31 Low-pass filter; band-pass filter that excludes all but the broadcasting band.
11-33 $R_C = 1.57\ \Omega$; $L = 25$ μh; $C = 20.7$ pf
11-35 5.5 v; 60.5 ma; 91 Ω
11-37 A non-linear volume control (variable resistance) with a characteristic that offsets the exponential curves.

Chapter 12

12-1 $15\ v - 4\ v + 4\ v - 15\ v = 0$ or $4\ v - 15\ v + 15\ v - 4\ v = 0$
12-3 $10\ A + 2\ A + 4\ A - 3\ A - 5\ A - 2\ A - 6\ A = 0$
(The sequence is arbitrary)

12-5 $I_1 = 1.5$ A; $I_2 = 0.25$ A; $I_3 = 1.25$ A; $I_1 R_1 = 24$ v; $I_2 R_2 = 4$ v; $I_3 R_3 = 12$ v

12-7   3 v; 7.2 Ω; 0.3 A

12-9 $E_{eq} = E_1 = 38$ v; $I_{RL} = 0.1$ A; $E_{RL} = 33.8$ v

12-11 $I_{R1} = 1.5$ A; $I_{R2} = 0.25$ A; $I_{R3} = 1.25$ A; $E_{R1} = 24$ v; $E_{R2} = 4$ v; $E_{R3} = 12$ v

12-13 -1.25 A; -5 v

12-15 -0.084 A; -28.5 v

12-17 $R_{TH} = 10$ Ω; $E_{TH} = 10$ v

12-19   0.084 A; 28.43 v

12-21   a. $Z_T = 450 - j48$;   b. $Z_T = 0 + j75$;   c. $Z_T = 42 + j12$;   d. $Z_T = 120 - j0$

12-23   a. 64 Ω, 38.7°;   b. 13.9 Ω, $-59.8°$

12-25 $57 + j3$, $23 + j47$

12-27   a. $125 \underline{/25°}$;   b. $1050 \underline{/-80°}$ c. $15 \underline{/-42°}$;   d. $0.12 \underline{/30°}$

12-29   a. $120 \underline{/-51°}$;   b. $6 \underline{/37°}$;   c. $5 \underline{/-62°}$

12-31   25 Ω; 0.4 A; $-36.8°$

12-33   89.5 ma; 63.4°

12-35 $Z_T = 43.4$ Ω; $i_T = 0.23 \underline{/68.9°}$

12-37   90 Ω; $0.222 \underline{/22°}$

Chapter 13

13-1   Plate, cathode and/or filament or heater.

13-3   To emit electrons which are the basis of electron-current flow.

13-5   Increasing the cathode temperature increases its thermionic emission which increases the current. Decreasing temperature has the opposite effect.

13-7   The control grid is placed between the cathode and plate, but nearer to the cathode.

13-9   5 kΩ

13-11   107 Ω

13-13   0.008 μf

13-15   4.5 ma

13-17   4,000 μ mhos

13-19 -6 v

13-21   14

13-23   When the plate voltage is lower than the screen voltage, and the screen current exceeds the plate current.

13-25 $I_f = I_C + I_S + I_P$

[Appendix E] Answers to the Odd Numbered Questions 427

13-27   Very small change in e_c results in a very large change in E_b because the grid voltage curves (see Figure 13-28) are nearly flat.
13-29 $E_b \approx 310$ v; $I_b = 2.5$ ma
13-31   120 v
13-33   Ratings that are below maximum values. A set of ratings for a particular (and typical) operating condition.
13-35   4.2 k ohms
13-37   Increased spacing between electrodes to prevent arcing; high voltage and current; high temperature due to high power; often utilizes a water cooled anode; often has a plate (anode) connection on end of glass envelope to further separate electrodes.
13-39   The cathode emits photoelectrons. Current flows with a positive plate potential.
13-41   The inside has a fluorescent coating which is illuminated when the electron beam contacts it. The illuminated screen is visible because the face is glass.

Chapter 14

14-1   A valence of four electrons.
14-3   Covalent bonds are established when atoms share their valence electrons. In terms of semiconductors, each atom acts as though it has a maximum valence of eight electrons.
14-5   Intrinsic is a pure crystal. Extrinsic crystals contain impurities. A doped crystal is an extrinsic crystal.
14-7   Gallium or indium
14-9   Electrons have about double the mobility of holes (refer to Table 14-3).
14-11   Forward bias decreases its width; reverse bias increases its width.
14-13   The flow of minority carriers.
14-15   The portion of a signal that causes forward bias is passed because current is allowed to flow. The portion causing reverse bias is blocked because the diode cannot conduct current, except for a small amount of leakage current.
14-17   Base-emitter is forward-biased, and collector-base is reverse-biased.
14-19   Base current flows from base to emitter. Collector current flows from collector to emitter. Emitter current is the sum of the base and collector current.
14-21 $I_B = 0.02$ ma; $I_C = 4$ ma

14-23 0.995
14-25 Saturation occurs at a point when increasing the base current will no longer increase the collector current.
14-27 -6.078 v
14-29 7.5 ma; -2.5 v
14-31 Common-base
14-33 Select an I_E value in the anticipated range of the circuit. Determine the I_C value which is horizontal from the I_E curve; $\alpha_{dc} = I_C/I_E$.
14-35 147
14-37 To define the operating characteristics of a transistor in terms of input resistance, output admittance, feedback voltage ratio, and current transfer ratio.
14-39 2 ma; 5 v

Chapter 15

15-1 To reduce its junction breakdown potential.
15-3 Zener breakdown occurs below about five volts and is caused by the electric field created by reverse voltage. Avalanche breakdown is above five volts and is the result of multiplying minority carriers that dislodge electrons.
15-5 -20 v and 5 ma
15-7 2 ma
15-9 By operating in the breakdown portion of the curve where a varying zener current maintains a constant voltage.
15-11 1.5 kΩ
15-13 200 Ω
15-15 $R_C = 12.5$ kΩ; 20 ma and 1 kΩ
15-17 10,000 Å
15-19 Negative to the drain, positive to the gate.
15-21 When the bias for a FET is varied, the depletion regions vary. While the regions are separated, drain current increases at a rate governed by the change in bias. When the bias potentials are sufficient to just close the depletion regions, the drain current stops increasing. This point is called **pinch-off**.
15-23 0.6 ma
15-25 $V_{DS} = -8$ v; $V_{GS} = 1.0$ v
15-27 $I_D = 2$ ma; $V_{GS} = 1.0$ v; $V_{DS} = -8$ v; $R_D = 7.5$ kΩ
15-29 Savings in weight and size; savings in cost; increased reliability; improved frequency response; low power.
15-31 Diodes are the base-emitter and collector-base junctions. Resistors are the resistivity of the doped silicon. Capacitors are the reverse-biased junction capacitances.

Chapter 16

16–1 Half-wave circuits rectify half of each cycle of the ac voltage. Full-wave circuits rectify both halves of each cycle.

16–3 Because of the *IR* drops across the rectifier and series choke.

16–5 More efficient rectification by using both halves of the input waveform; more efficient filtering.

16–7 Full use of secondary winding—does not require a center-tap; combines advantages of half- and full-wave rectifiers; requires four rectifiers.

16–9 12 h; 8 μf

16–11 The reactance of an *LC* type is higher than the resistance of an *RC* type in practical applications.

16–13 Higher current and inverse voltage ratings with smaller components.

16–15 Can operate without a power transformer; doubles the available ac voltage so that the dc output is twice what it would be if not doubled.

16–17 13.2 v to 16.0 v

16–19 Base reference has the advantage in that it regulates less current than the emitter reference.

16–21 200 kΩ

16–23 Replace the PNP with an NPN transistor; reverse the polarity of the zener diode.

Index

Abbreviations (*See* Symbols)
AC current, 111–13
Alpha, 339–41
Alternating voltage, 107–9
Ampere, defined, 41
Ampere-turns, 25–26, 31
Amplifier
   common-base, 353–54
   common-collector, 356–57
   common-emitter, 354–55, 357–58
   current (*See* Semiconductors, transistors)
   FET, 376–78
   FET, connections, 379
   grounded grid, vacuum tube, 319–20
   grounded plate, vacuum tube, 320–21
   pentode, vacuum tube, 315–16

Amplifier (Cont.)
   photoresistor, 369–70
   relay, 368 (*See also* Amplifier, photoresistor)
   special (*See* Vacuum tubes)
   triode, vacuum tube, 308–9
   voltage (*See* Vacuum tubes; Field effect transistors)
Amplification factor, 305–6, 309, 314
Amplitude measurements
   average, 113–15
   effective value, 114
   peak-to-peak, 113–15
   peak, 113–16
   rms, 113–16
   sine waves, 113–16
Angle
   amplitude measurements, 115–16

Angle (Cont.)
  cosine, 117
  phase, 116–17
  phase
    capacitive circuit, 152–54
    inductive circuit, 183–85
    rotation, 110–11
  sine, 117
Angstrom units, 372
Angular rotation, 109–11
Anode
  accelerating, 324
  collection, 324
Area, capacitor plates, 132–33
Atomic structure
  magnets, 17
  semiconductors, 329
Atoms
  basic, 1–8
  Bohr model, 1
  carbon, 2
  charges, 1–2
  electrons, 1–8
  germanium, 329
  hydrogen, 2–3
  neutrons, 1–8
  nucleus, 1–8
  protons, 1–8
  silicon, 2
  structure of magnets, 17
  valence band, 3
Autotransformer, 170–71
Avalanche breakdown, 362

Band, atoms, 2
Band-pass filter, 244–45
Band-stop filter, 244–45
Bandwidth, 236–38
Beam power tubes (*See* Vacuum tubes)
Beta, 339–41

Bias
  CRT, 325
  semiconductors, 334–35
  transistors, 341–44
Bohr, model 1
Bonding, covalent, 330
Bridge rectifiers (*See* Power supplies)
Bulbs, household, 11

Calculations
  amplification factor, 305–6, 309, 314
  amplitude values, sine waves, 113–16
  band-pass filter, 248–49
  capacitance, 131–33
  capacitive circuit, instantaneous current, 147–48
  capacitive reactance, 146–52
  capacitor plates, 132–33
  capacitors
    charges, series, 141–44
    parallel, 143–44
    plates, parallel, 143
    series, 138–43
    series formulas, 139
  circuit analysis (*See* Circuit analysis)
  coefficient of coupling, transformers, 164
  common-base amplifier, 353–54
  common-collector amplifier, 356–57
  common-emitter amplifier, 354–55, 357–58
  complex numbers, polar form, 278–80

INDEX

Calculations (Cont.)
  complex numbers, rectangular form, 273–78
  coulombs, 129
  current gain, transistor, 338–41
  current regulators, 401
  current, zener, 363–67
  dc current, 57–59
  duty cycle, 123–24
  FET
    bias, 377–78
    currents, 376–78
    load line, 376–77
    resistors, 378
  filters, 240–43
  filters, power supply, 392–94
  frequency, 117–19
  impedance, series RC, 190–94
  inductance
    parallel, 174–75
    series, 173–74
  inductance, defined, 159–60
  inductive reactance, 178–80
    parallel, 181–82
    series, 180–81
  inductor, coil, 161–62
  instantaneous current, 112–13
  instantaneous voltage, 111
  j factors, 273–78
  LR, parallel, 205–8
  LR, series, 200–205
  LR time constant, 208–11
  LRC
    parallel, 217–19
    series, 215–17
    series/parallel, 220–22
  metric, 120
  mutual inductance, 165–66
  mutual transconductance, 306, 314

Calculations (Cont.)
  non-linear circuits, 249–53
  parallel circuits
    current, dc, 70–72, 79–81
    power dissipation, 82–84
    resistance, 72–74, 76–77
    unknown quantities, dc, 84–87
  parallel/series resistance, 98–105
  phase, capacitive circuit, 152–54
  photoresistor, 368–70
  phototransistor, 370–71
  power dissipation 63–64
  pulses, 122–25
  RC, parallel, 194–97
  RC, series, 190–94
  RC time constant, 197–200
  resistance loads, 96–97, 103–5
  resonance
    parallel, 231–33
    series, 225–30
  series circuits
    current, dc, 57–59
    power dissipation, 63–64
    unknown quantities, 64–67
    voltage drops, 59–63
  series/parallel resistance, 90–97, 101–5
  thyrite, 250–52
  transistors
    bias resistors, 343–45
    common-base, 353–54
    common-collector, 356
    common-emitter, 354–55, 357–58
    current, 339–41, 347–50
    current gain, 338–41
    power dissipation, 351

Calculations (Cont.)
  vacuum tube
    amplification, 301
    bias voltage, 299
    cathode resistor, 302–3,
      315–16
    plate current, 300
    plate to cathode resistance,
      307, 309, 314
    plate voltage, 300
    voltage gain, defined, 301
    voltage gain, grounded grid
      amplifier, 319–20
    voltage gain, grounded
      plate amplifier, 320–21
    voltage gain, pentode,
      314–15
    voltage gain, triode, 309
  voltage, zener diode, 364–66
  voltage regulators
    series, 399–400
    shunt, 397–99
  wavelength, 119–21
Capacitance
  farad, defined, 131
  interelectrode, vacuum tube, 317
  measurement, 131–33
  parallel, 143–44
  RC time constant, 197–200
  series, 138–43
  series, chart, 418–19
  series/parallel, 145–46
  storage, 127–31
  stray, 134
  units, 131
Capacitive reactance, 146–52
Capacitive reactance chart, 416–17
Capacitor
  bypass, cathode, 303–4, 315
  charging, 127–31

Capacitor (Cont.)
  definition, 127–28
  dielectric constant, 128, 132–33
  dynes, 128–29
  plates, 127–34
Capacitors
  charging, series, 138
  dc and ac effects, 134–37
  electrolytics, 134
  electrons, 130–31
  losses (*See* Losses, capacitors)
  parallel
    area, plates, 143
    distance, plates, 143
  pulsating dc, 136–37
  series, 138–43
  series/parallel, 145–46
Cathode, 291–93
Cathode, cold, 322
Cathode, hot, 322
Cathode ray tubes, CRT (*See* Vacuum tubes)
Characteristics, FET, 374–75
Charge
  atoms, 1–2
  capacitors, series, 138, 141–44
  current, RC circuit, 197–98
Charges
  atoms, coulombs, 8–9
  capacitive reactance, 147
  capacitors, 127–34
  capacitors, series/parallel, 145–46
  carriers, 10
  dc/ac effects, 134–36
  diffusion, 334
Charging capacitors, 127–31, 138
Circuit analysis
  complex numbers
    applications, 280–85
    polar form, 278–80

INDEX

Circuit analysis (Cont.)
    j factors, 272–78
    loop equations, 257–62
    Norton's theorem, 266–69
    superposition theorem, 269–72
    Thévenin's theorem, 262–66
Circuits (*See also* Amplifiers)
    beam power pentode, 321
    capacitive, 130, 135–36, 138–46, 152
    cathode ray tube (CRT), 325
    common-base, 345, 353–54
    common-collector, 356–57
    common-emitter, 345–46, 354–55, 357–58
    current regulation, 401–3
    diode, vacuum tube, 295, 297–98
    FET, 373, 376–78
    filters, 239–45, 392–94
    gas tube, 322–23
    grounded grid amplifier, 319–20
    grounded plate amplifier, 320–21
    impedance, parallel RC, 194–97
    impedance, series RC, 190–94
    integrated, 379–85
    LC, parallel, 214
    LC, series, 212
    LR, parallel 205–8
    LR time constant, 209
    LRC, parallel, 218
    LRC, series, 216
    LRC, series/parallel, 220–22
    non-linear, 249–53
    pentode, 313, 315
    photoresistor, 367–69
    phototransistor, 370–71
    phototube, 323

Circuits (Cont.)
    power supplies, vacuum tube, 388–92
    Q, 233–35
    radio input, 238
    RC charge/discharge, 199
    resistance
        basic Ohm's law, 42–54
        combination, 101–5
        parallel, 69–87
        parallel/series, 98–101
        series, 57–67
        series/parallel, 90–97
    resonance
        parallel, 231
        series, 229
    superposition, ac/dc, 271–72
    tetrode, 311
    thyrite, 251
    transient, 245–48
    transistor
        common-base, 348
        common-collector, 346
        common-emitter, 349
    triode, 300, 302–4, 308–9
    voltage divider, 94–97
    voltage regulators, 397–400
    zener diode, 362–64
Class A operation, 354
Coefficient of coupling, transformer, 164
Configurations, transistors, 345–47
Coils
    inductance, calculation, 161–62
    induced current, 107–8
    magnet, 23–26
    variable inductance, 162–63
Color code, resistors, 405

Complex numbers
 applications, 280–85
 cosine functions, 279–80
 phase angle, polar form, 278–80
 polar form, 278–80
 rectangular form, 273–78
 sine functions, 279–80
Conductance, 11–12
Conductors
 copper, 10–11
 resistivity, 9–10
 silver, 10
Constants
 dielectric, 128, 132–33
 LR time, 208–11
 mathematics, 410
 RC time, 197–200
Conversions
 alpha, 340
 angstrom, 372
 area, 415–16
 beta, 340, 356
 coulombs, 129, 131–32
 gamma, 356
 length, 415–16
 metric, 120
 micron, 372
 Norton's theorem, 269
 polar form, 279–80
 rectangular form, 279–80
 tables, 415–16
 Thévenin's theorem, 269
 vacuum tube characteristics, 307
 volume, 415
Conversion units
 coulomb, 41
 esu, 129
 gilberts, 29–30
 magnetic flux, 18–20

Conversion units (Cont.)
 oersteds, 26–29
 volts, 37–38
Cores
 air, 173
 transformer, 171–73
Cosine, angle, 117
Cosine, table, 414
Cosine wave, 117
Cotangent, table, 414
Coulomb, 129, 131–32, 141–44
Coupling
 capacitive, 239
 coefficient, transformer, 164
 transformer, 238–39
Covalent bonding, 330
Crystals
 N-type, 332–33
 P-type, 332–33
Current
 ac, 111-13
 alternating, 111–13
 base, 338–41, 343–45
 calculation, 44–45
 capacitive, instantaneous, 147, 153–54
 capacitors, parallel, 144
 charge, RC circuit, 197-98
 collector, 338–41, 343–45
 complex numbers, 282, 284–85
 dc, 57–59
 drift, 9
 eddy, 172–73
 electric, 9
 electron, FET, 373–74
 electron flow, vacuum tube, 291–92, 295
 emitter, 338–41, 343–44
 exponential change, LR, 208–9

INDEX

Current (Cont.)
  FET, 374–78
  filament, 294, 316–17
  flow, 9
  induced, 34–36
  inductive, instantaneous, 184
  inductive, varying, 183–85
  inductors, 176–78
  instantaneous value, 112–13
  j factors, 275–76
  junction, 69–70
  LC, parallel, 213–14
  LC, series, 211–13
  LR, parallel, 205–8
  LRC, parallel, phase angle, 217–19
  LRC, series/parallel, 221
  Ohm's law, 42–44
  parallel circuit, dc, 70–72, 79–81
  phase, 113
  phase angle
    capacitive circuit, 152–54
    LR parallel, 205–7
  plate, 295–96
  RC, parallel, 194–97
  resonance
    parallel, 231–33
    Q, 235
    series, 227–30
  semiconductor, 338–41
    conventional, 331
    electron, 330–31
    leakage, 335
  series circuit, dc, 57–59
  shorted output, Nortons, 267–68
  transformer (*See* Transformers)
  transistor ratings, 351

Current (Cont.)
  transistors (*See* Current, semiconductor)
  varying, induction, 157–59
  varying, inductor, 176–78
  vacuum tube
    grids, 310–11
    maximum, 318
Current flow
  capacitor circuit, 134–36
  capacitors, series, 138
Current gain, 338–41
Current regulation, 401–3
  constant current generators, 401
  zener diode, 401–2
Curves
  B and *H*, 30–31
  characteristic
    common-base, transistor, 347–48
    common-emitter, transistor, 348–49
    FET, 374–75
    vacuum tubes, 304–9, 313
  diode, characteristic, 336
  diode, vacuum tube, 295–96
  FET, characteristic, 374–75
  hysteresis, 32
  non-linear, 249
  RC time constant, 198
  transistor characteristics, 347–49
  vacuum tube characteristics, 304–9, 313
Cutoff
  filter, 240–41
  remote, 316
  sharp, pentode, 313

Cycle
  duty, 123–24
  frequency, 117–19
  induced voltage, 107–9

Deflection
  electrostatic, CRT, 324–25
  magnetic, CRT, 325
Degrees, 110–11
Degrees, rotation, 110–11
Depletion region, 334–35, 337–38
DC current
  parallel circuit, 70–72
  series circuit, 57–59
Dielectric
  constant, 128, 132–33
  insulation, 138
  power factor, 137
Differentiating network, 246–47
Diffusion, integrated circuits, 381–83
Diode
  current, operating, 364–67
  zener
    breakdown voltage, 361–63
    depletion region, 361
    leakage current, 362
    load line, 363–66
Diode characteristic curve, 336
Diodes (*See also* Vacuum tubes)
  avalanche, 336
  clipping, 336–37
  semiconductor, 335–37
  sine waves, 337
  square waves, 336–37
  zener, 361–67
  zener, doping, 361
  zener region, 336
Distance, capacitor plates, 132–33

Drift current, 9
Duty cycle, 123–24

Eddy current, 172–73
Edge, pulse
  leading, 122
  trailing, 122
Electric current, 9
Electric field, 23–26
Electrical symbols, 47
Electromotive force, 8
Electron flow
  basic, 9
  CRT, 323–25
  FET, 373–74
  semiconductor, 330-31
  vacuum tube, 291–92, 295, 299, 310–13
Electronic symbols, 405–9
Electrons
  capacitors, 130–31, 138
  free, 330
  valence, 329
  valence band, 332
Emission
  secondary, 312
  thermionic, 291, 293

Fall time, 124
Farad, defined, 131
FET
  bias, 373, 377–78
  connections, 379
  construction, 372–73
  current flow, 373–74
  curves, characteristics, 374–75
  depletion regions, 373–74
  drain, 372–73
  gate, 372–73
  load line, 376-77
  N-channel, 372–73
  P-channel, 372–73

INDEX

FET (Cont.)
  pinch-off, 373–74
  self-bias, 377–78
  source, 372–73
  symbols, 375–76
Field
  electric, 23–26
  electrostatic, 127–28
  force, 16, 21
  force on a conductor, 33–34
  intensity, 26–29
  lines of force, 16
  magnetic, 17, 23–24
  strength, 37–38
Field Effect Transistors (FET), 372–79
Field intensity, defined, 26
Figure of merit, Q, 233–35
Filament, 291–94
Filament, materials, 293
Filters
  ac from dc, 238–39
  band-pass, 244–45
  band-pass design, 248–49
  band-stop, 244–45
  calculations, 393–94
  choke input, 392–93
  cutoff, 240–41
  dc from ac, 240
  high-pass, 241–42
  L-type, 392–93
  low-pass, 241–43
  Pi-type, 393
  power supplies, 392–94
  Q, 244
  Q, band-pass, 248
  RC, 393–94
  types, basic, 241
Flux, 29–30
  change, 34–38
  density, 19

Flux (Cont.)
  magnetic, 18–20
  magnetic lines, 168
  transformer, 164
Flux density, 24
Flux lines, capacitors, 127–28
Foot-candle, 367
Force
  capacitor charge, dynes, 128–29
  dynes, 128–29
  magnetomotive, 29
Frequency
  bandwidth, 236–38
  capacitive circuit, 147–52
  conversion, 117–18
  cycle, 117–19
  defined, 117–21
  Hertz, 117
  lambda, 119
  period, 119
  photodevices, 372
  resonance
    parallel, 231–32
    Q, 234–35
    series, 225–30
  resonance chart, 416–17
  time, 118–19
  transient signals, 247–48
  transistors, cutoff, 352
  wavelength, 119–21
Functions trigonometric, 410, 414

Gamma, 356
Gas tubes (*See* Vacuum tubes)
Gauss, 19–20
Germanium, 329

h parameters, 351–52
Heaters (*See* Filaments)
Hertz, 117

High-pass filter, 241–42
Hysteresis, 31–33
Henry, defined, 159
Hysteresis
  loss, capacitor, 137
  transformer, 172–73

Impedance
  complex numbers, 281–85
  defined, 189
  j factors, 272–78
  LC, parallel, 214–15
  LC, series, 212
  LRC, series/parallel, 221–22
  LRC series, phase angle, 215, 217
  parallel
    resistance/capacitive reactance, 194–97
    resistance/inductive reactance, 205–8
  phase angle
    LR, parallel, 205–8
    LR, series, 200–204
    RC, series, 190–94
  resistance/reactance, parallel, 217–19
  resistance/reactance, series, 215–17
  resonance
    parallel, 231, 233
    Q, 235
    series, 229–30
  series
    resistance/capacitive reactance, 189–94
    resistance/inductive reactance, 200–205
Induced voltage, 36–38, 107–9
Inductance
  defined, 159–63

Inductance (Cont.)
  LR time constant, 208–11
  mutual, 165–66
  parallel, 174–75
  parallel, chart, 418–19
  series, 173–74
  stray, 163
  variable, 162–63
Induction, current, 157–59
Induction, magnetic, 20–21
Inductive reactance
  chart, 416–17
  defined, 178–80
  parallel, 181–82
  series, 180–81
Inductor
  permeability, 161–62
  phase angle, 183–85
Instantaneous current, capacitive, 147
Insulators
  defined, 10–11
  glass, 10
  mica, 10
Integrated circuits
  advantages, 380
  capacitors, 383–85
  construction, general, 380, 385
  construction process, 381–83
  definition, 379–80
  diffusion, 381–83
  epataxial layer, 382
  hybrid, 380
  logic, 385
  memory, 385
  miniaturization, 385
  monolithic, 380
  multiple components, 383–85
  resistors, 383–85
  thinfilm, 380
  transistors, multiple, 383–84

INDEX

IR drops, 56–57, 59–63

j factors, 272–78

Kirchhoff's law
  basic, 56–57, 69–70
  loop equations, 257–58
  parallel circuits, 69–70
  series circuits, 56–57
  series/parallel circuits, 89–90
  transistor, current, 358
  transistors, 339
  zener diode, 364

Lambda, 119
Laws
  Kirchhoff's (*See* Kirchhoff's law)
  Lenz's 35–36, 161
L/C ratio, 234
Leakage current, transistor, 349
Leakage resistance, 134
Lenz's law, 35–36, 161
Linear functions, 42–43
Loading
  resistance, 96–97, 103–5
  transformer, 169
Load lines
  thyrite, 251–52
  vacuum tube, 308, 315
  zener diode, 363–66
Loop equations, 257–62
Loop, rotation, 110
Losses, capacitor
  absorption, 137
  heat, 137
  hysteresis, 137
  power factor, 137
Losses, hysteresis, 33
Low-pass filter, 241–43
LR time constant, 208–11

Magnet, demagnetizing, 33
Magnetic
  atomic structure, 17
  B and H curve, 30–31
  current, 33
  deflection, CRT, 325
  force per unit length, 26
  flux, 18–21, 29–30
  gauss, 19–20
  hysteresis, 31–33
  induced current, 34–36
  induction, 20–21
  lines of force, 23–24
  lines of force, direction, 23–24
  materials, 15–16, 20, 22–23
  maxwells, 18–19
  oersteds, 26–29
  saturation, 30–31
  theory, 17
  torque, 34
  weber, 18–20
Magnetism
  Flemming's rule, 35
  Lenz's law, 35–36
  permeability, 21–22
  reluctance, 29–30
  residual, 32
  shielding, 22
Magnetomotive force, 29
Magnets, 14–16
  air gaps, 21
  diamagnetic, 22
  electro, 22–23
  induction, 20–21
  lodestone, 14–15
  paramagnetic, 22
  permanent, 20–23
  poles, 14–16
Materials
  conductors, 10
  ferrite, 22

Materials (Cont.)
  insulators, 10
  magnetic, 15, 22–23
  non-magnetic, 15–16
  semiconductors, 10
Mathematic functions
  cubes, 411–13
  roots, 411–13
  squares, 411–13
Mathematic tables, 411–14
Mhos, 11–12
Miniaturization (*See* Integrated circuits)
Mutual inductance, 165–66
Mutual transconductance, 306, 314

Net reactance
  LC, parallel, 214
  LC, series, 212–13
  LRC, series, 215
Network, differentiating, 246–47
Non-linear circuits, 249–53
Non-linear functions, 43–44
Norton's theorem, 266–69

Ohms-centimeter, 10
Ohms, defined, 10–11
Ohm's law, 42–45
  ac, 111–12
  ac current, 114
  ac voltage, 114
  defined, 42
  examples, 48–50
  parallel circuits, unknown quantities, dc, 84–87
  parallel resistance, 72–74, 76–77
  series resistance, 57
  two power sources, 50–54

Pentodes (*See* Vacuum tubes)

Period, 119
Permeability, inductor, 161–62
Permeability, theory, 21–22, 24, 31
Phase, 113
Phase angle
  capacitive circuit, 152–54
  complex numbers, 281–85
  complex numbers, polar form, 278–80
  current, *LR*, parallel, 205–7
  defined, 116–17
  impedance
    RC, parallel, 194–97
    RC, series, 190–94
  inductive, 183–85
  in-phase, 116
  j factors, 272–73, 275–76
  LC, series, 211–12
  LR, series, 200–204
  LRC, parallel, 217–19
  LRC, series, 215–17
  LRC, series/parallel, 221–22
  out-of-phase, 116
  voltage, *RC*, series, 190–94
Phase inversion, vacuum tube, 300–301
Phase, power supplies, 389–90
Phase shift, filters, 239–40
Photo devices, 367–72
  angstrom units, 372
  foot-candle, 367
  micron units, 372
  photon, 367
  photoresistors, 367–70
  phototransistors, 370–72
  symbols, phototransistors, 371–72
Photoresistors, 367–70
Phototransistors, 370–72
Phototransistors, parameters, 371–72

INDEX

Phototube (*See* Vacuum tubes)
Picture tube, (CRT), 323
Pinch-off, 373–74
Plate, 291–92, 295
  capacitor, 127–34
  CRT, 324
  resistance (plate to cathode), 307, 309, 314
  wattage, 317
Plate characteristics, 296
P-N junction, 333–35
Potential difference, 8–9
Power, calculation, 46–47
Power consumption, transformer, 168–71
Power dissipation
  Ohm's law, 46–47
  parallel circuits, 82–84
  series circuits, 63–64
  series/parallel circuits, 92–94, 96, 102–3, 105
  transistors, 351
Power factor, 137
Power gain, 346–47
Power supplies
  bridge, solid state, 395
  filters, 392–94 (*See also* Filters)
  regulated, 402–3
  solid state
    full-wave, 395
    half-wave, 394–95
    voltage doubler, 395–96
  vacuum tube
    bridge, 391–92
    filament connections, 392
    filtering, 389–90
    full-wave, 390–92
    half-wave, 388–90
    waveforms, 389–90
  voltage regulation (*See* Voltage regulation)

Primary (*See* Transformers)
Pulsating dc
  capacitors, series 139
  duty cycle, 123–24
  general, 121–25
Pulses
  dc, 121–25
  definitions, 122
  duty cycle, 123–24
  fall time, 124
  repetition time, 122–24
  rise time, 124
  width, 125
Pulse width, 125

Q
  band-pass filter, 248
  bandwidth, 236–37
  filters, 244
  parallel circuit, 235
  series circuit, 234–35
Q, circuit, 233–35

Radians, 110, 148
Radio
  AM band, 238
  FM band, 120
Radio waves, 119
Ratings, vacuum tubes, 317–18
Ratio
  turns, 166–68
  voltage, transformer, 166–69
Reactance
  capacitive, 146–52
  chart, 416–17
  filters, 242–43
  inductive, 178–80
  inductive/capacitive, parallel, 213–15
  inductive/capacitive, series, 211–13

Reactance (Cont.)
  inductive
    defined, 178
    frequency, 178–80
    parallel, 181–82
    series, 180–81
  parallel, chart, 418–19
  resonance
    parallel, 230–33
    series, 225–26, 229–30
Reactance and resistance
  parallel, 217–19
  series, 215–17
  series/parallel, 220–22
RC time constant, 197–200
Rectangular form, j factors, 273–76
Rectification, 296–98
Rectification, diode, 336–37
Rectifiers (See Power supplies)
References, 420
Regulator, voltage
  gas tube, 322–23
  thyrite, 251–52
  zener diode, 364–66
Relay driver
  photoresistor, 368–69
  phototransistor, 370–71
Reluctance, 29–30
Remote cutoff, 316
Repetition time, 122–24
Residual, 32
Resistance
  ac, coil, 233–34
  calculation, 45
  cathode bias resistor, 302–3, 311, 315–16
  color code, 405
  defined, 10–11, 41–42
  j factors, 272–78
  leakage, 134

Resistance (Cont.)
  loads, 96–97, 103–5
  LR time constant, 208–11
  non-linear, 250–53
  ohms, 10–11
  Ohm's law, 42–44
  parallel, 69–87
  parallel, chart, 418–19
  parallel circuit, 72–79
  parallel/series, 98–101
  phase, voltage/current, 113
  photo device, 367–70
  plate to cathode, 307, 309, 314
  RC time constant, 197–200
  series, 57–67
  series/parallel, 90–97
  specific, 11
  thyrite, 250–52
  variable, non-linear, 252–53
  wire, 10–11
Resistance and reactance
  parallel, 217–19
  parallel, phase angle, 217–19
  series, 215–17
  series/parallel, 220–22
  series, phase angle, 215–17
Resistivity, 9–10
Resonance
  bandwidth, 236–38
  frequency, chart, 416–17
  parallel, 230–33
  series, 225–30
Rise time, 124
Root mean square (rms), 113–16
Rotation, angular, 109–11
Rules
  Flemming's left-hand, 35
  left-hand, magnetism, 24–25

Saturation, transistor, 341–42

INDEX

Secondary (*See* Transformers)
Secondary emission, 312
Screen grid, 310–12
Semiconductors, 10
  alpha, 339–41
  base, 338–41
  bias, diodes, 335–36
  bias effects, 341–44
  biasing, 334–35
  carbon, 10
  carriers
    majority, 332–33
    minority, 333
  characteristic curves, 347–49
  charge diffusion, 334
  charge potential, junction, 334
  collector, 338
  collector, voltage, 343–45
  configurations, comparison, 347
  configurations, transistors, 345–47
  construction, 333
  covalent bonding, 330
  crystal lattice, 330–32
  crystals
    N-type, 332–33
    P-type, 332–33
  current, 338–41
  current gain, 338–41, 345–50
  current, leakage, 335
  current ratings, 351
  current regulation (*See* Current regulation)
  depletion region, 334–35, 337–38
  diode
    characteristic curve, 336
    rectification, 336–37
  diodes, 335–37
  doping, 331–33

Semiconductors (Cont.)
  electrons, valence band, 332
  emitter, 338
  Field Effect Transistors (FET), 372–79
  free electron, 330
  germanium, 10, 329
  h parameters, 351–52
  hole movement, 333
  holes 330, 333
  integrated circuits (*See* Integrated circuits)
  intrinsic crystal, 331
  leakage current, 335, 349
  mobility, 333
  NPN junction, 337–39
  P-N junction, 333–35
  PNP junction, 337–39
  power dissipation, 351
  power gain, 346–47
  power supplies, 394–96
  recombination, 331
  resistivity, 329, 333
  saturation, 341–42
  silicon, 10, 329
  specifications, 350–52
  symbols, transistor, 338–39, 342–43, 352–53
  temperature, 333
  transistor action, 338
  transistor symbols, 338–39, 342–43, 352–53
  valence band, 329–30
  voltage gain, 346–47, 354–55
  voltage ratings, 351
  voltage regulation (*See* Voltage regulation)
Semiconductor theory, 329–33
Series regulators (*See* Voltage regulation)
Shell, atoms, 2

INDEX

Shunt regulators (*See* Voltage regulation)
Silicon, 329
Sine, 110–11
Sine, angle, 117
Sine, table, 414
Sine wave
   amplitude measurements, 113–16
   capacitive circuit, instantaneous current, 147–48
   capacitor circuit, 134–36
   defined, 111
   inductive, 184
   wavelength, 120
Solenoid, 27–28
Sound waves, 121
Special purpose tubes (*See* Vacuum tubes)
Specifications
   semiconductors, 350–52
   vacuum tubes, 316–18
Specific resistance, 11
Stray capacitance, 134
Stray inductance, 163
Superposition theorem, 269–72
Suppressor grid, 312–13
Symbols
   electrical, 47
   electronic, 405–9
   FET, 375–76
   transistor, 342–43, 352–53
   vacuum tube, 299, 302, 317

Tangent, table, 414
Temperature
   filament, 293–94
   semiconductors, 333
Tetrodes (*See* Vacuum tubes)
Thévenin's theorem, 262–66
Thyrites, 250–52

Time
   capacitive circuit, 147–48
   fall, 124
   rise, 124
Time constant
   LR, 208–11
   RC, 197–200
   transient circuit, 246–47
Torque, 34
Transconductance, mutual, 306, 314
Transformer
   autotransformer, 170–71
   cores, 171–73
   coupling, 238–39
   laminations, 173
   mutual inductance, 165–66
   ratios, 166–68
Transformer action, defined, 163
Transient circuits
   definition, 245
   RC time constant, 246–47
Transient response, 245–46
Transistor action, 338
Transistors (*See* Semiconductors)
Transistors
   integrated (*See* Integrated circuits)
   typical, 328
Transmitter tubes (*See* Vacuum tubes)
Trigonometric functions, 410, 414
Triodes (*See* Vacuum tubes)
Turns ratio, transformer, 166–68

Units
   CGS, 18, 29
   electronic, 405
   esu, 129
   mks, 18

INDEX

Universal chart, 418–19

Valence band, 329–30
Vacuum tubes
  amplification factor, 305–6, 309, 314
  beam power tubes, 321–22
  capacitance, interelectrode, 317
  cathode, cold, 322
  cathode, hot, 322
  cathode, photo, 323
  cathode ray tube, (CRT) 323–25
  control grid, 298–300
  current, maximum, 318
  curves, characteristic, 304–9
  diodes, 291–98
  filament current, 316–17
  filament voltage, 316–17
  gas, 322–23
  load line, 308, 315
  manuals, 316
  mutual transconductance, 306, 314
  peak inverse voltage, 318
  pentodes, 312–16
  pentodes, bias, 315–16
  pentodes, voltage gain, 314–15
  phototube, 323
  plate wattage, 317
  power supplies (*See* Power supplies)
  ratings, 317–18
  remote cutoff, 316
  screen grid, 310–12
  secondary emission, 312
  special amplifiers, 318–21
  special purpose, 321—25
  specifications, 316–18
  suppression grid, 312–13
  symbols, 299, 302, 317

Vacuum tubes (Cont.)
  tetrodes, 310–12
  tetrodes, bias, 311
  triodes, 298–304
  transmitter, 322
  voltage
    screen, 317
    plate, 317
  voltage gain, triode, 309
  voltage regulation (*See* Voltage regulation)
Vectors
  capacitive reactance, 152–53
  complex numbers, polar form, 280
  inductor, 184
  j factors, 272–75
  LC, parallel, 214
  LR, parallel, 206–7
  LR, series, 201–2
  LRC, parallel, 218
  LRC, series, 215–17
  RC parallel, 194–97
  RC, series, 193–97
Velocity, 119–20
Volt, 9
Volt, defined, 41–42
Voltage
  alternating, 107–9, 115
  angles, amplitude measurements, 115–16
  calculation, 45–46
  capacitor, 131–32
  capacitors, parallel, 143
  capacitors, series, 141–43
  collector, 343–45
  defined, 9
  doubler, 395–96
  drops, 59–63
  FET, circuit, 377–78
  filament, 292, 294, 316–17
  filter, 242–43

Voltage (Cont.)
  gain, vacuum tube, 301, 309
  induced
    basic, 36–38, 107–9
    coil, 178
    inductive, 157–59
    transformer, 163–64
  inductive, instantaneous, 183–85
  inductor, instantaneous, 175–76
  inductors, 175–77
  instantaneous, 111
  instantaneous, inductor, 175–76
  junction, 96–97, 353
  LC, parallel, 213
  LC, series, 212–13
  LR, parallel, 205
  LR, series, instantaneous, 202–4
  LRC, series/parallel, 221
  LRC, series phase angle, 216
  maximum alternating value, 111
  peak inverse, 318
  Ohm's law, 42–44
  open-circuit, Thévenin's, 263–66
  parallel circuits, dc, 79–82
  peak, 113–16
  phase, 113
  phase angle, capacitive circuit, 152–54
  plate, 295–96
  plate, maximum, 317
  power supplies, vacuum tube, 388–91
  pulsating dc, 121–25
  ratio, transformer, 166–69
  RC, parallel, 194
  RC, series, 192–93
  regulation, 396–400

Voltage (Cont.)
  regulation, zener diode, 364–66
  regulator, thyrite, 251–52
  resonance
    Q, 234–35
    series, 228–30
  series circuits, dc, 59–63
  screen, maximum, 317
  transistor ratings, 351
  transformer, induced, 165–66
  two sources, 60–61
  varying, capacitive circuit, 154
  zener breakdown, 361–63
Voltage divider, 94–97
Voltage gain
  grounded grid amplifier, 319–20
  grounded plate amplifier, 320–21
Voltage potential, 9
Voltage regulation
  circuits, basic, 397
  series, 399–400
  shunt, 397–99
  zener diode, 397–400

Waveform
  ac voltage vs. ac current, 112–13
  cosine, 117
  sine wave, 111
  sinusoidal, 111
  sinusoidal, inductive, 184
  transient, 245–47
Wavelength, 119–21
Waves
  radio, 119
  sound, 121

Zener breakdown, 361–62
Zener diodes, 361–67

DISCARDED

JUN 25 2025